Selected Titles in This Series

27 **Michel Théra, Editor,** Constructive, experimental, and nonlinear analysis
26 **Luis G. Gorostiza and B. Gail Ivanoff, Editors,** Stochastic models
25 **V. Jurdjevic and R. W. Sharpe, Editors,** Geometric control and non-holomorphic mechanics
24 **Idun Reiten, Sverre O. Smalø, and Øyvind Solberg, Editors,** Algebras and modules II
23 **Idun Reiten, Sverre O. Smalø, and Øyvind Solberg, Editors,** Algebras and modules I
22 **Vlastimil Dlab and László Márki, Editors,** Trends in ring theory
21 **S. W. Drury and M. Ram Murty, Editors,** Harmonic analysis and number theory: Papers in honour of Carl S. Herz
20 **J. Borwein, P. Borwein, L. Jörgenson, and R. Corless, Editors,** Organic mathematics
19 **Raymundo Bautista, Roberto Martínez-Villa, and José Antonio de la Peña, Editors,** Representation theory of algebras and related topics
18 **Raymundo Bautista, Roberto Martínez-Villa, and José Antonio de la Peña, Editors,** Representation theory of algebras
17 **V. Kumar Murty, Editor,** Seminar on Fermat's Last Theorem
16 **Bruce N. Allison and Gerald H. Cliff, Editors,** Representations of groups
15 **Karl Dilcher, Editor,** Number theory
14 **Vlastimil Dlab and Helmut Lenzing, Editors,** Representations of algebras
13 **R. A. G. Seely, Editor,** Category theory 1991
12 **A. Nicas and W. F. Shadwick, Editors,** Differential geometry, global analysis, and topology
11 **H. Tachikawa and V. Dlab, Editors,** Representations of finite dimensional algebras
10 **A. Bialynicki-Birula, J. Carrell, P. Russell, and D. Snow, Editors,** Group actions and invariant theory
9 **Joel S. Feldman and Lon M. Rosen, Editors,** Mathematical quantum field theory and related topics
8 **F. V. Atkinson, W. F. Langford, and A. B. Mingarelli, Editors,** Oscillation, bifurcation and chaos
7 **H. Kisilevsky and J. Labute, Editors,** Number theory
6 **J. Carrell, A. V. Geramita, and P. Russell, Editors,** Proceedings of the 1984 Vancouver Conference in algebraic geometry
5 **D. J. Britten, F. W. Lemire, and R. V. Moody, Editors,** Lie algebras and related topics
4 **I. Hambleton and C. R. Riehm, Editors,** Quadratic and Hermitian forms
3 **Zeev Ditzian, Amram Meir, Sherman D. Riemenschneider, and Ambikeshwar Sharma, Editors,** Second Edmonton conference on approximation theory
2 **Richard M. Kane, Stanley O. Kochman, Paul S. Selick, and Vector P. Snaith, Editors,** Current trends in algebraic topology (Parts 1 and 2)
1 **Carl Herz and R. Rigelhof, Editors,** 1980 Seminar on harmonic analysis

(Continued in the back of this publication)

CONSTRUCTIVE, EXPERIMENTAL, AND NONLINEAR ANALYSIS

Jonathan Borwein

Canadian Mathematical Society
Société mathématique du Canada

CONFERENCE PROCEEDINGS • Volume 27

CONSTRUCTIVE, EXPERIMENTAL, AND NONLINEAR ANALYSIS

Michel Théra
Editor

Published by the American Mathematical Society
for the Canadian Mathematical Society
PROVIDENCE, RHODE ISLAND

CMS CONFERENCE PROCEEDINGS
Series Editors
Anthony V. Geramita and Niky Kamran
Volume 27

Accepted for publication by the Canadian Mathematical Society
Published by the American Mathematical Society

2000 *Mathematics Subject Classification.* Primary 11Y16, 11J70, 11Y65, 35J20, 41A65, 44A12, 46B20, 47A15, 49J52, 49K24, 58C20, 65K10, 90C25, 90C48.

Library of Congress Cataloging-in-Publication Data

Constructive, experimental, and nonlinear analysis / Michel Théra, editor.
p. cm. — (Conference proceedings / Canadian Mathematical Society, ISSN 0731-1036 ; v. 27)
Includes bibliographical references.
ISBN 0-8218-2167-9 (alk. paper)
1. Number theory—Congresses. 2. Numerical analysis—Congresses. 3. Nonlinear functional analysis—Congresses. I. Théra, M. A. (Michel A.), 1946– II. Conference proceedings (Canadian Mathematical society) ; v. 27.
QA241 .C6895 2000
512′.7—dc21 00-038997

∞ The paper used in this book is acid-free and falls within the guidelines
established to ensure permanence and durability.
Visit the AMS home page at URL: http://www.ams.org/

10 9 8 7 6 5 4 3 2 1 05 04 03 02 01 00

Contents

PREFACE

On September 23, 1999, the University of Limoges awarded the degree of Doctor Honoris Causa to Professor Jonathan Borwein. In connection with this event an International Workshop took place on September 22 and 23. Indeed, thanks to the existence of LACO (the Laboratoire d'Arithmétique, Calcul Formel et Optimisation), Limoges is signal in Europe in having internationally recognized research groups in all the major areas in which Jonathan Borwein has worked.

This volume presents 20 original and refereed papers dedicated to these events. The topics reflect different aspects of modern analysis in which Jonathan Borwein has been engaged, along with contributions in varied areas of application including algorithmic number theory, as well as tomography. More precisely, this book contains papers on:

- **Constructive:** analytic and computational number theory
- **Experimental:** symbolic and numerical computation
- **Nonlinear analysis:** theoretical and computational optimization, as well recent development in nonsmooth and functional analysis with applications to control theory.

These articles provide a fine snapshot showing the vitality of present research in these different areas of modern analysis. They also suggest some of the directions mathematical research is following at the beginning of this new millenium.

Many people made the degree ceremony, and this volume, possible. I am indebted to the University of Limoges, especially to the University's International Relations staff, for providing both financial support and concrete help with the organization of the events. I wish also to thank the LACO for its support.

I further extend my gratitude to all the authors who agreed with such enthusiasm to contribute to this volume and, of course, to all the referees for their valuable comments and suggestions

Michel THÉRA, Limoges, January 2000

Canadian Mathematical Society
Conference Proceedings
Volume **27**, 2000

Fejér monotonicity and weak convergence of an accelerated method of projections

Heinz H. Bauschke, Frank Deutsch, Hein Hundal, and Sung-Ho Park

Dédié à Jonathan Borwein à l'occasion de la remise de son titre de Docteur Honoris Causa

Abstract. The classical method of alternating or cyclic projections is of great importance in practice, but known to converge sometimes quite slowly. An acceleration of this method was proposed by Gubin, Polyak, and Raik in 1965, and subsequently studied by Gearhart and Koshy in 1989. Very recently, the present authors provided new estimates on the rate of convergence. Furthermore, the accelerated method was extended to handle general nonexpansive linear operators rather than just compositions of projections, and several sufficient conditions for norm convergence were obtained.
In this note, we show that the sequence produced by the accelerated method is always Fejér monotone. This allows us to establish weak convergence of the accelerated method. The question on whether or not norm convergence can be achieved in general remains open.

1. Introduction

Throughout this note, we assume that

$$\boxed{X \text{ is a real Hilbert space with inner product } \langle\cdot,\cdot\rangle \text{ and norm } \|\cdot\|,}$$

and that

$$\boxed{T : X \to X \text{ is linear with } \|T\| \le 1 \text{ and } M := \operatorname{Fix} T := \{x \in X : Tx = x\}.}$$

Put differently, the operator T is *nonexpansive*: $\|Tx - Ty\| \le \|x - y\|$, for all $x, y \in X$. Let us denote the *(orthogonal) projection* (or nearest point mapping) onto a closed convex nonempty set C in X by P_C.

Our aim is to find $P_M x$, given a point $x \in X$.

The most straight-forward idea to achieve this is to iterate the mapping T (in the spirit of Banach contractions). The success of this approach depends on T itself:

1991 *Mathematics Subject Classification.* Primary 41A65; Secondary 47A05, 65J05, 90C55.
Key words and phrases. accelerating convergence, alternating projections, cyclic projections, Fejér monotone sequence.
Research of Heinz H. Bauschke supported by NSERC.

FACT 1.1. ([**4**, Theorem 2.2]) $T^n x \to P_M x$, for every $x \in X$ if and only if T is *asymptotically regular*: $T^n x - T^{n+1} x \to 0$, for every $x \in X$.

Note that $T = -I$ is not asymptotically regular albeit nonexpansive.

Fact 1.1 is instrumental in proving the following result. Recall that T is *self-adjoint* (resp. *positive semidefinite*), if $T = T^*$ (resp. if $T \geq 0$, i.e., $\langle Tx, x \rangle \geq 0$, for all $x \in X$).

FACT 1.2. ([**4**, Theorem 2.5]) Suppose $T_1, T_2, \ldots, T_k$ are finitely many self-adjoint, positive semidefinite, nonexpansive linear operators on X. Then the composition $T := T_k \cdots T_2 T_1$ is nonexpansive and asympotically regular; consequently, $T^n x \to P_M x$, for every $x \in X$.

The most important incarnation of this iteration arises when each T_i is the projection onto a closed subspace M_i of X. Fact 1.1 then becomes the celebrated *method of alternating (or cyclic) projections* (see [**12, 10, 6**] for further information and applications):

FACT 1.3 (von Neumann-Halperin). Suppose $M_1, M_2, \ldots, M_k$ are finitely many closed subspaces in X. Let $T := P_{M_k} \cdots P_{M_2} P_{M_1}$. Then $\operatorname{Fix} T = M := M_1 \cap M_2 \cap \cdots \cap M_k$ and T^n converges pointwise in norm to P_M.

Unfortunately — consider two lines in the Euclidean plane intersecting at a small angle — the convergence of $T^n x$ to $P_M x$ can be very slow. (In infinite dimensional spaces, the convergence can be *arbitrarily slow* in a sense made precise in [**7, 3**].)

To remedy this, an acceleration for the method of alternating projections was suggested by Gubin, Polyak, and Raik [**9**], and by Gearhart and Koshy [**8**]. A comprehensive analysis of this method, which covers our present setting where T is not necessarily the product of projection operators but merely a nonexpansive linear operator, can be found in [**4**]. Let us now briefly explain the accelerated method and summarize some of its properties.

DEFINITION 1.4. The *accelerated mapping A_T of T* is defined by

$$A_T(x) := (1 - t_x)x + t_x Tx,$$

where

$$t_x := \begin{cases} \frac{\langle x, x - Tx \rangle}{\|x - Tx\|^2}, & \text{if } Tx \neq x; \\ 1, & \text{otherwise.} \end{cases}$$

We also let

$$c(T) := \|TP_{M^\perp}\|;$$

in other words, $c(T)$ is the operator norm of T restricted to $M^\perp$.

FACT 1.5 (Acceleration). Fix $x \in X$ and consider *the sequence (x_n) generated by the accelerated method*: $x_n := A_T^{n-1} Tx$, for $n \geq 1$. Then:

(i) $\max\{\|x_n - P_M x\|, \|T^n x - P_M x\|\} \leq c(T)^n \|x - P_M x\|$, for every $n \geq 1$.
(ii) If $T^* = T \geq 0$, then $\|x_n - P_M x\| \leq \|T^n x - P_M x\|$, for every $n \geq 1$.
(iii) If $T = P_{M_2} P_{M_1}$, then $\|x_n - P_M x\| \leq \|T^n x - P_M x\|$, for every $n \geq 1$.

PROOF. [**4**, Theorems 3.16, 3.20, and 3.23]. □

REMARK 1.6.

(i) In the Euclidean plane, let T be the product of two projections onto lines. Then the sequence (x_n) converges to the projection onto the intersection *in finitely many steps* — regardless of the size of the angle between the lines. This example also shows that the one-time application of T is indeed necessary to achieve this behaviour, and it provides a good motivation for the accelerated method.
(ii) The proof of Fact 1.5 is surprisingly intricate, requiring the use of the Spectral Theorem and some nontrivial estimates.
(iii) By Fact 1.2 and Fact 1.3, the sequence (x_n) converges to $P_M x$ in the setting of Fact 1.5.(ii) and (iii).
(iv) In contrast, if $c(T) = 1$, then $(T^n x)$ need not converge to $P_M x$: indeed, let $T := -I$, so that $T^n x = (-1)^n x$, for every $n \geq 1$. Note that for this particular example, $A_T \equiv 0$ (use [**4**, Remarks following Lemma 3.9]), so that the accelerated method converges in at most two steps!

The main open problem on the accelerated method concerns norm convergence.

QUESTION 1.7. Given $x \in X$, does $A_T^{n-1} Tx$ converge in norm to $P_M x$?

By Fact 1.5, the answer is affirmative if T is self-adjoint and positive semidefinite or if T is the product of two projections. However, we do not know what happens even T is is the product of three (or more) projections!

REMARK 1.8. Surprisingly, it is *not* true that $\|A_T^{n-1}Tx - P_M x\| \leq \|T^n x - P_M x\|$, for every $n \geq 1$. (If this were true, then, by Fact 1.1, asymptotic regularity would suffice to guarantee norm convergence.) Indeed, [**4**, Example 3.24] contains the construction of a certain $T := P_{M_5} \cdots P_{M_2} P_{M_1}$, where each M_i is a two-dimensional subspace in $X := \mathbb{R}^5$, and a starting point $x \in X$, such that $\|A_T^{n-1}Tx - P_M x\| > \|T^n x - P_M x\|$, for every $n \geq 3$!

In this note, we establish at least weak convergence *of the accelerated method. The proof relies on properties of* Fejér monotone sequences.

We conclude this section with some comments on the important case when T is the product of projection operators.

FACT 1.9. [**3**] Suppose $T := P_{M_k} \cdots P_{M_1}$, where each M_i is a closed subspace of X, and $k \geq 2$. Then $c(T) = \|P_{M_k} \cdots P_{M_1} P_{M^\perp}\| = c(M_1, M_2, \ldots, M_k)$, where $c(M_1, M_2, \ldots, M_k)$ is the *cosine of the angle of the k-tuple* $(M_1, \ldots, M_k)$ in the sense of [**3**], and $c(T) < 1$ if and only if $M_1^\perp + \cdots + M_k^\perp$ is closed.

REMARK 1.10. Fact 1.9 in conjunction with Fact 1.5.(i) yields *linear convergence* of both the unaccelerated and the accelerated method provided that $M_1^\perp + \cdots + M_k^\perp$ is closed. Sufficient conditions for the closedness of the sum of the orthogonal complements are (see [**2**, Theorem 5.19 and Corollary 5.20]):
(i) at least one subspace is finite-dimensional; or
(ii) all subspaces except possibly one have finite codimension.
In particular, this applies in Euclidean spaces or when each M_i is a hyperplane.

2. The Main Result

DEFINITION 2.1. Suppose C is a closed convex nonempty set in X, and $(x_n)_{n \geq 0}$ is a sequence in X. Then (x_n) is *Fejér monotone with respect to* C, if

$$\|x_{n+1} - c\| \leq \|x_n - c\|, \quad \text{for all } n \geq 0 \text{ and every } c \in C.$$

The notion of Fejér monotonicity has turned out to be quite useful in the study of various algorithms. (See [**2, 5**] for further information.) The following properties will be needed.

FACT 2.2. Suppose $(x_n)_{n\geq 0}$ is Fejér monotone with respect to some closed convex nonempty set C in X. Then:

(i) (x_n) is bounded.
(ii) $(P_C x_n)$ is norm convergent, say to $c^* \in C$.
(iii) (x_n) converges weakly to some point in C if and only if every weak cluster point of (x_n) lies in C.
(iv) If (x_n) converges weakly to some point in C, then this weak limit coincides with c^*.
(v) If C is a closed subspace, then $P_C x_n = c^*$, for every $n \geq 0$.

PROOF. Parts (i)–(iii) are standard; see, e.g., [**2**, Theorem 2.16]. Slightly stronger versions of the remaining parts can be found in the first author's PhD thesis; see [**1**, Theorem 6.2.2]. For the sake of completeness, we include the reasoning here.

(iv): Suppose $x_n \rightharpoonup w \in C$ and $P_C x_n \to c^* \in C$. The well-known characterization of the projection yields

$$\langle c - P_C x_n, x_n - P_C x_n \rangle \leq 0, \quad \text{for every } c \in C \text{ and all } n \geq 0.$$

Taking limits yields $\langle c - c^*, w - c^* \rangle \leq 0$, for every $c \in C$. Choose $c = w \in C$ to conclude $\|w - c^*\|^2 \leq 0$, or $w = c^*$.

(v): Fix $n \geq 0$ and $t \in \mathbb{R}$ arbitrarily. Then $\tilde{c} := (1-t)P_C x_n + tP_C x_{n+1} \in C$, and Fejér monotonicity yields $\|x_{n+1} - \tilde{c}\| \leq \|x_n - \tilde{c}\|$. Squaring and simplifying this results in

$$(1-2t)\|P_C x_{n+1} - P_C x_n\|^2 + \|P_{C^\perp} x_{n+1}\|^2 \leq \|P_{C^\perp} x_n\|^2.$$

Since t was chosen arbitrarily, we conclude $P_C x_n = P_C x_{n+1}$. □

We are now in a position to prove weak convergence of the accelerated method.

THEOREM 2.3 (Weak Convergence). Fix $x \in X$ and consider the sequence (x_n) generated by the accelerated method: $x_n := A_T^{n-1} T x$, for every $n \geq 1$. Then the sequence (x_n) is Fejér monotone with respect to M and it converges weakly to $P_M x$.

PROOF. Set $x_0 := x$. Fix $y \in M$ arbitrarily. Then, since T is nonexpansive,

$$\|x_1 - y\| = \|T(x_0 - y)\| \leq \|x_0 - y\|.$$

Now pick $n \geq 1$ arbitrarily. By [**4**, Theorem 3.7] (with $t = 0$), we have

$$\begin{aligned}\|x_{n+1} - y\|^2 &= \|A_T x_n - y\|^2 \\ &= \|x_n - y\|^2 - t_{x_n}^2 \|x_n - Tx_n\|^2 \\ &= \|x_n - y\|^2 - \|x_n - A_T x_n\|^2.\end{aligned}$$

It follows that $\|x_{n+1} - y\| \leq \|x_n - y\|$, so that the sequence $(x_n)_{n\geq 0}$ is Fejér monotone with respect to M. Denote the norm limit of the sequence $(P_M x_n)_{n\geq 0}$ by m^* (see Fact 2.2.(ii)). Next, [**4**, Lemma 3.9.(1)] implies that $t_{x_n} \geq \frac{1}{2}$. Altogether,

$$\|x_n - y\|^2 - \|x_{n+1} - y\|^2 = t_{x_n}^2 \|x_n - Tx_n\|^2 \geq \tfrac{1}{4}\|x_n - Tx_n\|^2.$$

Summing this over n, we conclude that $\sum_n \|x_n - Tx_n\|^2 < \infty$; in particular, $x_n - Tx_n \to 0$. By Fact 2.2.(i), the sequence (x_n) is bounded and hence possesses weak cluster points, say $x_{k_n} \rightharpoonup w$. Note that $x_{k_n} - Tx_{k_n} \to 0$ and that $Tx_{k_n} \rightharpoonup Tw$ (since T is weakly continuous). It follows that $w = Tw$, i.e., $w \in \operatorname{Fix} T = M$. By Fact 2.2.(iii), (x_n) is weakly convergent to some point in M, and by Fact 2.2.(iv) this weak limit is, in fact, m^*. On the other hand, M is a closed subspace so that $P_M x_n = m^*$, for every $n \geq 0$ (Fact 2.2.(v)). It follows that $x_n \rightharpoonup m^* = P_M x_0 = P_M x$, as required. □

REMARK 2.4. The use of Fejér monotonicity, a very nice tool in the study of several algorithms, made the proof of Theorem 2.3 relatively simple. It is interesting to note that the sequence generated by the accelerated method does not fit the usual construction of Fejér monotone sequences via iterates of nonexpansive mappings. In fact (see [**4**, Lemma 3.8.(5) and the subsequent remark]), while continuous [**4**, Lemma 3.8.(5)], the accelerated mapping A_T may fail to be even uniformly continuous!

Acknowledgment. We wish to thank Michel Théra and an anonymous referee for drawing our attention to a related paper by Lemaire [**11**], and for pointing out that the conclusion of Theorem 2.3 holds under the following, more general hypothesis: T is nonexpansive (not necessarily linear) and $\operatorname{Fix} T = M$ is a linear subspace.

References

[1] Heinz H. Bauschke. *Projection algorithms and monotone operators.* PhD thesis, Department of Mathematics and Statistics, Simon Fraser University, 1996. Preprint 96:080. Available at www.cecm.sfu.ca/preprints/1996pp.html.

[2] Heinz H. Bauschke and Jonathan M. Borwein. On projection algorithms for solving convex feasibility problems. *SIAM Rev.*, 38(3):367–426, 1996.

[3] Heinz H. Bauschke, Jonathan M. Borwein, and Adrian S. Lewis. The method of cyclic projections for closed convex sets in Hilbert space. In *Recent developments in optimization theory and nonlinear analysis (Jerusalem, 1995)*, pages 1–38. Amer. Math. Soc., Providence, RI, 1997.

[4] Heinz H. Bauschke, Frank Deutsch, Hein Hundal, and Sung-Ho Park. Accelerating the convergence of the method of alternating projections. Preprint 99:138, Centre for Experimental and Constructive Mathematics, Simon Fraser University, Burnaby, Canada, 1999. Available at www.cecm.sfu.ca/preprints/1999pp.html.

[5] Patrick L. Combettes. Fejér-monotonicity in convex optimization. In C. A. Floudas and P. M. Paradalos, editors, *Encyclopedia of Optimization.* Kluwer Acad. Publ., 2000.

[6] Frank Deutsch. The method of alternating orthogonal projections. In *Approximation theory, spline functions and applications (Maratea, 1991)*, pages 105–121. Kluwer Acad. Publ., Dordrecht, 1992.

[7] C. Franchetti and W. Light. On the von Neumann alternating algorithm in Hilbert space. *J. Math. Anal. Appl.*, 114(2):305–314, 1986.

[8] William B. Gearhart and Mathew Koshy. Acceleration schemes for the method of alternating projections. *J. Comput. Appl. Math.*, 26(3):235–249, 1989.

[9] L. G. Gubin, B. T. Polyak, and E. V. Raik. The method of projections for finding the common point of convex sets. *USSR Computational Mathematics and Mathematical Physics*, 7(6):1–24, 1967.

[10] Israel Halperin. The product of projection operators. *Acta Sci. Math. (Szeged)*, 23:96–99, 1962.

[11] Bernard Lemaire. Which fixed point does the iteration method select? In *Recent advances in optimization (Trier, 1996)*, pages 154–167. Springer, Berlin, 1997.

[12] John von Neumann. *Functional Operators. II. The Geometry of Orthogonal Spaces.* Princeton University Press, Princeton, N. J., 1950. Annals of Mathematics Studies, no. 22.

Department of Mathematics and Statistics, Okanagan University College, Kelowna, British Columbia V1V 1V7, Canada.
E-mail address: bauschke@cecm.sfu.ca (Heinz H. Bauschke)

Department of Mathematics, Pennsylvania State University, University Park, Pennsylvania 16802, USA.
E-mail address: deutsch@math.psu.edu (Frank Deutsch)

355 Applied Research Lab, Pennsylvania State University, University Park, Pennsylvania 16803, USA.
E-mail address: hundal@psu.edu (Hein Hundal)

Department of Mathematics, Sogang University, Seoul, Korea.
E-mail address: shpark@ccs.sogang.ac.kr (Sung-Ho Park)

Canadian Mathematical Society
Conference Proceedings
Volume **27**, 2000

General "Squeeze Theorems" in Nonsmooth Analysis

J. Benoist and J.-B Hiriart-Urruty

At the occasion of the Doctorate Honoris Causa conferred to Prof. J. M. Borwein

ABSTRACT. In this note, we prove a general result concerning Clarke's normal cone to the intersection of a collection of sets in a Banach space. As a consequence we solve a problem raised by Lewis and Lucchetti ([**7**]) on Clarke's generalized subdifferentials of comparable locally Lipschitz functions; in doing so we generalize the answer to this problem given by Borwein and Fitzpatrick in [**2**]. We also provide a variant result in terms of generalized subdifferentials in the sense of Michel-Penot. Finally, possible refinements in terms of Clarke's generalized Jacobian sets of matrices are sketched.

1. Getting a non-null direction normal to nested subsets

Let E be a real Banach space, let $E^\star$ denote its topological dual space; we denote by $\langle p, x\rangle$ the image of $x \in E$ by $p \in E^\star$. Recall ([**3**, Section 2.4]) that, given $X \subset E$ and $x \in X$ an element d of E is said to be tangent to X at x (in Clarke's sense) when: for every sequence $\{x_k\} \subset X$ converging to x and sequence $\{t_k\} \subset]0, +\infty[$ decreasing to 0, there exists a sequence $\{d_k\}$ converging to d such that $x_k + t_k d_k \in X$ for all k. The set of all tangent directions to X at x is called the tangent cone to X at x and denoted $T_X(x)$. Indeed $T_X(x)$ is a closed convex cone of E (with apex 0).
A way of defining the normal cone to X at x (in Clarke's sense) is to do it with polarity with $T_X(x)$ [**3**, Section 2.4]:

$$N_X(x) := \{p \in E^\star : \langle p, d\rangle \leq 0 \text{ for all } d \in T_X(x)\}.$$

When $X_1 \subset X_2$ and $x \in (bdX_1) \cap (bdX_2)$, it is far from true that $N_{X_1}(x) \subset N_{X_2}(x)$. But in applications we often are content with having a non-null common p in $N_{X_1}(x)$ and $N_{X_2}(x)$. This is not even true, as simple examples show (see Example 2.3 below if necessary). The next theorem provides a general framework in which we ensure to get a non-null p normal to all the X_i of a collection $\{X_i\}_{i\in I}$ of subsets of E.

1991 *Mathematics Subject Classification.* Primary 58C20, 49J52, 26A27.
Key words and phrases. "Squeeze theorems", Clarke's normal cone, Clarke's subdifferential, Clarke's generalized Jacobian set of matrices.
The present work was prompted by lectures R. Lucchetti gave at seminars in Limoges and Toulouse. We also thank D. Azé, J. Borwein and A. Lewis for their constructive reactions to drafts of our results.

THEOREM 1.1. *Let $\{X_i\}_{i\in I}$ be a collection of subsets of E, let $x \in \bigcap_{i\in I} X_i$. We assume:*

(H_1) For all $i \neq j$ in I, either $X_i \subset X_j$ or $X_j \subset X_i$;
(H_2) There exists a nonempty open cone K (with apex 0) such that $(x+K)\cap X_i = \emptyset$ for all $i \in I$.

Then

$$\bigcap_{i\in I} N_{X_i}(x) \neq \{0\}.$$

The assumption (H_1) means the sets X_i $(i \in I)$ are "totally ordered" by inclusion. The assumption (H_2) ensures that x lies on the boundary of every X_i, $i \in I$; actually, should it not be the case, there would be no hope to find a non-null normal direction to X_i at x.
Since $N_{X_i}(x)$ is a "local" notion (in the sense that it depends on X_i only in a neighbourhood of x), it suffices to have (H_2) with $X_i \cap B(x,\epsilon)$, where $\epsilon > 0$ is arbitrarily small. A way of securing (H_2), as often in applications of Theorem 1.1, is to have some $X \supset \bigcup_{i\in I} X_i$ such that $(x+K)\cap X \neq \emptyset$ for all $i \in I$.

PROOF OF THEOREM 1.1.
(a) *Preliminaries.* When $B \subset E^\star$, B° is defined as the set of x in E such that $\langle p, x\rangle \leq 0$ for all $p \in B$. A classical result in Convex analysis, called the "bipolarization lemma", asserts the following: whenever A is a nonempty subset of E, $A^{\circ\circ}$ $(:= (A^\circ)^\circ)$ is the smallest closed convex cone containing A.
Now, if $\{A_i\}_{i\in I}$ is a collection of cones of E with $\bigcup_{i\in I} A_i \neq \emptyset$, we easily infer from the bipolarization lemma above

$$\left(\bigcap_{i\in I} A_i^\circ\right)^\circ = \overline{\text{conv}}\left(\bigcup_{i\in I} A_i\right), \tag{1}$$

where $\overline{\text{conv}}$ stands for the closed convex hull operation.
(b) *The proof itself.* We prove Theorem 1.1 ab absurdo. Suppose

$$\bigcap_{i\in I} N_{X_i}(x) = \{0\},$$

so that $\left(\bigcap_{i\in I} N_{X_i}(x)\right)^\circ = \{0\}^\circ = E$. Hence, according to the result recalled in Preliminaries above (with $A_i := T_{X_i}(x)$ for all $i \in I$),

$$\overline{\text{conv}}\left(\bigcup_{i\in I} T_{X_i}(x)\right) = E.$$

Consequently,

$$K \bigcap \overline{\text{conv}}\left(\bigcup_{i\in I} T_{X_i}(x)\right) \neq \emptyset, \tag{2}$$

where K is the cone from assumption (H_2). By considering a nonempty open convex subcone of K if necessary, we may assume that K (in (2)) is an open convex cone (still with apex 0). Therefore, (2) entails

$$K \bigcap \operatorname{conv}\left(\bigcup_{i\in I} T_{X_i}(x)\right) \neq \emptyset. \tag{3}$$

There thus exists $d \in K$, a finite subset $J \subset I$, $\{d_i\}_{i\in J}$ such that $d_i \in T_{X_i}(x)$ for all $i \in J$ and $d = \sum_{i\in J} d_i$. Thanks to Assumption (H_1), one can arrange the subsets X_i, $i \in J$, so that:

$$J = \{i_1, i_2, \dots, i_p\},\ X_{i_1} \subset X_{i_2} \subset \cdots \subset X_{i_p}.$$

We know that $d_{i_1} \in T_{X_{i_1}}(x)$; there then exists a sequence $\{d_{i_1,n}\}_n$ converging to d_{i_1} such that

$$x + \frac{1}{n} d_{i_1,n} \in X_{i_1}\ (\subset X_{i_2}) \quad \text{for all } n.$$

Similarly, $d_{i_2} \in T_{X_{i_2}}(x)$; there thus exists a sequence $\{d_{i_2,n}\}_n$ converging to d_{i_2} such that

$$(x + \frac{1}{n} d_{i_1,n}) + \frac{1}{n} d_{i_2,n} \in X_{i_2}\ (\subset X_{i_3}) \quad \text{for all } n.$$

We continue in this fashion. Finally, since $d_{i_p} \in T_{X_{i_p}}(x)$, there exists a sequence $\{d_{i_p,n}\}_n$ converging to d_{i_p} such that

$$x + \frac{1}{n}(d_{i_1,n} + d_{i_2,n} + \cdots + d_{i_p,n}) \in X_{i_p} \text{ for all } n.$$

But $d_{i_1,n} + d_{i_2,n} + \cdots + d_{i_p,n} \overset{n\to+\infty}{\longrightarrow} d_{i_1} + d_{i_2} + \cdots + d_{i_p} = d \in K$; since K is a cone (with apex 0) and open, we therefore have:

$$\frac{1}{n}(d_{i_1,n} + d_{i_2,n} + \cdots + d_{i_p,n}) \in K \text{ for } n \text{ large enough.}$$

Consequently, $x + \frac{1}{n}(d_{i_1,n} + d_{i_2,n} + \cdots + d_{i_p,n}) \in (x + K) \cap X_{i_p}$ for n large enough. This contradicts Assumption (H_2).
The hypothesis in the beginning of the proof is therefore absurd, we indeed have $\bigcap_{i\in I} N_{X_i}(x) \neq \{0\}$. □

An area of applications of techniques and results from Nonsmooth analysis is Mathematical economics; see [**1**] for a recent survey on the subject. Given a collection of (non-convex) production subsets, say $F_1, F_2, \dots, F_m$ in $\mathbb{R}^d$, a goal aimed at is to get a common non-null equilibrium price p, which mathematically finds expression in:

$$0 \neq p \in \bigcap_{i=1}^{m} N_{F_i}(x_i), \text{ where } x_i \in F_i \text{ for all } i = 1, \dots, m. \tag{4}$$

A result in that vein is derived from a simple application of Theorem 1.1.

THEOREM 1.2. *Let $F_1, F_2, \dots, F_m$ be subsets of $\mathbb{R}^d$, let $(x_1, x_2, \dots, x_m)$ in $F_1 \times F_2 \times \cdots \times F_m$ such that $F_1 + F_2 + \cdots + F_m$ satisfies the "outside cone property" (H_2'), i.e.*

(H_2') *There exists a nonempty open cone K (with apex 0) such that*

$$(x_1 + x_2 + \cdots + x_m + K) \cap (F_1 + F_2 + \cdots + F_m) \neq \emptyset.$$

Then, for every permutation σ of $\{1,2,\dots,m\}$, there exists a non-null p_σ such that

$$p_\sigma \in N_{F_{\sigma(1)}}(x_{\sigma(1)}),\ p_\sigma \in N_{F_{\sigma(1)}+F_{\sigma(2)}}(x_{\sigma(1)}+x_{\sigma(2)}),\dots$$
$$p_\sigma \in N_{F_1+F_2+\cdots+F_m}(x_1+x_2+\cdots+x_m). \tag{5}$$

This statement is somehow similar to the following one, due to Cornet and Rockafellar ([**4**]): *Let $F_1, F_2, \dots, F_m$ be subsets of $\mathbb{R}^d$, let $(x_1, x_2, \dots, x_m) \in F_1 \times F_2 \times \cdots \times F_m$ such that $x_1+x_2+\cdots+x_m$ lies on the boundary of $F_1+F_2+\cdots+F_m$; there then exists a non-null p such that*

$$p \in N_{F_i}(x_i) \quad \textit{for all} \quad i = 1,2,\dots,m. \tag{6}$$

When the production subsets F_i are convex, the result by Cornet-Rockafellar and our Theorem 1.2 indeed are equivalent; this is due to the following "decomposition rule" on the normal cones to convex subsets:

$$N_{F_{\sigma(1)}+F_{\sigma(2)}+\cdots+F_{\sigma(p)}}(x_{\sigma(1)}+x_{\sigma(2)}+\cdots+x_{\sigma(p)}) = \bigcap_{i=1}^{p} N_{F_{\sigma(i)}}(x_{\sigma(i)}).$$

In the case of non-convex production subsets F_i, to our best knowledge, the two results cannot be compared.

PROOF OF THEOREM 1.2.
Let $X_1 := F_{\sigma(1)} - x_{\sigma(1)}$, $X_2 := F_{\sigma(1)} + F_{\sigma(2)} - (x_{\sigma(1)} + x_{\sigma(2)})$, $\dots$, $X_m := F_1 + F_2 + \cdots + F_m - (x_1+x_2+\cdots+x_m)$. Now, $X_1 \subset X_2 \subset \cdots \subset X_m$ and $0 \in X_i$ for all $i = 1,2,\dots,m$. Moreover, by assumption , the "cone property" (H_2) is satisfied at 0. Thus, as an application of Theorem 1.1, we get $\bigcap_{i=1}^{m} N_{X_i}(0) \neq \{0\}$, which is the announced result since

$$N_{X_i}(0) = N_{F_{\sigma(1)}+F_{\sigma(2)}+\cdots+F_{\sigma(i)}}(x_{\sigma(1)}+x_{\sigma(2)}+\cdots+x_{\sigma(i)}).\square$$

2. Getting a common element to Clarke's generalized subdifferentials of ordered functions

Suppose one asks first-year students in a Calculus course the following question: if two real-valued functions f and g of the real variable are ordered, say $f \leq g$, and coincide at a point x where both are differentiable, does $f'(x) = g'(x)$? They are tempted to answer by using the very definition of the derivative of a function. The same thing happens at a more advanced level, when the differential of a real -valued function on a Banach space is defined. The answer is easy indeed once the "variational reflex" is acquired: $g - f \geq 0$ and $(g-f)(x) = 0$, thus x is a minimizer of $g-f$, whence $(g-f)'(x) = g'(x) - f'(x) = 0$. In the realm of Nonsmooth analysis, when the considered generalized subdifferentials $\partial\varphi$ of a real-valued functions φ are not "monotone with respect to the ordering of functions" (i.e. when $f \leq g$ and $f(x) = g(x)$ do not necessarily imply $\partial f(x) \subset \partial g(x)$), the case of two ordered functions is again handled via the variational approach: under some condition, $\partial(g-f)(x) \subset \partial g(x) - \partial f(x)$, so that the optimality condition $0 \in \partial(g-f)(x)$ gives rise to a common p to $\partial g(x)$ and $\partial f(x)$. When more than two functions are involved the situation is radically different: given m functions $f_1, f_2, \dots, f_m : E \to \mathbb{R}$ such that $f_1 \leq f_2 \leq \cdots \leq f_m$ and $f_1(x) = f_2(x) = \cdots = f_m(x)$, is it true that there exists a common p to all the $\partial f_i(x)$'s? The question was posed by Lewis and Lucchetti ([**7**, p. 19]), and positively answered by them under the following assumptions: $E = \mathbb{R}^d$, the functions f_i are locally Lipschitz, $\partial f_i(x)$ stands for Clarke's generalized subdifferential of f_i at x, and $m \leq 3$ or $d = 2$. This is a so-called "squeeze theorem". Later on, the question was considered by Borwein and Fitzpatrick ([**2**]); these authors proved there indeed is a common element to Clarke's generalized subdifferentials of the f_i's under the following assumptions: E is a Banach space and the functions f_i are locally Lipschitz ([**2**, pp. 9-10]). A partial generalization to the case where some of the f_i are not locally Lipschitz was also provided by them ([**2**, Theorem 4.3]), but this comes as an application of the "all locally Lipschitz" result to an ad hoc constructed framework. We generalized below the main result by Borwein and Fitzpatrick, using a completely different approach derived from the "general geometrical intersection theorem" obtained in Section 1.

Let E be a real Banach space, let $\varphi : E \to \overline{\mathbb{R}}$ be finite at $x \in E$. Then a geometric definition of the generalized subdifferential of φ at x in Clarke's sense (or Clarke and Rockafellar' sense) is the subset of $E^\star$ defined as follows ([**3**, p. 61]):

$$\partial\varphi(x) := \{s \in E^\star : (s, -1) \in N_{epi\varphi}(x, \varphi(x))\}. \tag{7}$$

We use the notation $\partial^{cl}\varphi(x)$ when another generalized subdifferential enters into the picture. When φ is Lipschitz around x, $\partial\varphi(x)$ is a nonempty convex set whose support function is the so-called Clarke generalized directional derivative $\varphi^\circ(x, .)$ of φ at x ([**3**, p.27]):

$$d \in E \longmapsto \varphi^\circ(x, d) = \limsup_{x' \to x, t \to 0^+} \frac{\varphi(x' + td) - \varphi(x')}{t}.$$

When φ is no longer Lipschitz around x, $\partial\varphi(x)$ may be empty; in any case, one may define the so-called asymptotic generalized subdifferential $\partial^\infty\varphi(x)$ of φ at x as follows:

$$\partial^\infty\varphi(x) := \{s \in E^\star : (s, 0) \in N_{epi\varphi}(x, \varphi(x))\}. \tag{8}$$

When φ is Lipschitz around x, $N_{epi\varphi}(x, \varphi(x))$ has no "horizontal half-line", which ensures that $\partial^\infty\varphi(x)$ reduces to the singleton $\{0\}$.

We now are ready to state our general "squeeze theorem".

THEOREM 2.1. *Let $\{f_i\}_{i\in I}$ be a collection of functions from E into $\overline{\mathbb{R}}$, let x be a point at which all the f_i coincide (i.e. $f_i(x) = c \in \mathbb{R}$ for all $i \in I$). We assume:*

($\mathcal{H}_1$) *For all $i \neq j$ in I, either $f_i \leq f_j$ or $f_j \leq f_i$;*

($\mathcal{H}_2$) *There is $\varphi : E \to \overline{\mathbb{R}}$ minorizing all the f_i, $i \in I$, with $\varphi(x) = c$, and a direction $d \in E$ such that*

$$\liminf_{v \to d, t \to 0^+} \frac{\varphi(x + tv) - \varphi(x)}{t} > -\infty. \tag{9}$$

Then:

$$\bullet \textit{either there is some } s \in \partial f_i(x) \textit{ for all } i \in I, \tag{10}$$

$$\bullet \textit{or there is a non-null } s \in \partial^\infty f_i(x) \textit{ for all } i \in I. \tag{11}$$

COMMENTS

- Assumption ($\mathcal{H}_1$) means the functions f_i ($i \in I$) are "totally ordered" pointwise.

- Clearly $(\mathcal{H}_2)$ is equivalent to having (9) with $\varphi = \inf_{i \in I} f_i$.
  $(\mathcal{H}_2)$ is nothing more than a "cone property from below" on φ at x, a mere expression of the "outside cone property" (H_2) of Theorem 1.1 for $epi\varphi$ at $(x, \varphi(x))$. This "outside cone property" for $epi\varphi$ at $(x, \varphi(x))$ can be expressed in various ways in terms of the behaviour of φ around x; the one given in (9) appears the most palpable to us.
- When one of the f_i turns out to be Lipschitz around x, then clearly it is (10) which holds true (still under the cone condition $(\mathcal{H}_2)$).
- If there is φ, Lipschitz around x with $\varphi(x) = c$, minorizing all the f_i, then $(\mathcal{H}_2)$ is secured and, again, it is (10) which holds true (since we could include φ in the collection). In particular one retrieves the following known result: if $f \geq \varphi$ on a neighbourhood of x, $f(x) = \varphi(x)$, and φ is Lipschitz around x, then $\partial f(x)$ is nonempty.

EXAMPLE 2.2. Let f_1 and f_2 be defined on $\mathbb{R}$ as follows:

$$f_1(x) = x^{\frac{1}{3}} \quad \text{for all} \quad x \in \mathbb{R},$$
$$f_2(x) = 0 \quad \text{if } x \leq 0, +\infty \text{ if not.}$$

Then f_1 does satisfy $(\mathcal{H}_2)$ at $x = 0$, while $\partial f_1(0)$ is empty. In this case, it is (10) which occurs: $\partial^\infty f_1(0) \cap \partial^\infty f_2(0) = \mathbb{R}_+$.

EXAMPLE 2.3. Let f_1 and f_2 be defined on $\mathbb{R}$ as follows:

$$f_1(x) = - \mid x \mid^{\frac{1}{2}} \quad \text{for all} \quad x \in \mathbb{R},$$
$$f_2(x) = 0 \quad \text{for all} \quad x \in \mathbb{R}.$$

Then there is no way to secure $(\mathcal{H}_2)$ at $x = 0$; indeed neither (10) nor (11) holds true here.

PROOF OF THEOREM 2.1 We apply Theorem 1.1 to the collection of subsets $X_i := epif_i$, $i \in I$, of $E \times \mathbb{R}$ at the point (x, c). Assumption $(\mathcal{H}_1)$ ensures (H_1), while $(\mathcal{H}_2)$ secures a local version of (H_2). Hence there is some non-null $p = (\sigma, r) \in E^\star \times \mathbb{R}$ in $\bigcap_{i \in I} N_{epif_i}(x, c)$. Since the involved sets X_i are epigraphs of functions, $r \leq 0$ for any $p = (\sigma, r)$ in $N_{epif_i}(x, c)$. Two eventualities then arise from the result above:

- either $r < 0$, whence $(\frac{\sigma}{-r}, -1)$ is still in $N_{epif_i}(x, c)$; as a result,
$$s := \frac{\sigma}{-r} \in \bigcap_{i \in I} \partial f_i(x).$$
- or $r = 0$, whence $\sigma \neq 0$ since $p \neq (0, 0)$; as a result,
$$0 \neq \sigma \in \bigcap_{i \in I} \partial^\infty f_i(x). \Box$$

COROLLARY 2.4. *Let $\{f_i\}_{i \geq 1}$ be a countable family of locally Lipschitz functions from E into $\mathbb{R}$, let x be a point at which all the f_i coincide. We assume that*

on a neighbourhood of x, $f_i \leq f_{i+1}$ for all $i \geq 1$.

Then:

$$\bigcap_{i=1}^{\infty} \partial f_i(x) \neq \emptyset. \tag{12}$$

3. Getting a common element to Michel-Penot generalized subdifferentials of ordered functions

When f admits a Gâteaux differential at x, denoted $D^G f(x)$, Clarke's generalized subdifferential of f at x may contain other elements than $D^G f(x)$. Still, a "squeeze theorem" holds true for Gâteaux-differentiable functions: if $f_1, f_2, \dots, f_m$ are Gâteaux-differentiable at x, and all the f_i coincide at x, and if $f_1 \leq f_2 \leq \dots \leq f_m$, then $D^G f_1(x) = D^G f_2(x) = \dots = D^G f_m(x)$ clearly.
There is an alternative to Clarke directional derivative and subdifferential, due to Michel-Penot ([**8, 9**]). The generalized directional derivative, in the sense of Michel-Penot, of the function φ at the point x is the sublinear lower-semicontinuous function

$$d \in E \longmapsto \varphi^\diamond(x, d) := \sup_{v\in E} \limsup_{t\to 0^+} \frac{\varphi(x + tv + td) - \varphi(x + td)}{t},$$

and the Michel-Penot' generalized subdifferential of φ at x is the set

$$\partial^{mp}\varphi(x) := \{s \in E^\star : \langle s, d\rangle \leq \varphi^\diamond(x, d) \text{ for all } d \text{ in } E\}.$$

If φ is Gâteaux-differentiable at x, then $\partial^{mp}\varphi(x) = \{D^G f(x)\}$, but on the other hand the set-valued mapping $\partial^{mp}\varphi(.)$ does not enjoy the various continuity properties of $\partial^{cl}\varphi(.)$.

We prove below an example of "squeeze theorem" for finitely many locally Lipschitz function f_i in terms of the $\partial^{mp} f_i(x)$. The way of proving it is an adaptation of techniques used in proving Theorem 1.1: this is a "functional" version of the "geometrical" approach seen in the proof of Theorem 1.1.

THEOREM 3.1. *Let $f_1, f_2, \dots, f_m$ be coincide at x. We assume that all the f_i are Lipschitz around x and satisfy $f_1 \leq f_2 \leq \dots \leq f_m$ in a neighbourhood of x. Then:*

$$\bigcap_{i=1}^{m} \partial^{mp} f_i(x) \neq \emptyset. \tag{13}$$

PROOF OF THEOREM 3.1 The notations are simpler in the proof if we assume instead $f_1 \geq f_2 \geq \dots \geq f_m$ in a neighbourhood of x.
Let $K_i := epi f_i^\diamond(x, .)$ for all $i = 1, 2, \dots, m$. Since f_i is assumed Lipschitz around x, $f_i^\diamond(x, .)$ is finite-valued. Consequently, the polar cone K_i° of K_i has no "horizontal half-line": $s = 0$ whenever $(s, 0) \in K_i^\circ$.

We want to prove that $\bigcap_{i=1}^{m} K_i^\circ$ does not reduce to $\{(0, 0)\}$ (in $E^\star \times \mathbb{R}$). We prove it ab absurdo.

Suppose $\bigcap_{i=1}^{m} K_i^\circ = \{(0, 0)\}$. Then $\left(\bigcap_{i=1}^{m} K_i^\circ\right)^\circ = E \times \mathbb{R}$, but

$$\left(\bigcap_{i=1}^{m} K_i^\circ\right)^\circ = \overline{\text{conv}}\left(\bigcup_{i=1}^{m} K_i\right) = \overline{K_1 + K_2 + \dots + K_m} \tag{14}$$

(cf. the "bipolarization lemma" in the proof of Theorem 1.1). Since $K_1 + K_2 + \dots + K_m$ is a convex cone with a nonempty interior, one infers from (14):

$$K_1 + K_2 + \dots + K_m = E \times \mathbb{R}. \tag{15}$$

As a particular choice, $(0,-1) \in K_1+K_2+\cdots+K_m$. Indeed there exist $d_1, d_2, \ldots, d_m$ in E such that:

$$d_1 + d_2 + \cdots + d_m = 0. \tag{16}$$

$$f_1^{\diamond}(x, d_1) + f_2^{\diamond}(x, d_2) + \cdots + f_m^{\diamond}(x, d_m) \leq -1. \tag{17}$$

For $t > 0$, let $e_0(t) := x$, and for $k = 1, 2, \ldots, m$

$$e_k(t) := x + t(d_1 + d_2 + \cdots + d_k) = e_{k-1}(t) + td_k.$$

Therefore:

$$\begin{aligned}
&\limsup_{t\to 0^+} \sum_{k=1}^{m} \frac{f_k(e_k(t)) - f_k(e_{k-1}(t))}{t} \\
&\quad\leq \sum_{k=1}^{m} \limsup_{t\to 0^+} \frac{f_k(e_k(t)) - f_k(e_{k-1}(t))}{t} \\
&\quad\leq \sum_{k=1}^{m} f_k^{\diamond}(x, d_k) \quad \text{by definition of } f_k^{\diamond}(x, d_k), \\
&\quad\leq -1 \quad \text{from (17).}
\end{aligned}$$

In the first expression of this string of inequalities, we have

$$\begin{aligned}
&[f_1(e_1(t)) - f_1(e_0(t))] + [f_2(e_2(t)) - f_2(e_1(t))] + \\
&\qquad \cdots + [f_m(e_m(t)) - f_m(e_{m-1}(t))],
\end{aligned}$$

which equals

$$\begin{aligned}
&-f_1(e_0(t)) + [f_1(e_1(t)) - f_2(e_1(t))] + [f_2(e_2(t)) - f_3(e_2(t))] + \\
&\qquad \cdots + [f_{m-1}(e_{m-1}(t)) - f_m(e_{m-1}(t))] + f_m(e_m(t)),
\end{aligned}$$

a sum of non-negative quantities since $f_k \geq f_{k+1}$ in a neighbourhood of x, plus $-f_1(e_0(t)) + f_m(e_m(t))$ which also is non-negative since $e_m(t) = e_0(t) + (d_1 + d_2 + \cdots + d_m)$ and $d_1 + d_2 + \cdots + d_m = 0$ (cf. (16)). We therefore have a contradiction. □

In some situations we have even more than (13). Let $\mathcal{S}_n(\mathbb{R})$ denote the Euclidean space of $n \times n$ real symmetric matrices, where the inner product of two matrices A and B in $\mathcal{S}_n(\mathbb{R})$ is $\langle A, B\rangle := \operatorname{tr}(AB)$. For $k = 1, 2, \ldots, n$, we write the k-th largest eigenvalue of A (counted with multiplicity) as $\lambda_k(A)$. Indeed, $\lambda_1 \geq \lambda_2 \geq \cdots \geq \lambda_n$ on $\mathcal{S}_n(\mathbb{R})$. Associated with $A \in \mathcal{S}_n(\mathbb{R})$ and a fixed index k are two other indices, $\underline{k}$ and $\overline{k}$, defined as follows:

$$\underline{k} := \min\{i : \lambda_i(A) = \lambda_k(A)\} \text{ and } \overline{k} := \max\{i : \lambda_i(A) = \lambda_k(A)\}.$$

We are in a situation of the "squeeze theorem" since:

$$\begin{aligned}
&\lambda_1 \geq \lambda_2 \geq \cdots \geq \lambda_n, \\
&\lambda_{\overline{k}}(A) = \cdots = \lambda_k(A) = \cdots = \lambda_{\underline{k}}(A).
\end{aligned}$$

Actually we have a much more precise result ([**6**, Theorem 5.3]): *the Clarke and Michel-Penot generalized subdifferential of λ_k at A are equal; these subdifferentials all coincide for k varying from $\underline{k}$ to $\overline{k}$.*

The eigenvalue functions λ_k, although Lipschitz on $\mathcal{S}_n(\mathbb{R})$, are very "bumpy" around a given A, they actually capture most of the nastinesses of a nonsmooth function;

the result recalled above made us believe that the question posed by Lewis and Lucchetti had good chances to be answered positively.

4. A "squeeze theorem" in terms of Clarke's generalized jacobian

Let $f_1, f_2, \ldots, f_m$ be real-valued locally Lipschitz functions on $\mathbb{R}^n$; we denote as $F = (f_1, f_2, \ldots, f_m)^\top : \mathbb{R}^n \to \mathbb{R}^m$ the vector-valued locally Lipschitz function on $\mathbb{R}^n$ whose component functions are $f_1, f_2, \ldots, f_m$. Instead of considering the generalized subdifferentials of the f_i separately, there is a way of handling all the f_i together: this is Clarke's generalized Jacobian. The so-called *the Clarke generalized Jacobian set* of F at x is the set of matrices (in $\mathcal{M}_{m,n}(\mathbb{R})$) defined as follows:

$$\mathcal{J}F(x) := \operatorname{conv}\{\lim JF(x_k) : x_k \to x, x_k \in D_F\}, \tag{18}$$

where D_F stands for the set of points where F is differentiable and $JF(x_k)$ the Jacobian matrix of F at x_k. This mathematical object is more precise than

$$\begin{aligned} \otimes_{i=1}^m \partial f_i(x) := \{ & M \in \mathcal{M}_{m,n}(\mathbb{R}) : \text{ the } j-\text{th line vector of } M \text{ is} \\ & \text{taken from } \partial f_j(x), \text{ for all } j = 1, 2, \ldots, m\}, \end{aligned} \tag{19}$$

since it takes into account the possible interdependence of the component functions f_i. For example, if

$$F : x \in \mathbb{R} \longmapsto F(x) := (\mid x \mid, - \mid x \mid)^\top (\in \mathbb{R}^2),$$

then

$$\mathcal{J}F(0) = \left\{ \begin{bmatrix} \alpha \\ -\alpha \end{bmatrix} : -1 \leq \alpha \leq +1 \right\},$$

while

$$\partial f_1(0) \otimes \partial f_2(0) = \left\{ \begin{bmatrix} \alpha \\ \beta \end{bmatrix} : -1 \leq \alpha \leq +1 \text{ and } -1 \leq \beta \leq +1 \right\}.$$

If $F = (f, g)^\top$ is Lipschitz around a point x where $f(x) = g(x)$, and if $f \leq g$ in a neighbourhood of x, there then exists *a matrix in $\mathcal{J}F(x)$ whose two line vectors are equal.* This is again due to the "variational approach" (cf. beginning of Section 2) which yields $0 \in \partial(g - f)(x)$, and the following exact calculus rule ([**3**, Theorem 2.6.6]):

$$\partial(g - f)(x) = \left\{ \ell_2 - \ell_1 : \begin{bmatrix} \ell_1 \\ \ell_2 \end{bmatrix} \in \mathcal{J}F(x) \right\}. \tag{20}$$

It is tempting to generalize by posing the following question: *if $f_1 \leq f_2 \leq \cdots \leq f_m$ and if $f_1(x) = f_2(x) = \cdots = f_m(x)$, is that true there exists a matrix M in the generalized Jacobian set of $F := (f_1, f_2, \ldots, f_m)^\top$ at x whose m line vectors are equal?*

We thought for some time that this "squeeze theorem" could be true; this is not the case for $m \geq 3$ as the next example shows it.

EXAMPLE 4.1. Let $F = (f_1, f_2, f_3)^\top : \mathbb{R} \to \mathbb{R}^3$ defined by

$$F(x) := (- \mid x \mid, x, \mid x \mid)^\top.$$

Then

$$\mathcal{J}F(0) = \left\{ \begin{bmatrix} -\beta \\ 1 \\ \beta \end{bmatrix} : -1 \leq \beta \leq +1 \right\},$$

There is no matrix in $\mathcal{J}F(0)$ with equal line vectors (while

$$\partial f_1(0) \cap \partial f_2(0) \cap \partial f_3(0) = \{1\}).$$

What we actually get in terms of the generalized Jacobian set $\mathcal{J}F(x)$ is a "squeeze theorem" of a different kind.

THEOREM 4.2. *Let $F = (f_1, f_2, \ldots, f_m)^\top$ be Lipschitz around x. Suppose that $f_1 \leq f_2 \leq \cdots \leq f_m$ in a neighbourhood of x and also $f_1(x) = f_2(x) = \cdots = f_m(x)$. Then, for all $v = (v_1, v_2, \ldots, v_m)$ satisfying*

$$v_1 \geq 0, v_1 + v_2 \geq 0, \ldots, v_1 + v_2 + \cdots + v_{m-1} \geq 0, v_1 + v_2 + \cdots + v_m = 0, \tag{21}$$

there exists $M_v = \begin{bmatrix} \cdots & \ell_1 & \cdots \\ \cdots & \ell_2 & \cdots \\ & \vdots & \\ \cdots & \ell_m & \cdots \end{bmatrix}$ *in $\mathcal{J}F(x)$ such that*

$$v_1\ell_1 + v_2\ell_2 + \cdots + v_m\ell_m = 0. \tag{22}$$

EXAMPLE 4.1 (revisited). Let a non-null $v = (v_1, v_2, v_3)$ satisfy (21); then necessarily either $v_1 > 0$ or $v_1 + v_2 > 0$. Then the matrix

$$M_v := \begin{bmatrix} -\beta \\ 1 \\ \beta \end{bmatrix}, \text{ with } \beta := \frac{v_2}{2v_1 + v_2},$$

lies in $\mathcal{J}F(0)$ and its line vectors do satisfy (22).

PROOF OF THEOREM 4.2. Let

$$K := \{(y_1, y_2, \cdots, y_m) \in \mathbb{R}^m : y_1 \leq y_2 \leq \cdots \leq y_m\}.$$

By assumption, there is a neighbourhood V of x such that:

$$F(x) \in K \text{ for all } x \in V, \tag{23}$$

this is a way of expressing that $f_1, f_2, \cdots, f_m$ are ordered in a neighbourhood of x. K is a closed convex cone whose polar cone K° is known (see [**5**, pp. 212-213] for example):

$$\begin{aligned} K^\circ = \{ & v = (v_1, v_2, \cdots, v_m) \in \mathbb{R}^m : v_1 \geq 0, v_1 + v_2 \geq 0, \cdots, \\ & v_1 + v_2 + \cdots + v_{m-1} \geq 0 \text{ and } v_1 + v_2 + \cdots + v_m = 0\}. \end{aligned}$$

Hence, since $(K^\circ)^\circ = K$, (23) can also be expressed as:

$$\langle F(x), v \rangle \leq 0 \text{ for all } x \in V \text{ and for all } v \in K^\circ, \tag{24}$$

where $\langle .,. \rangle$ denotes the usual inner product in $\mathbb{R}^m$.
Moreover, since $v_1 + v_2 + \cdots + v_m = 0$ for the $v = (v_1, v_2, \cdots, v_m)$ in K° and $F(\overline{x}) = (f_1(\overline{x}), f_2(\overline{x}), \cdots, f_m(\overline{x}))^\top$ has equal component values, $\langle F(\overline{x}), v \rangle = 0$. As a result we have the following "variational situation":

$$\text{For all } v \text{ in } K^\circ, \ \langle F(x), v \rangle \leq \langle F(\overline{x}), v \rangle \text{ in a neighbourhood of } x. \tag{25}$$

It then remains to exploit that. For $v \in K^\circ$, let θ_v be the real-valued function $x \mapsto \theta_v(x) := \langle F(x), v \rangle$. The result (25) above asserts that $\overline{x}$ is an unconstrained local maximizer of θ_v. Hence $0 \in \partial\theta_v(\overline{x})$, and the announced result follows from the exact calculus rule alluded to earlier ([**3**, Theorem 2.6.6]):

$$\partial\theta_v(\overline{x}) = \left\{ \sum_{i=1}^{m} v_i \ell_i : \begin{bmatrix} \cdots & \ell_1 & \cdots \\ \cdots & \ell_2 & \cdots \\ & \vdots & \\ \cdots & \ell_m & \cdots \end{bmatrix} \in \mathcal{J}F(\overline{x}) \right\}. \square$$

References

[1] BONNISSEAU, J.M., *Equilibre général dans une économie avec ensembles de production non convexes*, Actes des sixièmes journées du groupe MODE de la SMAI 1998, Editions de l'Actualité scientifique Poitou-Charentes, pp. 38-59, 1999.

[2] BORWEIN, J.M., FITZPATRICK, S.P., *Duality inequalities and sandwiched functions*, Preprint, Simon Fraser University at Burnaby, Canada, 1998.

[3] CLARKE, F.H., *Optimization and Nonsmooth analysis*, J. Wiley and Sons, New York, 1983. Reprinted as Vol. 5 of the series Classics in Applied Mathematics of SIAM, Philadelphia, 1990.

[4] CORNET, B., ROCKAFELLAR, R.T., *Separation theorems and supporting price theorems for nonconvex sets*, Unpublished manuscript, 1990.

[5] HIRIART-URRUTY, J.-B., *Optimisation et Analyse convexe (Exercices et problèmes corrigés)*, Collection Mathématiques, Presses Universitaires de France, 1998.

[6] HIRIART-URRUTY, J.-B., LEWIS , A.S., *The Clarke and Michel-Penot subdifferentials of the eigenvalues of a symmetric matrix*, Computational Optimization and Applications 13, pp. 13-23, 1999.

[7] LEWIS, A.S., LUCCHETTI, R.E., *Nonsmooth duality, sandwich and squeeze theorems*, to appear in SIAM J. Control and Optimization.

[8] MICHEL, P., PENOT, J.-P., *Calcul sous-différentiel pour les fonctions Lipschitziennes et non Lipschitziennes*, C. R. Acad. Sci. Paris, Vol. 298, pp. 269-272, 1984.

[9] MICHEL, P., PENOT, J.-P., *A generalized derivative for calm and stable functions*, Differential and Integral Equations, Vol. 5, pp. 433-454, 1992.

LACO, URA-CNRS 1586, FACULTÉ DES SCIENCES, UNIVERSITÉ DE LIMOGES, FRANCE
E-mail address: `benoist@unilim.fr`

MIP, UMR 5640, U.F.R. MATHÉMATIQUES, INFORMATIQUE, GESTION, UNIVERSITÉ PAUL SABATIER, TOULOUSE, FRANCE
E-mail address: `jbhu@cict.fr`

Canadian Mathematical Society
Conference Proceedings
Volume **27**, 2000

Ramanujan's Short Unpublished Manuscript on Integrals and Series Related to Euler's constant

Bruce C. Berndt[1] and Douglas C. Bowman[2]

ABSTRACT. Published with is his lost notebook is a fragment, less than two pages long, of a paper which Ramanujan began to write on Euler's constant and integrals of Frullanian type. We prove the theorems in this fragment as well as a few new related theorems.

1. Introduction

Included in the volume [**15**] featuring his lost notebook are some published and unpublished fragments of papers by Ramanujan. In particular, on pages 274 and 275 in [**15**], there is the beginning of a manuscript that probably was to focus on integrals related to Euler's constant γ and $\psi(s) := \Gamma'(s)/\Gamma(s)$, and to integrals and series related to Frullani's integral theorem [**3**, p. 313, eq. (2.15)], [**9**]. This fragment contains only two short sections comprising $1\frac{1}{2}$ pages. Afterwards, Ramanujan wrote "3." to indicate the beginning of a third section, but the manuscript ends abruptly at this point.

The purpose of this brief note is to present Ramanujan's work in this fragment and to relate it to other theorems in the literature. We also offer a couple new related theorems.

2. Ramanujan's Theorems on γ and $\psi(s)$

We first prove the primary theorem in Ramanujan's first section.

PROPOSITION 2.1 (Ramanujan). *Let p, q, and r be positive. Then*

$$\int_0^1 \left(\frac{x^{p-1}}{1-x} - \frac{rx^{q-1}}{1-x^r}\right) dx = \psi(q/r) - \psi(p) + \log r. \tag{2.1}$$

PROOF. (Ramanujan) Using the continuity of the integrand on the right side below for $0 \le x, s \le 1$, a well-known integral representation for the beta function,

1991 *Mathematics Subject Classification.* 40A99, 33B15.

Key words and phrases. Euler's constant, Frullani integral, gamma function, logarithmic derivative of the gamma function, Ramanujan's lost notebook.

[1]This research was conducted while the author was a Guggenheim Fellow.

[2]Research partially supported by NSF grant DMS 9705782.

the change of variable $t = x^r$ in the second part of the integrand, and L'Hospital's rule, we find that

$$\begin{aligned}
&\int_0^1 \left(\frac{x^{p-1}}{1-x} - \frac{rx^{q-1}}{1-x^r}\right) dx \\
&= \lim_{s\to 0+} \int_0^1 \left\{x^{p-1}(1-x)^{s-1} - r^{1-s}x^{q-1}(1-x^r)^{s-1}\right\} dx \\
&= \lim_{s\to 0} \left\{\frac{\Gamma(p)\Gamma(s)}{\Gamma(s+p)} - r^{-s}\frac{\Gamma(q/r)\Gamma(s)}{\Gamma(s+q/r)}\right\} \\
&= \lim_{s\to 0} \frac{\left\{\dfrac{\Gamma(p)}{\Gamma(s+p)} - r^{-s}\dfrac{\Gamma(q/r)}{\Gamma(s+q/r)}\right\}\Gamma(s+1)}{s} \\
&= \lim_{s\to 0} \left\{-\frac{\Gamma(p)\Gamma'(s+p)}{\Gamma^2(s+p)} + \Gamma(q/r)\left(\frac{r^{-s}\log r}{\Gamma(s+q/r)} + \frac{r^{-s}\Gamma'(s+q/r)}{\Gamma^2(s+q/r)}\right)\right\} \\
&= -\,\psi(p) + \log r + \psi(q/r),
\end{aligned}$$

which completes the proof. □

THEOREM 2.2 (Ramanujan). *Suppose that a, b, and c are positive with $b > 1$. Then*

$$\int_0^1 \left(\frac{x^{c-1}}{1-x} - \frac{bx^{bc-1}}{1-x^b}\right) \sum_{k=0}^{\infty} x^{ab^k}\, dx = \psi\left(\frac{a}{b} + c\right) - \log\frac{a}{b}.$$

PROOF. By Proposition 2.1 and the facts that $b > 1$ and $\psi(x) \sim \log x$, as x tends to ∞,

$$\begin{aligned}
\int_0^1 \left(\frac{x^{c-1}}{1-x} - \frac{bx^{bc-1}}{1-x^b}\right) \sum_{k=0}^{n} x^{ab^k}\, dx &= \sum_{k=0}^{n} \int_0^1 \left(\frac{x^{c+ab^k-1}}{1-x} - \frac{bx^{bc+ab^k-1}}{1-x^b}\right) dx \\
&= \sum_{k=0}^{n} \left(\psi\left(ab^{k-1}+c\right) - \psi\left(ab^k+c\right) + \log b\right) \\
&= \psi\left(\frac{a}{b}+c\right) - \psi\left(ab^n+c\right) + (n+1)\log b \\
&= \psi\left(\frac{a}{b}+c\right) - \log\left(ab^n+c\right) + (n+1)\log b + o(1) \\
&= \psi\left(\frac{a}{b}+c\right) - n\log b - \log a + (n+1)\log b + o(1) \\
&= \psi\left(\frac{a}{b}+c\right) - \log\frac{a}{b} + o(1),
\end{aligned}$$

as n tends to ∞. Letting $n \to \infty$, we complete the proof. □

COROLLARY 2.3 (Ramanujan). *We have*

(a) $$\int_0^1 \frac{1}{1+x} \sum_{k=1}^{\infty} x^{2^k}\, dx = 1-\gamma,$$

(b) $$\int_0^1 \frac{1+2x}{1+x+x^2} \sum_{k=1}^{\infty} x^{3^k}\, dx = 1-\gamma,$$

(c) $$\int_0^1 \frac{1+\frac{1}{2}\sqrt{x}}{(1+\sqrt{x})(1+\sqrt{x}+x)} \sum_{k=1}^{\infty} x^{(3/2)^k}\, dx = 1-\gamma.$$

PROOF. In Theorem 2.2, set, respectively, $c = 1, a = b = 2; c = 1, a = b = 3;$ and $c = 1, a = b = 3/2$. Use the fact that [**8**, p. 954]

$$\psi(2) = 1 - \gamma \tag{2.2}$$

to complete the proof. □

According to Bromwich [**5**, p. 526], (a) is due to Catalan. Parts (b) and (c) may be new.

Before discussing the very brief second section of Ramanujan's fragment, we offer some alternative proofs, references, and connections with further work of Ramanujan, as well as others.

LEMMA 2.4. *For $x > 0$ and any integer $n > 1$,*

$$\frac{1}{\log x} + \frac{1}{1-x} = \sum_{k=1}^{\infty} \frac{(n-1) + (n-2)x^{1/n^k} + (n-3)x^{2/n^k} + \cdots + x^{(n-2)/n^k}}{n^k(1 + x^{1/n^k} + x^{2/n^k} + \cdots + x^{(n-1)/n^k})}. \tag{2.3}$$

PROOF. It is easy to verify that

$$\frac{1}{1-x^n} = \frac{1}{n}\left(\frac{(n-1) + (n-2)x + (n-3)x^2 + \cdots + x^{n-2}}{1 + x + x^2 + \cdots + x^{n-1}} + \frac{1}{1-x}\right). \tag{2.4}$$

Replacing x by $x^{1/n}$ and iterating m times, we find that

$$\begin{aligned}\frac{1}{1-x} = &\sum_{k=1}^{m} \frac{(n-1) + (n-2)x^{1/n^k} + (n-3)x^{2/n^k} + \cdots + x^{(n-2)/n^k}}{n^k(1 + x^{1/n^k} + x^{2/n^k} + \cdots + x^{(n-1)/n^k})} \\ &+ \frac{1}{n^m(1 - x^{1/n^m})}.\end{aligned}$$

If we now let m tend to ∞ and apply L'Hospital's rule, we complete the proof. □

The special cases $n = 2, 3$ of Lemma 2.4 can be found in Ramanujan's third notebook [**13**, p. 364], and proofs can be found in Berndt's book [**4**, pp. 399–400]. Our proof generalizes these proofs.

LEMMA 2.5. *For every integer $n > 1$,*

$$\gamma = \int_0^1 \left(\frac{n}{1-x^n} - \frac{1}{1-x}\right) \sum_{k=1}^{\infty} x^{n^k - 1} dx. \tag{2.5}$$

PROOF. Integrate (2.3) over $0 \le x \le 1$ and employ the well-known integral representation [**5**, p. 507], [**8**, p. 955]

$$\gamma = \int_0^1 \left(\frac{1}{\log x} + \frac{1}{1-x}\right) dx.$$

Accordingly, replacing x by x^{n^k}, we find that

$$\begin{aligned}\gamma &= \int_0^1 \sum_{k=1}^{\infty} \frac{(n-1)+(n-2)x^{1/n^k}+(n-3)x^{2/n^k}+\cdots+x^{(n-2)/n^k}}{n^k(1+x^{1/n^k}+x^{2/n^k}+\cdots+x^{(n-1)/n^k})}dx \\ &= \sum_{k=1}^{\infty} \int_0^1 \frac{1}{n^k} \frac{(n-1)+(n-2)x^{1/n^k}+(n-3)x^{2/n^k}+\cdots+x^{(n-2)/n^k}}{1+x^{1/n^k}+x^{2/n^k}+\cdots+x^{(n-1)/n^k}}dx \\ &= \sum_{k=1}^{\infty} \int_0^1 \frac{(n-1)+(n-2)x+(n-3)x^2+\cdots+x^{n-2}}{1+x+x^2+\cdots+x^{n-1}} x^{n^k-1}dx \end{aligned}$$

$$= \int_0^1 \left(\frac{n}{1-x^n} - \frac{1}{1-x}\right) \sum_{k=1}^{\infty} x^{n^k-1}dx, \tag{2.6}$$

by (2.4). This completes the proof. □

Lemma 2.5 is equivalent to Theorem 2.2 in the case $c = 1, a = b = n$. To see this, first make these substitutions in Theorem 2.2 and use (2.2) to deduce that

$$1 - \gamma = \int_0^1 \left(\frac{1}{1-x} - \frac{nx^{n-1}}{1-x^n}\right) \sum_{k=1}^{\infty} x^{n^k} dx. \tag{2.7}$$

Adding (2.5) and (2.7) and simplifying, we readily find that

$$1 = (n-1) \int_0^1 \sum_{k=1}^{\infty} x^{n^k-1}dx,$$

which is trivially verified by termwise integration.

The arguments in the proof of Lemma 2.5 lead to another formula for γ.

THEOREM 2.6. *If b is an integer exceeding 1, let*

$$\epsilon_r = \begin{cases} b-1, & \text{if } b|r, \\ -1, & \text{if } b \nmid r. \end{cases} \tag{2.8}$$

Then

$$\gamma = \sum_{r=1}^{\infty} \frac{\epsilon_r}{r} \left[\frac{\log r}{\log b}\right],$$

where $[x]$ denotes the greatest integer $\leq x$.

PROOF. Return to the last equality in (2.6), with $n = b$, and rewrite it in the form

$$\gamma = \int_0^1 \sum_{k=1}^{\infty} \frac{b-1-x-x^2-\cdots-x^{b-1}}{1-x^b} x^{b^k-1}dx.$$

Now expand $1/(1-x^b)$ in a geometric series and integrate termwise to deduce that

$$\begin{aligned}\gamma &= \sum_{k=1}^{\infty}\sum_{j=0}^{\infty} \int_0^1 (b-1-x-x^2-\cdots-x^{b-1})x^{b^k+bj-1}dx \\ &= \sum_{k=1}^{\infty}\sum_{j=0}^{\infty} \left(\frac{b-1}{b^k+bj} - \frac{1}{b^k+bj+1} - \frac{1}{b^k+bj+2} - \cdots - \frac{1}{b^k+bj+(b-1)}\right).\end{aligned}$$

Now count the number of appearances of each harmonic term $1/r, 1 \leq r < \infty$. If r is a multiple of b, then the coefficient $b-1$ appears $[\log r/\log b]$ times, while if r is not a multiple of b, the coefficient -1 also appears $[\log r/\log b]$ times, and so the proof is complete. □

COROLLARY 2.7. *We have*

$$\gamma = \sum_{r=1}^{\infty} \frac{(-1)^r}{r} \left[\frac{\log r}{\log 2} \right]. \tag{2.9}$$

PROOF. Let $b = 2$ in Theorem 2.6. □

The representation for γ given in (2.9) was discovered in 1909 by G. Vacca [**17**] and is known as Dr. Vacca's series for γ. Corollary 2.7 was rediscovered by H. F. Sandham, who submitted it as a problem [**16**]. M. Koecher [**11**] obtained a generalization of (2.9) that includes a formula for γ submitted by Ramanujan as a problem [**12**], [**14**, p. 325] to the *Journal of the Indian Mathematical Society*, and found in his notebooks [**13**], [**3**, p. 196]. Further series in the spirit of those of Ramanujan and Koecher were found by F. L. Bauer [**2**]. A result similar to that of Bauer was found by A. W. Addison [**1**], with a simpler version later established by I. Gerst [**6**].

J. W. L. Glaisher [**7**] generalized Theorem 2.6. We give below a proof of a theorem which is apparently equivalent to his theorem. We again use Theorem 2.2, but we use a different approach from that in the proof of Theorem 2.6.

THEOREM 2.8. *Let a and b be positive integers with $b > 1$, and let ϵ_r be defined by (2.8). Then*

$$\log a + \gamma - \sum_{n=1}^{a-1} \frac{1}{n} = \sum_{r=a}^{\infty} \frac{\epsilon_r}{r} \left[\frac{\log(r/a)}{\log b} \right].$$

PROOF. From Theorem 2.2,

(2.10)
$$\int_0^1 \left\{ b\left(1 + \frac{x^{bc}}{1-x^b}\right) - 1 - \frac{x^c}{1-x} \right\} \sum_{k=0}^{\infty} x^{ab^k - 1} dx = \frac{b}{a} + \log\frac{a}{b} - \psi\left(\frac{a}{b} + c\right).$$

(That (2.10) is equivalent to Theorem 2.2 can be easily demonstrated by adding (2.10) to the equality of Theorem 2.2.) Setting $c = 1$ in (2.10) and recalling that

ϵ_j is defined by (2.8), we find that

$$\begin{aligned}
\frac{b}{a}+\log\frac{a}{b}-\psi\left(\frac{a}{b}+1\right) &= \int_0^1\left(\sum_{k=0}^{\infty}x^{ab^k-1}\right)\left(\sum_{j=0}^{\infty}\epsilon_j x^j\right)dx \\
&= \int_0^1 \sum_{n=a-1}^{\infty}\left(\sum_{\substack{ab^k-1+j=n\\ k,j\geq 0}}\epsilon_j\right)x^n dx \\
&= \int_0^1 \sum_{n=a}^{\infty}\left(\sum_{k=0}^{\left[\frac{\log(n/a)}{\log b}\right]}\epsilon_{n-ab^k}\right)x^{n-1}dx \\
&= \int_0^1 \sum_{n=a}^{\infty}\left(\epsilon_{n-a}+\sum_{k=1}^{\left[\frac{\log(n/a)}{\log b}\right]}\epsilon_n\right)x^{n-1}dx \\
&= \sum_{n=a}^{\infty}\frac{1}{n}\left(\epsilon_{n-a}+\left[\frac{\log(n/a)}{\log b}\right]\epsilon_n\right), \qquad (2.11)
\end{aligned}$$

where in the penultimate step we used the fact that $\epsilon_i=\epsilon_j$ if $i\equiv j \pmod b$. It remains to evaluate the sum $\sum_{n\geq a}\epsilon_{n-a}/n$. In the evaluation below, we use three familiar properties of $\psi(z)$ [**8**, formulas 8.362, no. 1, p. 952; 8.365, no. 6, p. 954; 8.366, no. 1, p. 954]. Accordingly, we find that

$$\begin{aligned}
\sum_{n=1}^{a-1}\frac{1}{n}-\sum_{n=a}^{\infty}\frac{\epsilon_{n-a}}{n} &= \sum_{k=0}^{\infty}\sum_{j=1}^{b}\left(\frac{1}{kb+j}-\frac{1}{kb+a}\right) \\
&= \frac{1}{b}\sum_{j=1}^{b}\sum_{k=0}^{\infty}\left\{\left(\frac{1}{k+\frac{j}{b}}-\frac{1}{k+1}\right)+\left(\frac{1}{k+1}-\frac{1}{k+\frac{a}{b}}\right)\right\} \\
&= \frac{1}{b}\sum_{j=1}^{b}\left\{\left(-\psi\left(\frac{j}{b}\right)-\gamma\right)+\left(\psi\left(\frac{a}{b}\right)+\gamma\right)\right\} \\
&= \frac{1}{b}\sum_{j=1}^{b}\left(\psi\left(\frac{a}{b}\right)-\psi\left(\frac{j}{b}\right)\right) \\
&= \psi\left(\frac{a}{b}\right)-\frac{1}{b}\sum_{j=0}^{b-1}\psi\left(\frac{j+1}{b}\right) \\
&= \psi\left(\frac{a}{b}\right)-\psi(1)+\log b \\
&= \psi\left(\frac{a}{b}\right)+\gamma+\log b. \qquad (2.12)
\end{aligned}$$

Substituting (2.12) into (2.11) and simplifying slightly, we complete the proof. □

We have found an analogue of Theorem 2.8 in the case that $c>1$. Since the proof again uses Theorem 2.2, and since our result is not as elegant as Theorem 2.8, we state the theorem but do not prove it.

THEOREM 2.9. *Let a, b, and c be positive integers with $b > 1$. Define*

$$\epsilon_{r,c} = \begin{cases} 1-b, & \text{if } r = b\ell, \quad \ell \geq c, \\ 1, & \text{otherwise.} \end{cases}$$

Then

$$\psi(a+c) - \log a - \sum_{n=a+c+2}^{ab+cb-1} \frac{\left[\frac{\log((n-c)/a)}{\log b}\right]}{n}$$
$$= \sum_{n=ab+cb}^{\infty} \frac{\left[\frac{\log((n-cb)/a)}{\log b}\right](\epsilon_{n,c}-1) + \left[\frac{\log((n-c)/a)}{\log b}\right]}{n}.$$

We complete this section with a remark about Proposition 2.1. After replacing x by e^{-x} in (2.1), we obtain an integral of Frullani-type. In his third quarterly report, Ramanujan found a beautiful generalization of Frullani's theorem. In particular, the formula

$$\int_0^\infty \frac{(1+ax)^{-p} - (1+bx)^{-q}}{x} dx = \psi(q) - \psi(p) + \log \frac{b}{a}, \tag{2.13}$$

where $a, b, p, q > 0$, is a special instance of Ramanujan's theorem [**3**, p. 314]. In view of the right sides of (2.1) and (2.13), one might surmise that (2.1) can be derived from (2.13), or Ramanujan's generalization of Frullani's theorem, but we have been unable to do this.

3. Integral Representations of $\log x$

Section 2 in Ramanujan's unpublished fragment is devoted solely to the statements of the following theorem and (3.1) below.

THEOREM 3.1 (Ramanujan). *If a, b, and c are positive with $b > 1$, then*

$$\int_0^1 \frac{x^{c-1} - x^{bc-1}}{\log x} \sum_{k=0}^{\infty} x^{ab^k} dx = \log\left(1 + \frac{bc}{a}\right).$$

PROOF. As indicated by Ramanujan, we begin with the equality [**8**, p. 575]

$$\int_0^1 \frac{x^{p-1} - x^{q-1}}{\log x} dx = \log \frac{q}{p}, \tag{3.1}$$

where $p, q > 0$. Thus, since $b > 1$,

$$\begin{aligned}\int_0^1 \frac{x^{c-1}-x^{bc-1}}{\log x}\sum_{k=0}^{n} x^{ab^k}\,dx &= \sum_{k=0}^{n}\int_0^1 \frac{x^{c+ab^k-1}-x^{bc+ab^k-1}}{\log x}dx\\ &=\sum_{k=0}^{n}\log\frac{bc+ab^k}{c+ab^k}\\ &=\sum_{k=0}^{n}\left(\log b+\log(c+ab^{k-1})-\log(c+ab^k)\right)\\ &=(n+1)\log b+\log(c+a/b)-\log(c+ab^n)\\ &=(n+1)\log b+\log(c+a/b)-n\log b-\log a+o(1)\\ &=\log(1+bc/a)+o(1),\end{aligned}$$

as n tends to ∞. Letting n tend to ∞, we complete the proof. □

COROLLARY 3.2. *We have*

$$\int_0^1 \frac{1-x}{\log x}\sum_{k=1}^{\infty} x^{2^k} = \log 2.$$

PROOF. Set $c = 1$ and $a = b = 2$ in Theorem 3.1. □

Observe that if x is replaced by e^{-x} in (3.1), we obtain an example of Frullani's integral theorem. Ramanujan's ideas can be extended to other examples of Frullani-type integrals found by, among others, Ramanujan in his Quarterly Reports [**3**] and Hardy [**9**], [**10**, pp. 195–226]. For example, consider the integral [**9**, eq. (29)], [**10**, p. 200]

$$\int_0^\infty \frac{e^{-ax}\cos(\alpha x)-e^{-bx}\cos(\beta x)}{x}dx = -\frac{1}{2}\log\frac{a^2+\alpha^2}{b^2+\beta^2}, \tag{3.2}$$

where $a, b, \alpha, \beta > 0$. We conclude with a brief application of (3.2) which we do not prove, as the proof is similar to those above.

THEOREM 3.3. *If $a, b, c, \alpha > 0$ with $b > 1$, then*

$$\begin{aligned}\int_0^1 &\frac{x^{c-1}\cos(\alpha\log x)-x^{bc-1}\cos(\alpha b\log x)}{\log x}\sum_{k=0}^{\infty} x^{ab^k}\,dx\\ &= -\frac{1}{2}\log\left\{\left(1+\frac{bc}{a}\right)^2+\frac{\alpha^2b^2}{a^2}\right\}.\end{aligned}$$

The authors are grateful to Kenneth S. Williams for finding an error in our original formulation and proof of Theorem 2.8.

References

[1] A. W. Addison, *A series representation for Euler's constant*, Amer. Math. Monthly **74** (1967), 823–824.

[2] F. L. Bauer, *Eine Bemerkung zu Koechers Reihen für die Eulersche Konstante* Sitz. Bayer. Akad. Wiss . Math. –Natur. Kl. (1990), 27–33.

[3] B. C. Berndt, *Ramanujan's Notebooks, Part I*, Springer–Verlag, New York, 1985.

[4] B. C. Berndt, *Ramanujan's Notebooks, Part IV*, Springer–Verlag, New York, 1994.

[5] T. J. I'a. Bromwich, *An Introduction to the Theory of Infinite Series, 2nd ed.*, Macmillan, London, 1926.

[6] I. Gerst, *Some series for Euler's constant*, Amer. Math. Monthly **76** (1969), 273–275.
[7] J. W. L. Glaisher, *On Dr. Vacca's series for* γ, Quart. J. Pure Appl. Math. **41** (1909–10), 365–368.
[8] I. S. Gradshteyn and I. M. Ryzhik, *Table of Integrals, Series, and Products, 5th ed.*, Academic Press, New York, 1994.
[9] G. H. Hardy, *On the Frullanian integral* $\int_0^\infty \{[\varphi(ax^m) - \psi(bx^n)]/x\}(\log x)^p dx$, Quart. J. Math. **33** (1902), 113–144.
[10] G. H. Hardy, *Collected Papers, Vol. V*, Clarendon Press, Oxford, 1972.
[11] M. Koecher, *Einige Bemerkungen zur Eulerschen Konstanten*, Sitz. Bayer. Akad. Wiss. Math.–Natur. Kl. (1990), 9–15.
[12] S. Ramanujan, *Question 327*, J. Indian Math. Soc. **3** (1911), 209.
[13] S. Ramanujan, *Notebooks (2 Volumes)*, Tata Institute of Fundamental Research, Bombay, 1957.
[14] S. Ramanujan, *Collected Papers*, Chelsea, New York, 1962.
[15] S. Ramanujan, *The Lost Notebook and Other Unpublished Papers*, Narosa, New Delhi, 1988.
[16] H. F. Sandham, *Advanced Problem No. 4353; Solution by D. F. Barrow*, Amer. Math. Monthly **58** (1951), 116–117.
[17] G. Vacca, *A new series for the Eulerian constant* $\gamma = \cdot 577\ldots$, Quart. J. Pure Appl. Math. **41** (1909–10), 363–364.

Department of Mathematics, University of Illinois, 1409 W. Green St., Urbana, IL 61801, USA
E-mail address: berndt@math.uiuc.edu
E-mail address: bowman@math.uiuc.edu

Canadian Mathematical Society
Conference Proceedings
Volume **27**, 2000

An Efficient Algorithm for the Riemann Zeta Function

P. Borwein

This paper is for the occassion of the Honorary Doctorate of Dr. Jonathan M. Borwein

ABSTRACT. A very simple class of algorithms for the computation of the Riemann-zeta function to arbitrary precision in arbitrary domains is proposed. These algorithms compete with the standard methods based on Euler-Maclaurin summation, are easier to implement and are easier to analyze.

1. Introduction

We propose some very simple algorithms for the arbitrary precision calculation of the Riemann-zeta function which is the analytic continuation of

$$(1.1)\qquad \zeta(s) := \sum_{n=1}^{\infty} \frac{1}{n^s} \qquad \text{re(s)} > 1$$

$$= \frac{1}{(1-2^{1-s})} \sum_{n=1}^{\infty} \frac{(-1)^{n+1}}{n^s} \qquad \text{re(s)} > 0$$

These algorithms do not compete with the Riemann-Siegel formula based algorithms for computations concerning zeros on the critical line (Im(s) $= 1/2$) where multiple low precision evaluations are required. (See [2,6].) They can however improve on the standard algorithms for arbitrary precision computation of the zeta function in the major symbolic algebra packages (all of Maple, Mathematica and Pari use Euler-Maclaurin based algorithms [2,5,7]). They are easier to implement and far easier to analyse.

2. Algorithms

We commence by presenting the algorithm in generic form and then offer two specializations.

1991 *Mathematics Subject Classification.* Primary 11Y16; Secondary 11M99.
Key words and phrases. Riemann zeta function, computation, high-precision, algorithm .
Research supported by NSERC.

ALGORITHM 1. Let $p_n(x) := \sum_{k=0}^n a_k x^k$ be an arbitrary polynomial of degree n that does not vanish at -1. Let

$$c_j := (-1)^j \left(\sum_{k=0}^{j}(-1)^k a_k - p_n(-1)\right) \tag{2.1}$$

then

$$\zeta(s) = \frac{-1}{(1-2^{1-s})p_n(-1)} \sum_{j=0}^{n-1} \frac{c_j}{(1+j)^s} \quad + \quad \xi_n(s) \tag{2.2}$$

where

$$\xi_n(s) = \frac{1}{p_n(-1)(1-2^{1-s})} \frac{1}{\Gamma(s)} \int_0^1 \frac{p_n(x)|\log x|^{s-1}}{1+x} dx. \tag{2.3}$$

Here Γ is the gamma function.

Note that the c_j are (up to sign) just the coefficients of $\frac{p_n(x)-p_n(-1)}{1+x}$ which is a polynomial of degree $n-1$.

PROOF. We use the standard formulae.

$$\zeta(s) = \frac{1}{(1-2^{1-s})\Gamma(s)} \int_0^1 \frac{|\log x|^{s-1}}{1+x} dx \qquad \text{re(s)} > 0 \tag{2.4}$$

and

$$\frac{1}{(m+1)^s} = \frac{1}{\Gamma(s)} \int_0^1 x^m |\log x|^{s-1} dx \qquad re(s) > 0 \tag{2.5}$$

See [1,7], though both follow easily from

$$\Gamma(s) = \int_0^\infty u^{s-1} e^{-u} du = \int_0^1 |\log x|^{s-1} dx \qquad re(s) > 0 \tag{2.6}$$

which is just the definition of Γ and (1) .

We now write

$$\begin{aligned} \xi_n(s) : &= \frac{1}{p_n(-1)(1-2^{1-s})} \frac{1}{\Gamma(s)} \int_0^1 \frac{p_n(x)|\log x|^{s-1}}{1+x} dx \\ &= \frac{1}{p_n(-1)(1-2^{1-s})} \frac{1}{\Gamma(s)} \int_0^1 \frac{p_n(-1)|\log x|^{s-1}}{1+x} dx \\ &- \frac{1}{p_n(-1)(1-2^{1-s})} \frac{1}{\Gamma(s)} \int_0^1 \frac{p_n(-1)-p_n(x)}{1+x} |\log x|^{s-1} dx \end{aligned}$$

The first term above gives $\zeta(s)$ by (2.4) and the last term expands with (2.5) to give the series expansion in (2.2). □

The trick now is to choose p_n so that the error in the integral for ξ_n divided by $p_n(-1)$ is as small as possible.

The Chebychev polynomial, shifted to $[0, 1]$, and suitably normalized maximizes the value $p_n(-1)$ over all polynomials of comparable supremum norm on [0,1]. So the Chebychev polynomials are one obvious choice for p_n and give the next result.

ALGORITHM 2. Let

$$d_k := n\sum_{i=0}^{k} \frac{(n+i-1)!4^i}{(n-i)!(2i)!}$$

then

$$\zeta(s) = \frac{-1}{d_n(1-2^{1-s})}\sum_{k=0}^{n-1} \frac{(-1)^k(d_k - d_n)}{(k+1)^s} + \gamma_n(s)$$

where for $s = \sigma + it$ with $\sigma \geq \frac{1}{2}$

$$\begin{aligned} |\gamma_n(s)| &\leq \frac{2}{(3+\sqrt{8})^n}\frac{1}{|\Gamma(s)|}\frac{1}{|(1-2^{1-s})|} \\ &\leq \frac{3}{(3+\sqrt{8})^n}\frac{(1+2|t|)e^{\frac{|t|\pi}{2}}}{|(1-2^{1-s})|} \end{aligned}$$

PROOF. The formula we need for the nth Chebychev polynomial on $[0,1]$ is

$$T_n(x) = (-1)^n n\sum_{k=0}^{n}(-1)^k\frac{(n+k-1)!}{(n-k)!(2k)!}4^k x^k$$

from which the expression for d_k is deduced. To estimate the error we observe that, by Algorithm 1,

$$\begin{aligned} |\gamma_n(s)| &= \left|\frac{1}{d_n(1-2^{1-s})}\frac{1}{\Gamma(s)}\int_0^1 \frac{T_n(x)|\log x|^{s-1}}{1+x}dx\right| \\ &\leq \frac{2}{(3+\sqrt{8})^n}\frac{1}{|(1-2^{1-s})\Gamma(s)|}\int_0^1 \frac{|\log x|^{s-1}}{1+x}dx \end{aligned}$$

since on $[0,1]$, $|T_n(x)|$ is bounded by 1 and $|T_n(-1)| \geq \frac{1}{2}(3+\sqrt{8})^n$. We now compute that

$$\int_0^1 \frac{|\log x|^{\frac{1}{2}}}{1+x}dx \leq .68$$

to deduce that

$$|\gamma_n(s)| \leq \frac{1.36}{(3+\sqrt{8})^n}\frac{1}{|(1-2^{1-s})\Gamma(s)|}.$$

Now for $s = \sigma + it$ with $\sigma \geq \frac{1}{2}$

$$\left|\frac{\Gamma(\sigma)}{\Gamma(\sigma+it)}\right|^2 = \prod_{n=0}^{\infty}(1+\frac{t^2}{(\sigma+n)^2}).$$

So

$$\begin{aligned}\frac{1}{|\Gamma(s)|} &= \frac{\left(\prod_{n=0}^{\infty}(1+\frac{t^2}{(\sigma+n)^2})\right)^{\frac{1}{2}}}{|(\Gamma(\sigma)|} \\ &\leq \frac{\left(\prod_{n=0}^{\infty}(1+\frac{t^2}{(\frac{1}{2}+n)^2})\right)^{\frac{1}{2}}}{|\Gamma(\sigma)|} \\ &\leq \frac{(\frac{1+4t^2}{|t|\pi})^{\frac{1}{2}}\,(\sinh(t\pi))^{\frac{1}{2}}}{|\Gamma(\sigma)|}.\end{aligned}$$

Since $|\Gamma(\sigma)|^{-1} \leq 1.5$ on $[\frac{1}{2}, \infty)$ we are done. □

Since $(3+\sqrt{8}) = 5.828\ldots$ and this is the driving term in the estimate, we see that we require roughly (1.3)n terms for n digit accuracy, provided we are close to the real axis.

An even simpler algorithm, though not quite as fast, can be based on taking $p_n(x) := x^n(1-x)^n$.

ALGORITHM 3. Let

$$e_j = (-1)^j[\sum_{k=0}^{j-n} \frac{n!}{k!(n-k)!} - 2^n]$$

(where the empty sum is zero). Then

$$\zeta(s) = \frac{-1}{2^n(1-2^{1-s})} \sum_{j=0}^{2n-1} \frac{e_j}{(j+1)^s} + \gamma_n(s)$$

where for $s = \sigma + it$ with $\sigma > 0$

$$|\gamma_n(s)| \leq \frac{1}{8^n} \frac{(1+|\frac{t}{\sigma}|)e^{\frac{|t|\pi}{2}}}{|1-2^{1-s}|}.$$

If $-(n-1) \leq \sigma < 0$ then

$$|\gamma_n(s)| \leq \frac{1}{8^n|1-2^{1-s}|} \frac{4^{|\sigma|}}{|\Gamma(s)|}$$

(Note that the $\gamma_n(s) = 0$ for $s = -1, -2, \ldots, -n+1$.)

The details of this are very similar to those of Algorithm 2 on using $p_n(x) := x^n(1-x)^n$ and we omit them. The fact that convergence persists into the part of the half plane $\{re(s) < 0\}$ is a consequence of the fact that

$$\int_0^1 \frac{x^n(1-x)^n}{1+x} |\log x|^{s-1}\, dx$$

converges provided re(s) > −n. Thus Algorithm 3 gives another proof of the analytic continuation of $\zeta(s)(1-s)$. (Note that $|e_j/2^n| = 1$ for $j = 0, \ldots, n$ and $|e_j/2^n| \leq 1$ for all j.)

Because $1/\Gamma(s) = 0$ for s a negative integer we have that $\gamma_n(s) = 0$ for $s = -1, -2, \ldots, -n+1$. However since

$$\zeta(-2n+1) = -\frac{\beta_{2n}}{2n}$$

the sum in Algorithm 2 computes Bernoulli numbers, for $s = -1, \ldots, -n+1$, exactly.

In order to make comparisons some care has to be taken. For an Euler-Maclaurin based computation, Bernoulli numbers have to be computed. If they are then stored a second evaluation will be much faster than an initial evaluation. Part of what makes Euler-Maclaurin unattractive for very large precision computations is that it is storage intensive and computationally expensive to compute the Bernoulli numbers, at least by usual methods. Roughly speaking, order of n Bernoulli numbers are required for n-digit precision and this requires in excess of order of n^2 storage.

The Binomial-like coefficients of Algorithms 2 and 3 are much easier to compute and if done sequentially require only one additional binomial coefficient per term which computes by a single multiplication and division.

3. Optimality

Algorithms 2 and 3 are nearly optimal in the following sense. There is no sequence of n-term exponential polynomials that can converge to $\zeta(s)$ on an interval $[a, b], a > 1$ very much faster than those of the algorithms. Precisely we have.

THEOREM 3.1. *Let $1 < \alpha < \beta$ and let n be fixed. Then*

$$||\zeta(s) - \sum_{k=1}^{n} \frac{a_k}{b_k^s}||_{[\alpha,\infty)} \geq \frac{1}{\left(2^\alpha(3+\sqrt{8})^2\right)^n}$$

and

$$||\zeta(s) - \sum_{k=1}^{n} \frac{a_k}{b_k^s}||_{[\alpha,\beta)} \geq (D(\alpha,\beta))^n$$

for any real (a_k) and (b_k). Here $D(\alpha,\beta)$ is a positive constant that depends only on α and β and $||.||_{[\alpha,\beta]}$ denotes the supremum norm on $[\alpha,\beta]$.

PROOF. The proof follows the method of [4]. Under the change of variables $s \to -\log(x)/\log(2)$ for some real (c_k), (d_k) and (e_k)

$$\begin{aligned} &||\zeta(s) - \sum_{k=1}^{n} \frac{a_k}{b_k^s}||_{[\alpha,\beta]} \\ &= ||\sum_{k=1}^{\infty} x^{log(k)/log(2)} - \sum_{k=1}^{n} a_k x^{c_k}||_{[2^{-\beta},2^{-\alpha}]} \\ &\geq ||\sum_{k=0}^{n} x^k - \sum_{k=1}^{n} d_k x^{e_k}||_{[2^{-\beta},2^{-\alpha}]} \end{aligned}$$

where the last inequality follows by the comparison theorem (Corollary 2 of [4]). Now Theorem 8 of [4] gives the explicit estimate

$$||\sum_{k=0}^{n} x^k - \sum_{k=1}^{n} d_k x^{e_k}||_{[2^{-(\beta-\alpha)},1]} \geq \frac{1}{(C+\sqrt{C^2-1})^{2n}}$$

where $C := (3+2^{-(\beta-\alpha)})/(1-2^{-(\beta-\alpha)})$ and the result follows with the aid of Corollary 2 of [4] again. □

Another way in which Algorithms 2 and 3 are (somewhat) near optimal is the following. At even integers the algorithms generate rational approximations that satisfy, for each positive integer N,

$$||\zeta(2N) - \frac{p_n}{q_n}|| < \frac{1}{q_n^{\epsilon}}$$

for infinitely many integers $(p_n), (q_n)$ and some positive $\epsilon := \epsilon(N)$. But results of Mahler show that no such inequalities exist with arbitrarily large ϵ and it is expected that in fact ϵ can be no greater than two. (See Chapter 11 of [3].)

References

[1] M. Abromowitz and I. Stegun, *Handbook of Mathematical Functions*, Dover, N.Y., 1972.

[2] M. V. Berry and J. P. Keating, *A new asymptotic representation for $\zeta(\frac{1}{2}+it)$ and quantum spectral determinants*, Proc. R. Soc. Lond. A, **437** (1992), 151–173.

[3] J. Borwein and P. Borwein, *Pi and the AGM*, Wiley N.Y., 1987.

[4] P. Borwein, *Uniform approximation by polynomials with variable exponents*, Canadian J. Math. **35** (1983), 547–557.

[5] H. Cohen and M. Olivier, *Calcul des valeurs de la function zêta de Riemann en multiprécision*, C.R. Acad. Sci. Paris, Série 1 **314** (1992), 427–430.

[6] A. Odlyzko and A. Schönhage, *Fast algorithms for multiple evaluations of the Riemann zeta-function*, Trans. A.M.S. **309** (1988), 797–809.

[7] E. C. Titchmarsh, *The Theory of the Riemann Zeta-function*, Second ed. Oxford Scientific. Publication, 1986.

Department of Mathematics and Statistics, Simon Fraser University, Burnaby, BC V5A 1S6, Canada

E-mail address: pborwein@cecm.sfu.ca

Canadian Mathematical Society
Conference Proceedings
Volume **27**, 2000

On Periods of Elements from Real Quadratic Number Fields

Edward B. Burger and Alfred J. van der Poorten

To Jonathan Borwein, on the occasion of his award of a Dhc

Preliminaries

Anyone meeting a continued fraction expansion

$$a_0 + \cfrac{1}{a_1 + \cfrac{1}{a_2 + \cfrac{1}{a_3 + \cfrac{1}{a_4 + \cfrac{1}{a_5 + \ddots}}}}}$$

will see that a less wasteful notation, say $[\,a_0\,,a_1\,,a_2\,,\ldots]$, is needed to represent it. Anyone attempting to compute the truncations $[\,a_0\,,a_1\,,\ldots,a_h\,]=p_h/q_h$ will be delighted to notice that the definition $[\,a_0\,,a_1\,,\ldots,a_h\,]=a_0+1/[\,a_1\,,\ldots,a_h\,]$ immediately implies by induction on h that there is a correspondence

$$\begin{pmatrix}a_0 & 1\\ 1 & 0\end{pmatrix}\begin{pmatrix}a_1 & 1\\ 1 & 0\end{pmatrix}\cdots\begin{pmatrix}a_h & 1\\ 1 & 0\end{pmatrix}=\begin{pmatrix}p_h & p_{h-1}\\ q_h & q_{h-1}\end{pmatrix}\longleftrightarrow[\,a_0\,,a_1\,,\ldots,a_h\,]=p_h/q_h$$

between products of certain two by two matrices and continued fractions.

1. Introduction

Different explanations of a phenomenon often provide new insights and greater understanding. Thus our purpose here is to re-explore the phenomena reported and proved in [**1**].

Consider the two purely periodic continued fraction expansions

$$\alpha=[\,\overline{1\,,14\,,1\,,4\,,1\,,3\,,12\,,3\,,1\,,4}\,]\text{ and }\beta=[\,\overline{1\,,1\,,4\,,2\,,55\,,2\,,4\,,1\,,1\,,2\,,2}\,].$$

1991 *Mathematics Subject Classification.* Primary 11J70, 11A65, 11J68.

Key words and phrases. unit, periodic continued fractions.

The first author was a Visiting Fellow at Macquarie University during the writing of this paper and wishes to thank the members of the ceNTRe for their warm hospitality. The second author was supported in part by a grant from the Australian Research Council.

There's nothing immediately hitting the eye that suggests that the two expansions represent quadratic irrationals in the same quadratic number field. So if such a thing is claimed, there's then nothing for it but to do some computation, at least to the extent of evaluating the two expansions for an entire period. The respective tableaux

h			0	1	2	3	4	5	6	7	8	9
$a_{\alpha,h}$			1	14	1	4	1	3	12	3	1	4
$p_{\alpha,h}$	0	1	1	15	16	79	95	364	4463	13753	18216	86617
$q_{\alpha,h}$	1	0	1	14	15	74	89	341	4181	12884	17065	81144

and

h			0	1	2	3	4	5	6	7	8	9	10
$b_{\beta,h}$			1	1	4	2	55	2	4	1	1	2	2
$p_{\beta,h}$	0	1	1	2	9	20	1109	2238	10061	12299	22360	57019	136398
$q_{\beta,h}$	1	0	1	1	5	11	610	1231	5534	6765	12299	31363	75025

then invite us to study the matrices

$$M_\alpha = \begin{pmatrix} 86617 & 18216 \\ 81144 & 17065 \end{pmatrix} \quad \text{and} \quad M_\beta = \begin{pmatrix} 136398 & 57019 \\ 75025 & 31363 \end{pmatrix}.$$

The rule of the game now is that each matrix

$$M = \begin{pmatrix} x & x' \\ y & y' \end{pmatrix}$$

is of a special shape; namely

$$M_\gamma = \begin{pmatrix} x & -Ny \\ y & x - Ty \end{pmatrix},$$

where $T = T_\gamma$ is the trace $\gamma + \overline{\gamma}$, and where $N = N_\gamma$ is the norm $\gamma\overline{\gamma}$, of the element γ whose period has produced M.

Specifically, $86617 - 17065 = (\alpha + \overline{\alpha}) \cdot 81144$ and $18216 = -\alpha\overline{\alpha} \cdot 81144$. Plainly, we can now compute the trace T_α, and norm N_α, of α. If we recall that a real quadratic irrational has a purely periodic continued fraction expansion if and only if it is greater than 1 and its conjugate lies between -1 and 0, then we know we need the positive zero of the defining polynomial. Pressing buttons on our calculator yields $\alpha = 1.067447993571368484\ldots$; and we're little the wiser. Moreover, since our real object is to decide whether α and β belong to the same quadratic number field it might have demonstrated greater wisdom had we computed the discriminant Δ_α of α. It is $T_\alpha^2 - 4N_\alpha = 1.6326530612244897959183673 46\ldots$. Perhaps, we're still in trouble!

How about we also compute Δ_β. When we do that, we find that $T_\beta = 7/5$, $N_\beta = -19/25$, and $\Delta_\beta = 5$.

Recovering our senses, we now notice that

$$1.632653061224489795918367346\ldots = [\,1\,,1\,,1\,,1\,,2\,,1\,,1\,,2\,] = 80/49\,,$$

so that $\Delta_\alpha = (4^2/7^2) \cdot 5$, and plainly the quadratic number field containing both α and β is $\mathbb{Q}(\sqrt{5})$. In fact, backtracking a little, it's now clear that $\alpha = (3 + 2\sqrt{5})/7$ and $\beta = (7 + 5\sqrt{5})/10$. Checking the continued fractions of those numbers confirms that conclusion.

However, all this is not quite to the point. In [**1**] it is remarked, in effect, that the two matrices

$$M_\alpha^{25} = \begin{pmatrix} p_{\alpha,249} & -N_\alpha q_{\alpha,249} \\ q_{\alpha,249} & p_{\alpha,249} - T_\alpha q_{\alpha,249} \end{pmatrix} \quad \text{and} \quad M_\beta^{24} = \begin{pmatrix} p_{\beta,263} & -N_\beta q_{\beta,263} \\ q_{\beta,263} & p_{\beta,263} - T_\beta q_{\beta,263} \end{pmatrix}$$

have the same trace

$$2469358527651528622763891388578931265566414510770004830269847 \backslash$$
$$8395289566538179507389432113883234418865101546019834683808080800002\,.$$

Given, moreover, that the integers $250 = 25 \cdot 10$ and $264 = 24 \cdot 11$ have the same parity, it is then suggested that this pair of coincidences of itself entails that α and β belong to the same quadratic number field. Worse, it is then *proved* in [**1**] that such coincidence of trace and parity is necessary and sufficient for two given pure periodic elements to belong to the same quadratic number field.

All this is rather surprising, sufficiently so as to make one confident that the result must be entirely natural. Indeed, we will show it to be a manifestation of readily recalled properties of the group of units of the real quadratic number field.

2. What is going on

2.1. The core fact is

THEOREM 1. *Suppose that* $[\,a_0\,, a_1\,, \ldots\,, a_{r-1}\,, a_r\,] = x/y$ *and that the previous convergent* $[\,a_0\,, a_1\,, \ldots\,, a_{r-1}\,]$ *is denoted by* x'/y'. *Then the claim*

$$[\,\overline{a_0\,, a_1\,, \ldots\,, a_r}\,] = \gamma\,, \quad \textit{where} \quad \gamma^2 - T\gamma + N = 0\,,$$

is equivalent to the allegation

$$\begin{pmatrix} x & x' \\ y & y' \end{pmatrix} = \begin{pmatrix} x & -Ny \\ y & x - Ty \end{pmatrix}.$$

PROOF. Suppose $[\,\overline{a_0\,, a_1\,, \ldots\,, a_r}\,] = \alpha$, that is, $\alpha = [\,a_0\,, a_1\,, \ldots\,, a_r\,, \alpha\,]$. By the correspondence between continued fractions and two by two matrices we must have

$$\begin{pmatrix} x & x' \\ y & y' \end{pmatrix} \begin{pmatrix} \alpha & 1 \\ 1 & 0 \end{pmatrix} = \begin{pmatrix} x\alpha + x' & x \\ y\alpha + y' & y \end{pmatrix} \longleftrightarrow \alpha.$$

That is, $(x\alpha + x')/(y\alpha + y') = \alpha$, or $y\alpha^2 - (x - y')\alpha - y' = 0$.

Thus if $x' = -Ny$ and $y' = x - Ty$ we have $y(\alpha^2 - T\alpha + N) = 0$, so $\alpha = \gamma$. Conversely, if $\alpha = \gamma$ *a priori*, then $(x - y')/y = T$ and $y'/y = -N$. □

Moreover, the kernel of this core is that the matrix $M = \left(\begin{smallmatrix} x & -Ny \\ y & x-Ty \end{smallmatrix}\right)$ must be unimodular , because

$$\begin{pmatrix} x & -Ny \\ y & x - Ty \end{pmatrix} = \begin{pmatrix} x & x' \\ y & y' \end{pmatrix} = \begin{pmatrix} a_0 & 1 \\ 1 & 0 \end{pmatrix} \begin{pmatrix} a_1 & 1 \\ 1 & 0 \end{pmatrix} \cdots \begin{pmatrix} a_r & 1 \\ 1 & 0 \end{pmatrix}$$

plainly has determinant $(-1)^{r+1}$.

COROLLARY. The element $x - \gamma y$ is a unit in the ring $\mathbb{Z}[\gamma]$.

PROOF. The integer $x - \gamma y$ is an eigenvalue of the unimodular matrix M. □

2.2. Of course $\gamma = (A+B\sqrt{D})/C$ for some squarefree positive integer $D > 1$ and integers A, B and C, where $B \neq 0$ and, without loss of generality, $C > 0$. Thus γ is an element of the quadratic number field $\mathbb{K} = \mathbb{Q}(\sqrt{D})$. It is usual then to abuse language by referring to $x - \gamma y$ as 'a unit of $\mathbb{K}$'.

Next, we should recall that if $D \equiv 1 \pmod 4$ then $\delta = \frac{1}{2}(1+\sqrt{D})$ is an integer; if $D \not\equiv 1 \pmod 4$ we set $\delta = \sqrt{D}$. We will also write $n = \operatorname{Norm} D = \delta\overline{\delta}$ and $t = \operatorname{Trace}\delta = \delta + \overline{\delta}$. Thus if $D \equiv 1 \pmod 4$ then $n = -\frac{1}{4}(D-1)$, $t = 1$, while if $D \not\equiv 1 \pmod 4$ we have $n = -D$, $t = 0$.

We come now to an important rule: One *must* write γ as $(P + f\delta)/Q$, where Q divides $\operatorname{Norm}(P + f\delta) = P^2 + ftP + f^2 n$; here P, $f \neq 0$ and $Q > 0$ are integers, which we may suppose not to have any redundant common factor. It is a simple matter to transform $(A+B\sqrt{D})/C$ appropriately. Other than for a possible redundant common factor, $\gamma = (CA+CB\sqrt{D})/C^2$ will certainly do; the additional adjustment to change $\sqrt{D}$ to δ is straightforward.

Our insistence that Q divide $\operatorname{Norm}(P + f\delta)$ is more than just a convenient convention. One needs it to validate a correspondence between $\mathbb{Z}$-modules and ideals.

REMARK 2. The $\mathbb{Z}$-module $\mathcal{I} = \langle Q, P + f\delta\rangle$ is an ideal of $\mathbb{Z}[f\delta]$.

PROOF. It suffices to check that $f\delta(P + f\delta)$ is in $\mathcal{I}$. But

$$f\delta(P + f\delta) = -(P^2 + ftP + f^2 n) + (P + ft)(P + f\delta),$$

and then $Q \mid (P^2 + ftP + f^2 n)$ is essential to complete the verification. □

2.3. Plainly, we are obliged to think of the example elements $\alpha = (3+2\sqrt{5})/7$ and $\beta = (7 + 5\sqrt{5})/10$ as $\alpha = (7 + 28\varphi)/49$ and $\beta = (10 + 50\varphi)/50$, with $\varphi = (1 + \sqrt{5})/2$. The corresponding eigenvalues of the respective matrices M_α and M_β are, as noted, $\eta_\alpha = x_\alpha - y_\alpha\alpha$ and $\eta_\beta = x_\beta - y_\beta\beta$. That is

$$\eta_\alpha = 86617 - 81144\alpha \quad\text{and}\quad \eta_\beta = 136398 - 75025\beta.$$

In better terms,

$$\eta_\alpha = 86617 - (11592 + 46368\varphi) \quad\text{and}\quad \eta_\beta = 136398 - (15005 + 75025\varphi).$$

We will find it useful to note that $46368 = 2^5 \cdot 3^2 \cdot 7 \cdot 23$, and $75025 = 5^2 \cdot 3001$.

2.4. Finally, we confirm that the quantities

$$\eta_\alpha = 75025 - 46368\varphi \quad\text{and}\quad \eta_\beta = 121393 - 75025\varphi$$

both *are* units, lest we have blundered and are about to waste our effort pointlessly. Indeed, they are; with respective norms $\operatorname{Norm}\eta_\alpha = 1$ and $\operatorname{Norm}\eta_\beta = -1$. We should recognise, however that the example is degenerate — not quite general — in that $D = 5$ is the unique D so that δ is a unit. Moreover, we also notice that $\eta_\beta = 46368 + 75025\overline{\varphi}$, so $-\varphi\eta_\beta = \eta_\alpha$.

3. Units in quadratic number fields

3.1. For nonsquare positive D there always are units $x - \delta y$, that is, positive integers x, y satisfying "Pell's equation" $x^2 - txy + ny^2 = \pm 1$. That follows readily from several applications of the box principle. One first proves Dirichlet's result that there are infinitely many integers q so that the distance to the nearest integer $\|q\delta\|$ of $q\delta$ satisfies $\|q\delta\| < 1/q$. It follows that for each such q there is an integer p

so that $|p^2 - tpq + nq^2| < (p - q\overline{\delta})/q < (\delta - \overline{\delta})$. Again by the box principle, there is then an integer k, with $|k| < (\delta - \overline{\delta})$, so that the equation $p^2 - tpq + nq^2 = k$ has infinitely many pairs of solutions (p, q) and, once again by the box principle, two distinct pairs of solutions (p, q) and (p', q') so that $p \equiv p'$ and $q \equiv q' \pmod{k}$. Then $x = |pp' - Dqq'|/k^2$, $y = |pq' - p'q|/k^2$ displays positive integers x and y satisfying $x^2 - txy + ny^2 = 1$. We add that there may be a smaller solution (in that x or y is smaller than the minimal x, y just shown to be a solution to $x^2 - txy + ny^2 = 1$) for the equation $x^2 - txy + ny^2 = -1$.

3.2. The following is peculiarly poorly known.

REMARK. One may say that a two by two matrix $\left(\begin{smallmatrix} a & b \\ c & d \end{smallmatrix}\right)$ is *column dominant* if $a > b$ implies $c \geq d$, or if $a < b$ implies $c \leq d$. A matrix that is not column dominant will be said to be *column balanced.* Dually, by transposition, we similarly have the notions of *row dominant* and *row balanced.* A matrix that is both column and row dominant may be said to be *dominant.* Similarly if it is both column and row balanced it may be said to be *balanced.* Happily, it is plain that

PROPOSITION. A two by two positive integer unimodular matrix is dominant.

Even if one were to allow zero entries as well, the only exceptions that occur are

$$J := \begin{pmatrix} 0 & 1 \\ 1 & 0 \end{pmatrix} \quad \text{and} \quad I = \begin{pmatrix} 1 & 0 \\ 0 & 1 \end{pmatrix}; \text{ notice that } J^2 = I.$$

THEOREM. A positive integer unimodular matrix $\left(\begin{smallmatrix} p & r \\ q & s \end{smallmatrix}\right)$ has a unique decomposition

$$\begin{pmatrix} p & r \\ q & s \end{pmatrix} = \begin{pmatrix} a_0 & 1 \\ 1 & 0 \end{pmatrix} \begin{pmatrix} a_1 & 1 \\ 1 & 0 \end{pmatrix} \cdots \begin{pmatrix} a_m & 1 \\ 1 & 0 \end{pmatrix},$$

in positive integers a_i, other than perhaps for a_0 or a_m which may be zero.

PROOF. Apply the Euclidean algorithm to the rows of the matrix. Thus step 0 of the decomposition is

$$\begin{pmatrix} p & r \\ q & s \end{pmatrix} = \begin{pmatrix} a_0 & 1 \\ 1 & 0 \end{pmatrix} \begin{pmatrix} q & s \\ p - a_0 q & r - a_0 s \end{pmatrix},$$

with a_0 selected maximally so that the remainders $p - a_0 q$ and $r - a_0 s$ both are nonnegative. Plainly one is left with a matrix whose first row dominates its second row. Thus on repeating the process the succeeding a_i are positive, at any rate until one is left with a balanced matrix.

Dually, one may apply the Euclidean algorithm to the columns of the matrix, with step 0 then being

$$\begin{pmatrix} p & r \\ q & s \end{pmatrix} = \begin{pmatrix} r & p - a_m r \\ s & q - a_m s \end{pmatrix} \begin{pmatrix} a_m & 1 \\ 1 & 0 \end{pmatrix}.$$

□

It follows that $a_0 = 0$ is possible only if $p \leq q$, and $a_m = 0$ only if $p \leq r$. Moreover, $p/q = [a_0, a_1, \ldots, a_s]$ and $r/s = [a_0, a_1, \ldots, a_{s-1}]$. Furthermore, the preliminary description of the notion 'continued fraction', as well as the matrix correspondence, entails that $[a_0, a_1, \ldots, a_{s-1}, 0] = [a_0, a_1, \ldots, a_{s-2}]$.

3.3. Recall, see §§2.2, that we have set $\delta = \sqrt{D}$ or $\frac{1}{2}(1+\sqrt{D})$ according as $D \not\equiv 1$ or $D \equiv 1 \pmod 4$; with $n = \delta\overline{\delta}$ and $t = \delta + \overline{\delta}$.

Applying the peculiarly poorly known remark to the case of units $x - \delta y$, with x and y positive integers, one checks readily that

$$M = \begin{pmatrix} x & -ny \\ y & x-ty \end{pmatrix} = \begin{pmatrix} a_0 & 1 \\ 1 & 0 \end{pmatrix}\begin{pmatrix} a_1 & 1 \\ 1 & 0 \end{pmatrix}\cdots\begin{pmatrix} a_r & 1 \\ 1 & 0 \end{pmatrix}\begin{pmatrix} 0 & 1 \\ 1 & 0 \end{pmatrix},$$

for positive integers $a_0, a_1, \ldots, a_{r-1}$. Moreover, the near identity of the elements x and $x - ty$ in the main diagonal (recall that $t = 0$ or $t = 1$, and in particular that t is an integer) entails $a_r = a_0 - t$ and that the word $a_1 a_2 \ldots a_{r-1}$ is a palindrome[1].

In the immediate sequel x, y is the minimal pair of positive integers satisfying $x^2 - txy + ny^2 = \pm 1$, and M_δ is the matrix

$$M_\delta = \begin{pmatrix} x & -ny \\ y & x-ty \end{pmatrix}$$

corresponding to the *fundamental* unit $\eta = x - \delta y$ of $\mathbb{K} = \mathbb{Q}(\sqrt{D})$.

It is fairly easy to see naïvely that $\pm\eta^h$, $h \in \mathbb{Z}$ yields all the units of $\mathbb{K} = \mathbb{Q}(\sqrt{D})$. For if there were another unit ε, say, one could readily construct a unit $\pm\eta^k\varepsilon$, some $k \in \mathbb{Z}$, 'smaller than' the fundamental unit.

REMARK. In the case $D = 5$ the *fundamental* unit is $1 - \varphi$ (recall that above we set $\varphi = \frac{1}{2}(1+\sqrt{5})$). Then $x = 1$, $y = 1$ and $n = -1$, $t = 1$, so $a_r = a_0 - 1 = 0$ and

$$M_\varphi = \begin{pmatrix} 1 & 1 \\ 1 & 0 \end{pmatrix} = \begin{pmatrix} 1 & 1 \\ 1 & 0 \end{pmatrix}\begin{pmatrix} 0 & 1 \\ 1 & 0 \end{pmatrix}\begin{pmatrix} 0 & 1 \\ 1 & 0 \end{pmatrix},$$

as claimed.

3.4. It is a manifestation of the Cayley-Hamilton Theorem that

$$M_\delta^{h+2} = (2x - ty)M_\delta^{h+1} - (x^2 - txy + ny^2)M_\delta^h\,, \quad h \in \mathbb{Z}. \tag{3.1}$$

If we now set $M_\delta^h = M_h$ then (3.1) becomes the recurrence relation

$$M_{h+2} = (2x - ty)M_{h+1} - (x^2 - txy + ny^2)M_h \quad h = 0,\, 1\, \ldots,$$

reporting that the units $\eta^h = x_h - \delta y_h$ are given by the recurrence relations

$$x_{h+2} = (2x - ty)x_{h+1} \mp x_h \quad \text{and} \quad y_{h+2} = (2x - ty)y_{h+1} \mp y_h \quad h = 0,\, 1,\, \ldots.$$

Here $\eta^0 = 1 - \delta \cdot 0$ and $\eta^1 = x - \delta y$. The sign $\mp$ is chosen according as $\eta\overline{\eta} = \pm 1$.

In the example case $D = 5$ the recurrence relation is

$$\eta^{h+2} = \eta^{h+1} + \eta^h \quad \text{with } \eta^0 = 1 \text{ and } \eta = 1 - \varphi,$$

so the sequence $(y_h)_{h\ge 0}$ is 0, 1, 1, 2, 3, 5, …, $y_{24} = 46368$, $y_{25} = 75025$, ….

By the way, this calculation is computationally easy. One does *not* have to wade stepwise all the way up to $h = 24$. Nor is this ease just a consequence of our happening to be looking at the Fibonacci numbers.

In general, it is a familiar fact from the theory of recurrence sequences, and is in any case easy to verify directly, that

$$y_h = y(\eta^h - \overline{\eta}^h)/(\eta - \overline{\eta})\,. \tag{3.2}$$

Since, moreover, always $-1 < \overline{\delta} < 0$, it is plain that $y_h \approx y\eta^h/(\eta - \overline{\eta})$, and it *is* easy to compute h given y_h.

[1] Recall that a palindrome is never even, indeed, it's never odd or even. It's a toyota.

3.5. Now it is all very well to talk blithely about the fundamental unit η of $\mathbb{K} = \mathbb{Q}(\sqrt{D})$ but, rather more precisely, η is the fundamental unit of the domain $\mathbb{Z}[\delta]$ (recall that if D is squarefree then $\mathbb{Z}[\delta]$ is the domain of all integers of $\mathbb{K}$). This 'precision' is of importance. While $1 - \varphi = \overline{\varphi}$ is the fundamental unit of $\mathbb{Z}[\frac{1}{2}(1+\sqrt{5})]$, the fundamental unit of the domain $\mathbb{Z}[\sqrt{5}]$ cannot be $1-\varphi$ because φ does not even belong to $\mathbb{Z}[\sqrt{5}]$. Indeed, from the expansion $\sqrt{5} = [\,2\,,\overline{4}\,]$ we learn immediately that the fundamental unit of $\mathbb{Z}[\sqrt{5}] = \mathbb{Z}[2\varphi]$ has trace 4 and norm -1, and thus is $2 - \sqrt{5} = \overline{\varphi}^3$. Just so, $\alpha = \overline{\varphi}^{24}$ is the fundamental unit of $\mathbb{Z}[92736\varphi]$, while $\beta = \overline{\varphi}^{25}$ is the fundamental unit of the domain $\mathbb{Z}[15050\varphi]$.

REMARK. We should confess that any one of the four units $\pm\eta$, $\pm\overline{\eta}$ has the right to step forward and to insist *it* generates all the units of $\mathbb{Z}[\delta]$. Our selecting $x - \delta y$ for the qualification 'fundamental' if both x and y are positive integers, is just a convention convenient for the present context.

3.6. One might well wonder how, given the fundamental unit, η_g say, of the domain $\mathbb{Z}[g\delta]$ one finds the fundamental unit, η_f say, of $\mathbb{Z}[f\delta]$, where $g\,|\,f$. Recall, however, that to compute η_f our remarks at §§3.3 point out that it suffices to find the period of $f\delta$. So the point of the following remarks is to explain the relationship between η_g and η_f. Indeed, if we have not insisted that D be squarefree it suffices to ask what power of the fundamental unit $\eta = x - \delta y$ has y divisible by f.

The first fact we recall is that if p is a rational prime and F is a polynomial with integer coefficients then $F(X)^p \equiv F(X^p) \pmod p$. Hence if γ is a zero of F it follows that $\gamma^p \equiv \gamma' \pmod p$, where γ' is some zero of F. Thus if F is the defining polynomial of $\eta = x - \delta y$ then $\eta^p \equiv \eta$ or $\eta^p \equiv \overline{\eta} \pmod p$.

Let $p \neq 2$. Suppose first that $p\,|\,D$. It is then immediate by the binomial expansion that η^p is equivalent to a rational integer modulo p; so η^p is in $\mathbb{Z}[p\delta]$. If $p \nmid D$ then similarly $\eta^{p-1} \equiv 1 \pmod p$, respectively $\eta^{p+1} \equiv \overline{\eta}\eta = \pm 1 \pmod p$, entail that η^{p-1}, respectively η^{p+1}, is in $\mathbb{Z}[p\delta]$. The last two cases are p splits, respectively p is inert, in $\mathbb{K} = \mathbb{Q}(\delta)$. So if $\epsilon = \left(\frac{p}{D}\right)$ denotes the Kronecker symbol then all three cases speak of $\eta^{p-\epsilon}$.

Suppose now that $p = 2$. Plainly $(x - y\sqrt{D})^2$ is then always in $\mathbb{Z}[2\sqrt{D}]$. A trifle less obviously, if $D \equiv 1 \pmod 4$ then also $(x - y\delta)^3$ is in $\mathbb{Z}[2\delta]$.

3.7. Returning to the example, it happens that $46368 = 2^5 \cdot 3^3 \cdot 7 \cdot 23$ and $75025 = 5^2 \cdot 3001$. By our previous computations we already know that $\alpha^{25} = \beta^{24}$, but it also seems worthwhile to check this in the light of the immediately preceding remarks. By those remarks, the smallest power of α that can have a 5 in its 'y' is α^5 and one notes that α^5 is not in $\mathbb{Z}[5^2\varphi]$. Thus we need α^{25} and it, indeed, is in $\mathbb{Z}[5^2 \cdot 3001\varphi]$. Had we endeavoured to get that 3001 as such, we should have considered α^m for m dividing 3000, discovering once again that $m = 25$ will do.

3.8. In particular it must be that the two matrices M_α^{25} and M_β^{24} determine the same unit. In other words, they have the same eigenvalues. In particular, they have the same 'norm' — that is, determinant — and the same trace. Since the only two possibilities for the determinant are ± 1, the coincidence of parities chatted about above is simply there to ensure the coincidence of norms.

3.9. We have said enough to make plain that (Theorem 1 of [**1**])

THEOREM 3. *Given a pure periodic element* $\alpha = [\,\overline{a_0\,,\,a_1\,,\,\ldots\,,\,a_r}\,]$ *set*

$$[\,a_0\,,\,a_1\,,\,\ldots\,,\,a_r\,] = p_{\alpha,r}/q_{\alpha,r} \quad \textit{and} \quad [\,a_0\,,\,a_1\,,\,\ldots\,,\,a_{r-1}\,] = p_{\alpha,r-1}/q_{\alpha,r-1}\,;$$

equivalently

$$M_\alpha = \begin{pmatrix} a_0 & 1 \\ 1 & 0 \end{pmatrix}\begin{pmatrix} a_1 & 1 \\ 1 & 0 \end{pmatrix}\cdots\begin{pmatrix} a_r & 1 \\ 1 & 0 \end{pmatrix} = \begin{pmatrix} p_{\alpha,r} & p_{\alpha,r-1} \\ q_{\alpha,r} & q_{\alpha,r-1} \end{pmatrix}.$$

Let $\beta = [\,\overline{b_0\,,\,b_1\,,\,\ldots\,,\,b_s}\,]$. *Then* α *and* β *belong to the same quadratic number field if and only if there are positive integers* u *and* v *so that the matrices* M_α^u *and* M_β^v *are similar, that is, have the same determinant and trace; equivalently,*

$$p_{\alpha,u(r+1)-1} + q_{\alpha,u(r+1)-2} = p_{\beta,v(s+1)-1} + q_{\beta,v(s+1)-2}\ ,,$$

and $u(r+1)$, $v(s+1)$ *have the same parity.*

REMARK. One can readily compute traces of powers of the matrices M^h featuring above by noticing that those traces satisfy a second order recurrence relation; to wit that satisfied by the M^h.

4. Pure periodicity and reduction

It is difficult to avoid feeling that our story is incomplete because we deal only with purely periodic continued fraction expansions.

4.1. Strictly speaking, there is nothing to add. The argument sketched at §§3.1 is readily adjusted to show that, given an arbitrary generator γ of a quadratic number field, there are integers a and b so that $\mathrm{Norm}(a-\gamma b) = 1$. A review of Theorem 1 shows that its core story only tacitly supposes that the partial quotients a_0, a_1, ... all are positive integers. So we may consider finite decompositions

$$M_\gamma = \begin{pmatrix} a & -N_\gamma b \\ b & a - T_\gamma b \end{pmatrix} = \begin{pmatrix} c_0 & 1 \\ 1 & 0 \end{pmatrix}\begin{pmatrix} c_1 & 1 \\ 1 & 0 \end{pmatrix}\cdots\begin{pmatrix} c_t & 1 \\ 1 & 0 \end{pmatrix}$$

in not necessarily positive integers c_i and deduce from Theorem 1 that γ has a purely periodic expansion $[\,\overline{c_0\,,\,c_1\,,\,\ldots\,,\,c_t}\,]$. One then works with that period, albeit that it is not unique. Then every generator of a quadratic number field has a purely periodic expansion.

There's little harm in working *en passant* with nonpositive partial quotients. Indeed, at §§3.3 we admit 0 as a partial quotient, presuming that the reader will recognise that $[\,\ldots\,,\,a\,,\,0\,,\,b\,,\,\ldots] = [\,\ldots\,,\,a+b\,,\,\ldots]$. In similar spirit, it is also not difficult to get rid of negative partial quotients; for example see the remarks at p363 of [**3**].

4.2. If, on the other hand, one *insists* that continued fraction expansions have all their partial quotients (other perhaps than the 0-th) positive — in other words that the expansions be 'admissible', then there's nothing for it but to deal only with *reduced* elements ρ. From here on our expansions *are* admissible.

Suppose one is given a sequence $(a_h)_{h\geq 1}$ of positive integers, and $a_0 \in \mathbb{Z}$. Then there is a real number α with continued fraction expansion $[a_0, a_1, a_2, \ldots\ldots]$. Given an arbitrary irrational complex number $\beta = \beta_0$ one may define the sequence $(\beta_h)_{h\geq 1}$ by $\beta_{h+1} = (\beta_h - a_h)^{-1}$. Vincent's Theorem (1836) points out that unless $\beta = \alpha$ there is some $k = k(\beta, \alpha)$ so that for $h \geq k$ necessarily β_h is in the left hand half of the unit circle.

Accordingly, consider the sequence of polynomials $(f_h)_{h\geq 0}$ obtained from $f(Y) = f_0(Y) = A_{0,0}Y^r + A_{0,1}Y^{r-1} + \cdots + A_{0,r}$ by the rule

$$f_{h+1}(Y) = \pm Y^r f_h(a_h + Y^{-1}). \tag{4.1}$$

The choice of sign in (4.1) is not particularly important, but morally should be made so that the f_h have positive leading coefficient; after a while the $-$ sign will always do. One sees readily that any zero β of f is transformed into a zero β_h of f_h. In particular, if α is itself a non-multiple zero of a polynomial f without rational zeros then there is an $k = k(f)$ so that for $h \geq k$ all the zeros β_h of f lie in the left hand half of the unit circle, other than for the zero α_h which is real and greater than 1. In [**2**] the polynomials f_h $(h \geq k)$ are said to be 'reduced', in analogy with the case $r = 2$ — where that terminology is standard.

Thus the continued fraction process eventually reduces all algebraic elements. Plainly $\alpha_k = -(q_{\alpha,k-1}\alpha - p_{\alpha,k-1})/(q_{\alpha,k}\alpha - p_{\alpha,k})$ generates the same number field over $\mathbb{Q}$ as does α, so it suffices to work just with reduced elements $\rho = \alpha_k$.

4.3. Specifically, a quadratic irrational ρ is reduced if $\rho > 1$ but its conjugate $\overline{\rho}$ satisfies $-1 < \overline{\rho} < 0$. It is a theorem attributed to Galois that ρ has a purely periodic continued fraction expansion if and only if ρ is reduced.

To see that clearly, suppose r denotes the integer part of ρ. Then the tableau yielding the continued fraction expansion of ρ begins with the line

$$\rho = r - (r - T_\rho + \overline{\rho}) .$$

We know that the said tableau is eventually periodic. Now consider the conjugate of the tableau, and in particular the conjugate

$$r - T_\rho + \rho = r - \overline{\rho}$$

of the given line. Because ρ is reduced, this too is a line in some continued fraction tableau.

The preceding lines of this conjugate tableau are, albeit in reverse order, the conjugates of the succeeding lines of the original tableau. So they are periodic because the original tableau is periodic. But a period with its order reversed is a period. We invite the reader to see it now manifest that the conjugate tableau, and therefore also the original tableau, must be *purely* periodic.

References

[1] Edward B. Burger, 'On real quadratic number fields and simultaneous diophantine approximation', Monatshefte Math., **128** (1999), 201–209.

[2] Enrico Bombieri and Alfred J van der Poorten, 'Continued fractions of algebraic numbers', (*Computational Algebra and Number Theory, Sydney* 1992, Wieb Bosma and Alf van der Poorten eds., Kluwer 1995, 138–154.

[3] Alf van der Poorten, 'Formal power series and their continued fraction expansion', in Joe Buhler ed., *Algorithmic Number Theory* (Proc. Third International Symposium, ANTS-III, Portland, Oregon, June 1998), Springer Lecture Notes in Computer Science **1423** (1998), 358–371.

Department of Mathematics, Williams College, Williamstown, Massachusetts 01267
E-mail address: `eburger@williams.edu (Edward B. Burger)`

ceNTRe for Number Theory Research, Macquarie University NSW 2109, Australia
E-mail address: `alf@mpce.mq.edu.au (Alf van der Poorten)`

Canadian Mathematical Society
Conference Proceedings
Volume **27**, 2000

Vector-valued perturbed minimization principles

Robert Deville and Catherine Finet

To Jonathan Borwein, on the occasion of his award of Dhc

ABSTRACT. We give a vector-valued extension of Deville–Godefroy–Zizler perturbed minimization principle. We also get a vector-valued extension of Borwein-Preiss smooth perturbed minimization principle.

Introduction

The lack of compactness in infinite dimensional Banach spaces results in functions not attaining their bounds and some perturbations are thus needed. The idea of perturbed optimization is simple : one starts with a real-valued function f which is, say, lower semi-continuous and bounded from below on a nice set and shows that there exist arbitrarily small perturbations g such that $f + g$ attains a (strong) minimum on the set (every minimizing sequence converges). A most convenient approach to these problems is through smooth perturbed minimization principles (also called smooth variational principles). This approach can be chosen in spaces that admit non trivial smooth functions. A smooth perturbed minimization principle substitutes for the Hahn-Banach theorem in non convex analysis. Many authors have studied variational principles (see [**1**], [**3**], [**4**]). Instead of considering real-valued functions, we can look at vector-valued functions. For example, Göpfert and Tammer have studied extensions of Ekeland's variational principle to vector optimization [**5**], [**10**] and [**11**]. In this context it is natural to look at some extension of Deville–Godefroy–Zizler perturbed minimization principle [**3**]. This principle extends Borwein-Preiss smooth perturbed minimization principle [**1**].
We want to dedicate this article to the memory of P. Loridan.

1991 *Mathematics Subject Classification.* 46B20, 49N.

Key words and phrases. Lower semi-continuous function, Vector-valued perturbed minimization principles.

The authors want to thank the FNRS for the grant they got. The paper was written when the first author visited the Department of Mathematics of the University of Mons-Hainaut (Belgium).

Preliminaries

Let X and Z be real Banach spaces and Z be partially ordered by a closed convex pointed cone K with non empty interior. Let f be a function from X to Z. The vector minimization problem (P) under consideration is :

$$\text{Find } \overline{x} \in X \text{ such that } (f(\overline{x}) - K) \cap f(X) = \{f(\overline{x})\}.$$

In other words :

$$\text{Find } \overline{x} \in X \text{ such that } (f(\overline{x}) - K \setminus \{0\}) \cap f(X) = \emptyset.$$

Such an $\overline{x}$ is called a minimal solution of (P) or Pareto solution or efficient solution of (P). We denote by $E(f)$ the set of all minimal solutions and by $\operatorname{Min} f$ the set of all minimal elements of $f(X)$ (with respect to K), that is $\operatorname{Min} f = f(E(f))$; for vector optimization in partially ordered spaces, see for example [**6**].
Starting with an arbitrary function $f : X \to Z$, there is no reason why the set $\operatorname{Min} f$ should be non empty. We want to find a "nice" perturbation h of f as close as we want to f and such that $\operatorname{Min} h$ is not empty. In the scalar case, usual conditions on the function f are boundedness from below and lower semi-continuity. These notions have already been extended in the vector case [**7**], [**11**].
A function $f : X \to Z$ is said *bounded from below* if there exists $z \in Z$ such that for all $x \in X$, $z \leq f(x)$.
A function $f : X \to Z$ is said *lower semi-continuous* (ℓsc) if, for all $z \in Z$, the set $\{x \in X, f(x) \leq z\}$, also denoted $\{f \leq z\}$, is closed.
In what follows k^o will be a fixed point in the interior of K.
A function $f : X \to Z$ is said to be k^o lower semi-continuous k^o-ℓsc if, for all $r \in \mathbb{R}$, $\{f \leq rk^o\}$ is closed. Of course, a continuous function is ℓsc, and a ℓsc function is k^o-ℓsc.
Let us remark that a notion of ℓsc function was also introduced by J.M. Borwein, J.P. Penot and M. Théra, see [**2**], [**8**] and [**9**]. As an ℓsc mapping in the sense of [**2**] has a closed epigraph (Proposition 1.4 of [**2**]) it follows that such a mapping is also ℓsc in our sense.
Let us start with an easy useful lemma :

LEMMA 1. *The sum of two k^o-ℓsc (respectively ℓsc) functions is a k^o-ℓsc (respectively ℓsc) function.*

PROOF. Let u_1 and $u_2 : X \to Z$ be k^o-ℓsc. We show that $u_1 + u_2$ is a k^o-ℓsc function by proving that, for all $r \in \mathbb{R}$,

$$\{u_1 + u_2 \leq rk^o\} = \bigcap_{\substack{s,t \in \mathbb{R} \\ s+t=r}} \{u_1 \leq sk^o\} \cup \{u_2 \leq tk^o\}$$

Let $x \in X$ be such that $(u_1 + u_2)(x) \leq rk^o$. Suppose that there exist $s, t \in \mathbb{R}$ such that $s + t = r$ and

$$\{sk^o - u_1(x), tk^o - u_2(x)\} \cap K = \emptyset.$$

Then by Hahn-Banach theorem, there exist $z^\star \in Z^\star$, $z^\star \notin 0$, and $\alpha \in \mathbb{R}$ such that for all $k \in K$, $z^\star(k) \leq \alpha$ and $z^\star(sk^o - u_1(x)) > \alpha$, $z^\star(tk^o - u_2(x)) > \alpha$.
Thus $z^\star((s + t)k^o - (u_1 + u_2)(x)) > 2\alpha$. But, $(s + t)k^o - (u_1 + u_2)(x) \in K$. This gives us that $\{u_1 + u_2 \leq rk^o\} \subset \bigcap_{\substack{s,t \in \mathbb{R} \\ s+t=r}} \{u_1 \leq sk^o\} \cup \{u_2 \leq tk^o\}$. The other inclusion is trivial. □

Following [**5**] and [**11**], let us consider the functional $\tilde{z} : Z \to \mathbb{R}$ defined for $z \in Z$ by

$$\tilde{z}(z) = \inf\{t \in \mathbb{R}, z \in tk^o - K\}.$$

That is $\tilde{z}(z)$ is the smallest abscisse of a vector u proportional to k^o such that the shifted cone $u - K$ contains z. The functional $\tilde{z}$ is well defined because k^o belongs to the interior of K.
Let us recall some properties of $\tilde{z}$ we will use [**10**] and [**11**] :

(1) Let $r \in \mathbb{R}$, $(\tilde{z}(z) > r) \Leftrightarrow (z \notin rk^o - K)$,
(2) $\tilde{z}(0) = 0$; $\tilde{z}(k^o) = 1$; $\forall \alpha > 0$, $\tilde{z}(\alpha k^o) = \alpha$, $\tilde{z}(-\alpha k^o) = -\tilde{z}(\alpha k^o) = -\alpha$,
(3) $\tilde{z}$ is continuous sublinear,
(4) $\tilde{z}$ is strictly monotone [that is : if $z_1, z_2 \in Z$ and $z_1 \in z_2 + K \setminus \{0\}$ then $\tilde{z}(z_1) > \tilde{z}(z_2)$],
(5) If $f : X \to Z$ is bounded from below, it follows, from the monotonicity of $\tilde{z}$ that $\tilde{z} \circ f : X \to \mathbb{R}$ is also bounded from below,
(6) If $f : X \to Z$ is k^o-ℓsc then $\tilde{z} \circ f : X \to \mathbb{R}$ is ℓsc.

Let us also remark that if $\tilde{z} \circ f$ attains its minimum at $x_0 \in X$ then by the strict monotonicity of $\tilde{z}$, we have that $x_0 \in E(f)$. Let us denote for $f : X \to Z$,

$$\|f\|_\infty = \sup_{x \in X} \|f(x)\|$$

and

$$\|f\|_{\tilde{z}} = \sup_{x \in X} \left[\max\left\{\tilde{z}(f(x)), \tilde{z}(-f(x))\right\}\right].$$

PROPOSITION 2. *The norm $\|f\|_\infty$ is finite if and only if the norm $\|f\|_{\tilde{z}}$ is finite.*

PROOF. We can suppose $f \not\equiv 0$. Let us first suppose that $\|f\|_\infty < \infty$. As $k^o \in \text{int } K$, there exists $r > 0$ such that the ball $B(k^o, r)$ is contained in K. Then, for each $x \in X$, $k^o \pm r\frac{f(x)}{\|f\|_\infty} \in K$ and as K is a cone,

$$\{f(x), -f(x)\} \subset -K + k^o \frac{\|f\|_\infty}{r} \quad ,$$

and

$$\|f\|_{\tilde{z}} \leq \frac{1}{r}\|f\|_\infty.$$

On the other hand, suppose $\|f\|_{\tilde{z}} < \infty$. Then $\pm f(x) \in -K + \|f\|_{\tilde{z}} k^o$, i.e. $k^o \pm \frac{f(x)}{\|f\|_{\tilde{z}}} \in K$.
The expected result will follow from the fact that the set $\{z \in Z, k^o \pm z \in K\}$ is bounded. For that we prove that this set is weakly bounded. Let us suppose that it is not the case. Then there exist $z^\star \in Z^\star$ and $(z_n) \subset Z$ such that, for all n, $k^o \pm z_n \in K$ and $|z^\star(z_n)| > n$. It follows that, for all n, $z^\star(k^o \pm z_n) \in z^\star(K)$, which is a convex cone in $\mathbb{R}$. Then, $(z^\star(k^o \pm z_n))_n \subset \mathbb{R}^+$ or $z^\star(k^o \pm z_n)_n \subset \mathbb{R}^-$. Let us suppose for example that, for all n, $z^\star(k^o) \pm z^\star(z_n) \geq 0$. Then, for all n, $|z^\star(z_n)| \leq z^\star(k^o)$, which contradicts the fact that $|z^\star(z_n)| > n$. □

Vector-valued Deville–Godefroy–Zizler perturbed minimization principle

In the following result, we call a bump function on a Banach space X a real valued function on X with bounded non empty support. Let $f : X \to \mathbb{R} \cup \{+\infty\}$ be a function and $y \in X$ be such that $f(y) < \infty$, then we say that f attains its strong minimum on X at y if $f(y) = \inf\{f(x); x \in X\}$ and $\|y_n - y\| \to 0$, whenever $y_n \in X$ are such that $f(y_n) \to f(y)$.

THEOREM 3. *Let $(Y, \|\cdot\|_Y)$ be a complete convex cone of bounded, k^o-ℓsc functions from X to Z such that*

(i) *for all $g \in Y$, $\|g\|_\infty \le \|g\|_Y$,*
(ii) *for every $g \in Y$ and every $u \in X$, $\|\mathbf{T}_u g\|_Y = \|g\|_Y$, where $\mathbf{T}_u g(x) = g(u+x)$ for all x in X,*
(iii) *for every $g \in Y$ and every $a > 0$, the function $h : X \to Z$ defined by $h(x) = g(ax)$ belongs to Y,*
(iv) *There exists a bump function $b : X \to \mathbb{R}$ such that $b(0) > 0$ and $\tilde{b} = -bk^o$ belongs to Y.*

Let $f : X \to Z \cup \{+\infty\}$ be a k^o-ℓsc, bounded from below function such that

$$D(f) = \{x \in X, f(x) \in Z\} \neq \emptyset.$$

Then the set of all $g \in Y$ such that $\tilde{z} \circ (f + g)$ attains its strong mimimum on X is a dense G_δ subset G of Y.
[We put $\tilde{z}(+\infty) = +\infty$].

PROOF. The proof follows the ideas used by Deville–Godefroy–Zizler ([**3**]) and also the properties of the functional $\tilde{z}$. We consider the set :

$$\begin{aligned} U_n \quad &= \{g \in Y, \quad \exists x_0 \in X, \tilde{z}(f+g)(x_0) \\ &\qquad\qquad < \inf\{\tilde{z}(f+g)(x), x \notin B(x_0, 1/n)\}\}. \end{aligned}$$

where for convenience, we denote $\tilde{z}(f+g)(x) = \tilde{z}\big((f+g)(x)\big)$. Since f and g are bounded from below, by monotonicity of $\tilde{z}$, we have $\tilde{z} \circ (f+g)$ is bounded from below.

Let us prove that U_n is an open set. Let g be in U_n, then there exists $x_0 \in X$ such that $\epsilon := \inf\{\tilde{z}(f+g)(x), \ , x \notin B(x_0, 1/n)\} - \tilde{z}(f+g)(x_0) > 0$. So there by Proposition 2 and property (i), there exists $C > 0$ such that for all $h \in Y$, $\|h\|_{\tilde{z}} \le C\|h\|_\infty \le C\|h\|_Y$. Let $k \in Y$ such that $\|g - k\|_Y < \epsilon/4C$. By using the sublinearity of $\tilde{z}$, we get that

$$\tilde{z}(f+k)(x_0) \le \tilde{z}(f+g)(x_0) + \tilde{z}(k-g)(x_0)$$

By definition of ϵ, for $x \notin B(x_0, 1/n)$,

$$\tilde{z}(f+k)(x_0) \le \tilde{z}(f+g)(x) + \tilde{z}(k-g)(x_0) - \epsilon$$

and again by sublinearity of $\tilde{z}$,

$$\begin{aligned} \tilde{z}(f+k)(x_0) \quad &\le \quad \tilde{z}(f+k)(x) + \tilde{z}(g-k)(x) + \tilde{z}(k-g)(x_0) - \epsilon \\ &\le \quad \tilde{z}(f+k)(x) + 2C\|g-k\|_Y - \epsilon \end{aligned}$$

Thus, for $x \notin B(x_0, 1/n)$, $\tilde{z}(f+k)(x_0) \le \tilde{z}(f+k)(x) - \epsilon/2$. This proves that $k \in U_n$.
To prove that U_n is a dense set, let $g \in Y$ and $\epsilon > 0$, we need to find $h \in Y$, $\|h\|_Y < \epsilon$ and $x_0 \in X$ such that

$$\tilde{z}(f+g+h)(x_0) < \inf\{\tilde{z}(f+g+h)(x), x \notin B(x_0, 1/n)\}.$$

There exists a bump function $b : X \to \mathbb{R}$ such that $b(0) > 0$, $b(x) = 0$, whenever $\|x\| \ge 1/n$, $-bk^o \in Y$ and $\|bk^o\|_Y < \epsilon$. Since $\tilde{z} \circ (f+g)$ is bounded from below, there exists $x_0 \in X$ such that

$$\tilde{z}(f+g)(x_0) < \inf_{x \in X} \tilde{z}(f+g)(x) + b(0),$$

We consider $h(x) = -b(x-x_0)k^o$. Then, by translation invariance, $h \in Y$, $\|h\|_Y < \epsilon$ and since $\tilde{z}(k^o) = 1$, by using the sublinearity of $\tilde{z}$,

$$\tilde{z}(f+g+h)(x_0) \leq \tilde{z}(f+g)(x_0) + \tilde{z}(h)(x_0) = \tilde{z}(f+g)(x_0) - b(0) < \inf_{x\in X} \tilde{z}(f+g)(x)$$

and, if $x \in X \setminus B(x_0, 1/n)$, then

$$\tilde{z}(f+g+h)(x) = \tilde{z}(f+g)(x) \geq \inf_{x\in X} \tilde{z}(f+g)(x)$$

Hence $g + h \in U_n$ and U_n is dense in Y.
Consequently, $G = \cap_n U_n$ is a dense G_δ subset of the complete cone Y. We now claim that if $g \in G$, then $\tilde{z} \circ (f+g)$ attains its strong minimum on X. Indeed, let $x_n \in X$ be such that

$$\tilde{z}(f+g)(x_n) < \inf\{\tilde{z}(f+g)(x), x \notin B(x_n, 1/n)\}.$$

Following [**3**], it is easy to see that (x_n) is a Cauchy sequence, converging to some $x_\infty \in X$. As f and g are k^o-ℓsc then by Lemma 1, $f+g$ is also k^o-ℓsc and it follows that $\tilde{z} \circ (f+g)$ is ℓsc (property (6) of $\tilde{z}$) then

$$\tilde{z}(f+g)(x_\infty) \leq \liminf_n \tilde{z}(f+g)(x_n).$$

And following again [**3**] we get that $\tilde{z} \circ (f+g)$ attains its strong minimum at x_∞. Conversely, if $\tilde{z} \circ (f+g)$ attains its strong minimum then $g \in G$. □

COROLLARY. *Let $f : X \to Z$ be k^o-ℓsc and bounded from below. Then for each $\epsilon > 0$, there exists $g \in Y$ such that $\|g\|_Y < \epsilon$ and $\mathrm{Min}(f+g) \neq \emptyset$.*

REMARK 4. As $f : X \to Z$ is k^o-ℓsc and bounded from below, it follows that $\tilde{z} \circ f : X \to \mathbb{R}$ is ℓsc and bounded from below. If you take for Y a Banach space of bounded continuous real valued functions on $\mathbb{R}$ satisfying the usual conditions of the scalar Deville–Godefroy–Zizler principle [**3**], you get that $\{g \in Y, (\tilde{z} \circ f) + g$ attains its strong minimum on $X\}$ is a G_δ subset of Y. But of course there is no reason why g should be of the form $\tilde{z} \circ \ell$ with $\ell : X \to Z$ and moreover as $\tilde{z}$ is not linear we do not get any information on $\mathrm{Min}(f+\ell)$. So Theorem 3 cannot be deduced directly from the scalar perturbed minimization principle.

REMARK 5. In the theorem, the points where $\tilde{z} \circ (f+g)$ attains its strong minimum can be localized. Indeed, for every $\epsilon > 0$, there exists $a_\epsilon > 0$ such that whenever $y_0 \in X$ satisfies

$$(\star) \qquad (\tilde{z} \circ f)(y_0) < \inf(\tilde{z} \circ f) + a_\epsilon$$

there exist $g \in Y$, $x_0 \in X$ such that

(a) $\tilde{z} \circ (f+g)$ attains its minimum at x_0,
(b) $\|g\|_Y \leq \epsilon$ and $\|x_0 - y_0\| \leq \epsilon$.

You follow the scalar proof [**3**] in the following way : let $b : X \to \mathbb{R}$ be a bump function such that $b(0) = 1$ and $\tilde{b} = -bk^o$ belongs to Y. Let us denote by $b_\epsilon(x) = b(x/\epsilon)$. By Proposition 2 and property (i) of Y, there exists $C > 0$ such that for all $\ell \in Y$, $\|\ell\|_{\tilde{z}} \leq C\|\ell\|_\infty \leq C\|\ell\|_Y$. Then define : $a_\epsilon = \frac{\epsilon}{3\|b_\epsilon\|_Y + 1/C}$, $h(x) = -3a_\epsilon b_\epsilon(x-y_0)k^o$. Then by property (iv) of Y, h belongs to Y. By Theorem 3, there exist $k \in Y$ and $x_0 \in X$ such that $\|k\|_Y \leq a_\epsilon/C$ and $\tilde{z} \circ (f+h+k)$ attains its minimum at x_0. Let us define $g = h + k$. Then condition (a) is satisfied. Moreover, $\|g\|_Y \leq \|h\|_Y + \|k\|_Y \leq \epsilon$.

On the other hand, by sublinearity of $\tilde{z}$,

$$\begin{aligned}\tilde{z}(f+g)(y_0) &\le (\tilde{z}\circ f)(y_0) - 3a_\epsilon + (\tilde{z}\circ k)(y_0) \le (\tilde{z}\circ f)(y_0) - 3a_\epsilon + C\|k\|_Y \\ &< \inf_X(\tilde{z}\circ f) - a_\epsilon\end{aligned}$$

and, whenever $\|y - y_0\| \ge \epsilon$, by sublinearity of $\tilde{z}$,

$$\begin{aligned}\tilde{z}(f+g)(y) &= \tilde{z}(f+k)(y) \ge (\tilde{z}\circ f)(y) - C\|k\|_Y \\ &\ge \inf_X(\tilde{z}\circ f) - a_\epsilon.\end{aligned}$$

This implies that $\|x_0 - y_0\| < \epsilon$ and shows (b).

Let us remark that if y_0 satisfies $(\star)$ then y_0 is an a_ϵ-approximatively efficient point, that is a point satisfying

$$f(X) \cap (f(y_0) - a_\epsilon k^o - K \setminus \{0\}) = \emptyset.$$

We do not know if we can get the same localization property with a_ϵ-approximatively efficient points.

We now give some applications of our theorem. As a Lipschitz continuous function $g : X \to Z$ is k^o-ℓsc, we can take for Y the Banach space of all bounded Lipschitz continuous functions from X to Z equipped with the norm :

$$\|g\|_Y = \|g\|_\infty + \sup\left\{\frac{\|g(x) - g(y)\|}{\|x - y\|}, x \ne y\right\}.$$

COROLLARY. Let $f : X \to Z$ be k^o-ℓsc, bounded from below. Then, for every $\epsilon > 0$, there exists $g : X \to Z$ bounded Lipschitz continuous function such that

$$\|g\|_Y < \epsilon\,,\ \mathrm{Min}(f+g) \ne \emptyset.$$

Take now for Y the Banach space of all $g : X \to Z$ that are bounded, Lipschitz continuous and Fréchet-differentiable (resp. Gateaux-differentiable) equipped with the norm : $\|g\|_Y = \max(\|g\|_\infty, \|g'\|_\infty)$. We then get a vector-valued version of the Borwein-Preiss smooth perturbed minimization principle.

COROLLARY. Let X be a Banach space that admits a Lipschitz continuous bump function which is Fréchet-differentiable (resp. Gateaux-differentiable). Then for every $f : X \to Z$ k^o-ℓsc, bounded from below, and for every $\epsilon > 0$, there exists a function $g : X \to Z$ which is Lipschitz continuous and Fréchet-differentiable (resp. Gateaux-differentiable) and such that $\|g\|_\infty < \epsilon$, $\|g'\|_\infty < \epsilon$, $\tilde{z}\circ(f+g)$ attains its strong minimum and $\mathrm{Min}(f+g) \ne \emptyset$.

References

[1] J. Borwein and D. Preiss, Smooth variational principle with applications to subdifferentiability of convex functions, *Trans. Amer. Math. Soc. 303, 1987, 517-527.*

[2] J. Borwein, J. Penot and M. Théra, Conjugate convex operators, *J. Math. Anal. Appl. 102, 1984, 399-414.*

[3] R. Deville, G. Godefroy and V. Zizler, Smoothness and renormings in Banach spaces, *Pitman monographs and Surveys in Pure and Appl. Math..* Longman Scientific & Technical, 1993.

[4] I. Ekeland, On the variational principle, *J. Math. Anal. Appl. 47, 1974, 324-353.*

[5] A. Göpfert and Chr. Tammer, A new maximal point theorem, *Zeitschr. f. Analysis und Anwendungen*, 1994.

[6] J. Jahn, *Mathematical vector optimization in partially ordered spaces*, Lang Verlag Frankfurt, Bern, New York, 1986.

[7] P. Loridan, Well-posedness in vector optimization, *Cermsem 95.25.*

[8] J. Penot and M. Théra, Applications sous-linéaires à valeurs dans un espace de fonctions continues, *Annali di Matematica Pura ed Applicata, 136, 1984, 133-151.*

[9] J. Penot and M. Théra, Semi-continuous mappings in general topology, *Archiv der Math, 38, 1982, 158-166.*

[10] Chr. Tammer, A generalization of Ekeland's variational principle, *Optimization 25, 1992, 129-141.*

[11] Chr. Tammer, A variational principle and applications for vectorial control approximation problems, Reports of the *Inst. of Optimization and Stochastics.* Martin-Luther Universität Halle - Wittenberg, 1996.

UNIVERSITÉ DE BORDEAUX I, MATHÉMATIQUES PURES DE BORDEAUX, COURS DE LA LIBÉRATION, 351, F 33405 TALENCE CEDEX, FRANCE

E-mail address: deville@math.u-bordeaux.fr

UNIVERSITÉ DE MONS-HAINAUT, INSTITUT DE MATHÉMATIQUE ET D'INFORMATIQUE, "LE PENTAGONE", AVENUE DU CHAMP DE MARS, 6, B 7000 MONS, BELGIQUE

E-mail address: catherine.finet@umh.ac.be

Canadian Mathematical Society
Conference Proceedings
Volume **27**, 2000

Continued Fractions, Comparison Algorithms, and Fine Structure Constants

Philippe Flajolet and Brigitte Vallée

To Jonathan Borwein, on the occasion of his award of a Doctorate Honoris Causa

ABSTRACT. There are known algorithms based on continued fractions for comparing fractions and for determining the sign of 2×2 determinants. The analysis of such extremely simple algorithms leads to an incursion into a surprising variety of domains. We take the reader through a light tour of dynamical systems (symbolic dynamics), number theory (continued fractions), special functions (multiple zeta values), functional analysis (transfer operators), numerical analysis (series acceleration), and complex analysis (the Riemann hypothesis). These domains all eventually contribute to a detailed characterization of the complexity of comparison and sorting algorithms, either on average or in probability.

Motivations

The topic of this paper is the study of one of the simplest possible algorithms for one of the simplest conceivable tasks—the comparison of two fractions. The algorithm has been proposed in the celebrated "Hackers' Memorandum" known as HAKMEM [**4**], an amazing bag of tricks for computational mathematics that was collected at M.I.T. in 1972 by M. Beeler, R.W. Gosper, and R. Schroeppel. Amongst many gems relative to continued fraction algorithms and written by William Gosper, we find:

> **Item 101A (Gosper):** Numerical comparison of continued fractions is slightly harder than in decimal, but much easier than with rationals – just invert the decision as to which is larger whenever the first discrepant terms are even-numbered. Contrast this with the problem of comparing the rationals 113/36 and 355/113.

The algorithm suggested here compares two rational numbers a/b and c/d by means of a continued fraction expansion algorithm applied simultaneously to the two numbers, and stopped as soon as a discrepancy of quotients is encountered. For instance,

1991 *Mathematics Subject Classification.* Primary 11Y65, 68Q25, 65U05, 37A45, 11Y60.
Key words and phrases. Continued fraction, algorithms, sorting, constants.
This work was supported in part by the ALCOM-FT Project of the European Union.

we discover in three "short division" steps that

$$\alpha = \frac{224}{97} = \mathbf{2} + \cfrac{1}{\mathbf{3} + \cfrac{1}{\underline{\mathbf{4}} + \cfrac{1}{\mathit{3} + \cfrac{1}{\mathit{2}}}}}, \qquad \beta = \frac{95}{41} = \mathbf{2} + \cfrac{1}{\mathbf{3} + \cfrac{1}{\underline{\mathbf{6}} + \cfrac{1}{\mathit{2}}}}$$

satisfy $\alpha < \beta$ since the discrepant digits $\underline{\mathbf{4}}, \underline{\mathbf{6}}$ occur in the third position and satisfy $4 < 6$. In turn-of-the-millennium hacker's parlance, one might as well describe the algorithm as two "lazy" parallel executions of the continued fraction algorithm.

For rational numbers, $x = a/b$ and $y = c/d$, the algorithm estimates the quantity $\operatorname{sign}(a/b - c/d)$. The purest mathematician may well deem this discussion senseless. Don't we have

$$\operatorname{sign}\left(\frac{a}{b} - \frac{c}{d}\right) = \operatorname{sign}(ad - bc), \tag{1}$$

or cannot we plainly compute with floating point numbers,

$$\operatorname{sign}\left(\frac{a}{b} - \frac{c}{d}\right) \approx \operatorname{sign}\left(\operatorname{float}(\frac{a}{b}) - \operatorname{float}(\frac{c}{d})\right), \tag{2}$$

and easily determine the answers? Here is the main point. In the computational world, numbers are not "known" in the abstract, but rather accessible through finite or imperfect information only, like integral fraction forms or floating point approximations. In such a context, both *complexity* and *accuracy* of computations matter. The great advantage of the HAKMEM algorithm arises precisely from the fact that it operates within the *set precision* of data. For instance the straight evaluation (1) implies handling double precision numbers, that is, as much as 8 single-precision multiplications by naïve methods and 6 at least, when Karatsuba's technique is used [**27**]; at any rate, the bit complexity is superlinear (e.g., quadratic, if standard multiplication is used), while our results imply that *the HAKMEM algorithm has linear bit complexity* on average. On another register, if divisions are performed in floating point, then wrong decisions may occur: for instance, one has

$$\frac{312689}{99532} - \frac{833719}{265381} \approx 3 \times 10^{-11},$$

and the sign is not even computed correctly with 10 digit arithmetics, although the fractions only involve 6-digit integers. To the contrary, the HAKMEM algorithm always provides the correct answer within the set precision. Such reasons explain for instance the multiple uses of the algorithm (under the name of ab_vs_cd) in Knuth's design of the Metafont system: see [**28**].

The HAKMEM algorithm has surfaced in a diversity of contexts of which we now discuss a few. First, there are general-purpose computer arithmetics systems that are entirely based on continued fractions, a notable case being the one developed by Vuillemin around 1987 [**49**]; comparison of numbers in this context is likely to involve a version of the HAKMEM algorithm. Second, issues of correctness and robustness are central in the design of computational geometry systems: there, the comparison problem is identical to the problem of deciding the sign of 2×2 determinants (which means deciding the orientation of triangles); in this range of application, the HAKMEM algorithm has been rediscovered and extended by Avnaim et al. [**1**]. Finally, the HAKMEM algorithm is structurally very similar to an optimal algorithm proposed by Gauss for lattice reduction in dimension 2 that is of special importance in various areas of computational number theory; the two analyses entertain very close ties as may be seen by examining references [**14, 42**] on which the present paper is partly based.

Our primary goal is a characterization, both on average and in probability, of the behaviour of the simple comparison algorithm whose principles have just been sketched. We also discuss in the paper an interesting generalization to sorting based on continued fraction digits. First, it turns out that the algorithms are best placed within the general framework of symbolic dynamics and expanding maps of the interval (Section 1). A whole variety of algorithms result, including two versions of the basic sign algorithm, one dependent on basic (standard) continued fractions, the other one relying on centred continued fractions. Fundamental intervals that are familiar from the elementary theory of continued fraction then conduce to slowly convergent sums, called "moment sums", that express the average case complexity of the sign algorithms (Section 2) and of the corresponding (radix) sorting algorithms (Section 3). In the basic continued fraction case, moment sums appear to be related to multiple zeta values and Euler-Zagier constants, so that they are surrounded by a strong set of identities and fast convergent sum or integral representations (Section 4). In contrast, the centred case leads to non-analytic sums arising from the floor function and irrational multipliers so that specific series acceleration need to be developed that rely on lattice generating functions (Section 5).

Functional analysis is then brought into play in Section 6. It turns out that transfer operators of the Ruelle-Mayer type have dominant eigenvalues that precisely describe the geometric rate of decay in the probability distribution of costs of the sign-algorithms (Sections 6). As a final surprise (Section 7), the cost of sorting n numbers exhibits an asymptotic dependency on n whose character is dictated by the position of the nontrivial zeros of the Riemann zeta function, that is to say, by the Riemann hypothesis.

One of the themes of this paper is the effective computability of the constants involved in the analysis. A real number α is said to be *polynomial time computable* if an approximation of α to accuracy 10^{-d} can be computed in time $\mathcal{O}(d^r)$ for some integer r. We let **P** denote the class of such numbers. (See [**8**] for an introduction to this aspect of computational mathematics.) Effective numerical procedures usually go along with proofs of membership in **P**.

1. Expanding maps, sign algorithms, and continued fractions

Symbolic dynamics concerns itself with the interplay between properties of a continuous transformation and discrete properties of trajectories of points under iteration of the transformation. A common framework in symbolic dynamics is the one of *expanding maps of the interval* and their associated *number representation systems*. The definition involves a triplet $(\mathcal{I}, U, \mathcal{M})$ that satisfies the following conditions:

(E_1) The set $\mathcal{M}$, whose elements are called digits, is a subset of the integers; there is a partition $\{I_j\}_{j\in\mathcal{M}}$ of the interval $\mathcal{I}$ satisfying additionally the condition below.

(E_2) The function $U(x)$ maps monotonically each I_j *onto* $\mathcal{I}$ and is expanding, meaning that $|U'(x)| > 1$ on the closure $\overline{I}_j$ of I_j.

See [**38**] for some of the main properties. An expanding map thus consists of finitely or denumerably many branches indexed by some set $\mathcal{M}$. We let $m(x) \in \mathcal{M}$

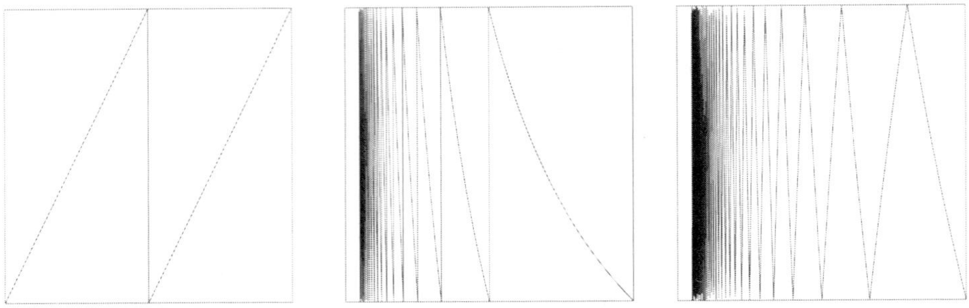

FIGURE 1. The three maps corresponding to binary representations ($\{2x\}$, left), basic continued fractions ($\{1/x\}$, middle), and centred continued fractions ($\{\{1/x\}\}$, right).

represent the branch of $U(x)$ in the domain of which x falls. A coding of a real number x is then obtained by the sequence

$$m(x),\ m(U(x)),\ m(U^2(x)),\ m(U^3(x)),\ \ldots \tag{3}$$

where powers denote iteration, $U^2 = U \circ U$, etc. The sequence of *digits* of x, $\mathbf{m} = (m_1, \ldots, m_k, \ldots) = (m(x), \ldots, m(U^{(k)}(x)), \ldots)$ is then produced by the following simple algorithm.

procedure $(\mathcal{I}, U, m)$-**expansion**$(x : \mathcal{I})$
  for $k := 1$ **to** $+\infty$ **do**
    $m_k := m(x);\ x := U(x);$

A special rôle is played by the set $\mathcal{H}_1$ of branches of the inverse function $U^{(-1)}$ of U that are also naturally numbered by the index set $\mathcal{M}$. If x_k and the sequence $m = (m_1, \ldots, m_k)$ is known, the algorithm can be run backwards, and the original x_0 is recovered by

$$x_0 = h_{\mathbf{m}}(x_k) \quad \text{where} \quad h_{\mathbf{m}}(y) = h_{m_1} \circ h_{m_2} \circ \cdots \circ h_{m_k}(y).$$

The set $\mathcal{H}$ of all compositions $h_{\mathbf{m}}$ is referred to as the set of *branches of the inverse function* or *inverse branches* for short; the index k in $h_{(m_1,\ldots,m_k)}$ is called the depth of h and denoted by $|h|$. Clearly, the stochastic behaviour of a numbering system is closely related to the dynamics of $U(x)$ on the interval $\mathcal{I}$; this dynamics is itself isomorphic to the dynamics of the semigroup of contractions $\mathcal{H}$ that is generated by the depth-one inverse branches $\mathcal{H}_1$.

Let $\lfloor y \rfloor$ be the integer part function of y and $\{y\} = (y \bmod 1)$ the fractional part. The scheme (3) generalises the usual binary representation of real numbers (see Fig. 1) that corresponds to fixing the choice

$$\mathcal{I} = [0, 1], \quad U(x) = \{2x\}, \quad m(x) = \lfloor 2x \rfloor.$$

As we shall explain in some detail, this constitutes also a very convenient framework for a discussion of continued fraction algorithms.

Sign algorithms. The basic numeration scheme relative to an abstract triplet $(\mathcal{I}, U, m)$ gives rise to a semi-decision algorithm that determines if two numbers are distinct: in order to detect distinctness of x and x' simply expand "lazily" in parallel x and x' until the first discrepant digit is detected. The scheme is then

procedure $(\mathcal{I}, U, m)$-**sign**$(x, x' : \mathcal{I})$
  for $k := 1$ **to** $+\infty$ **do**
    $m_k := m(x)$; $m'_k := m(x)$; $x := U(x)$; $x' = U(x')$;
    if $m_k \neq m'_k$ **then exit**.

Since the branches of U are by assumption each monotonic, the knowledge of the partial codings $(m_1, \ldots, m_{k-1}, m_k)$ and $(m_1, \ldots, m_{k-1}, m'_k)$, where k denotes the position of the first discrepant digit, fully determine the sign of the difference $x - x'$. For instance, binary representations, where every branch is monotone increasing, have $\text{sign}(x - x') = \text{sign}(m_k - m'_k)$.

In the sequel, we examine two numeration systems relative to continued fractions: the basic continued fraction system (where characteristic quantities are indicated by a bar sign $(\overline{\cdot})$) and the centred continued fraction system (indicated by a hat $(\widehat{\cdot})$).

Basic continued fraction expansions. The *basic continued fraction* (BCF) system fits into the general framework. It is defined by the triplet $(\overline{\mathcal{I}}, \overline{U}, \overline{m})$

$$\overline{I} = [0,1],\ \overline{U}(x) = \left\{\frac{1}{x}\right\},\ \overline{m}(x) = \left\lfloor \frac{1}{x} \right\rfloor.$$

By convention, we take $\overline{U}(0) = 0$, $\overline{m}(0) = 0$.

To one step of the BCF expansion algorithm, there corresponds the set of inverse branches of depth 1,

$$\overline{\mathcal{H}}_1 := \{h(z) = \frac{1}{m+z} \quad m \geq 1\},$$

and the set of digits is $\overline{\mathcal{M}} = \{1, 2, \ldots\}$. The inverse branches of arbitrary depth are then *linear fractional transformations* (LFT's for short). To a real number $x \in \overline{\mathcal{I}}$, the BCF algorithm thus associates the sequence of iterates $x_0 = x, x_1, x_2, \ldots, x_k$, and (with the convention $1/0 = 0$)

$$x_0 = h_{\mathbf{m}}(x_k) \qquad \text{where} \quad h(y) = \cfrac{1}{m_1 + \cfrac{1}{m_2 + \cfrac{1}{\ddots \, \overline{m_k + y}}}}.$$

The BCF-sign algorithm dependent on basic continued fraction expansions is exactly the HAKMEM algorithm described in the introduction. Here all branches are monotone decreasing, so that

$$\text{sign}(x - x') = (-1)^k \text{sign}(m_k - m'_k),$$

since each stage of the iteration is orientation-reversing.

Centred continued fraction expansions. Centred continued fractions are obtained when one replaces truncation (integer part function) by rounding to the nearest integer. First introduce the notations

$$[[y]] = \left\lfloor x + \frac{1}{2} \right\rfloor,\ \{\{y\}\} = |y - [[y]]|,\ \varepsilon(y) = \text{sign}(y - [[y]]),$$

so that the following identity holds:

$$y = [[y]] + \varepsilon(y)\{\{y\}\}.$$

Then, the *centred continued fraction* (CCF) system is defined by the triplet $(\widehat{\mathcal{I}}, \widehat{U}, \widehat{m})$ given by

$$\widehat{\mathcal{I}} = [0, \frac{1}{2}], \ \widehat{U}(x) = \{\{\frac{1}{x}\}\}, \ \widehat{m}(x) = \varepsilon(\frac{1}{x})[[\frac{1}{x}]],$$

with, conventionally, $\widehat{U}(0) = 0$, $\widehat{m}(0) = 0$, $1/0 = 0$.

The CCF expansion algorithm operates on the interval $\widehat{\mathcal{I}} = [0, \frac{1}{2}]$. By design, the allowed indices m must satisfy the condition $|m| \geq 2$, and, furthermore, the value $m = -2$ does not occur since $m(1/x) = -2$ could only arise from an x larger than $\frac{1}{2}$. In this way, the set of LFT's of depth 1 for the CCF–algorithm is

$$\widehat{\mathcal{H}}_1 := \left\{ h(z) = \frac{1}{m+z} \quad m \geq 2 \right\} \cup \left\{ h(z) = \frac{1}{m-z} \quad m \geq 3 \right\},$$

and the set of digits is $\widehat{\mathcal{M}} = \{+2, -3, +3, -4, +4, \dots\}$.

The relation between x_0 and x_k is described by a linear fractional transformation of depth k, which is associated to the k-tuple of signed integers $\mathbf{m} = (m_1, \dots, m_k)$

$$h_{\mathbf{m}}(z) = \cfrac{1}{|m_1| + \cfrac{\varepsilon_1}{|m_2| + \cfrac{\varepsilon_2}{\ddots \, \overline{|m_k| + \varepsilon_k z}}}}, \qquad \varepsilon_j = \mathrm{sign}(m_j). \tag{4}$$

The CCF-sign algorithm based on centred continued fractions is such that an iteration is increasing if the corresponding sign is -1 and decreasing otherwise. In that case, one has

$$\mathrm{sign}(x - x') = \sigma \prod_{j=1}^{k-1} (-\mathrm{sign}(m_j)),$$

where $\sigma \in \{-1, 1\}$ is $+1$ iff $m_k \succ m'_k$ in the nonstandard order over on $\mathbb{Z}$,

$$0 \prec -1 \prec +1 \prec -2 \prec +2 \prec -3 \prec +3 \cdots,$$

that is also the lexicographic order on $\mathbb{N} \times \{-1, +1\}$.

Note. A variety of practical implementations result from the sign algorithms based on continued fraction representations. For instance, when applied to rational numbers, inverses are effected plainly by exchanging components of integer pairs, and the sign algorithm should be halted as soon as one of the continued fraction expansions terminates, etc. When applied to real numbers in fixed precision, the number of iterations can be bounded in advance by a quantity that depends on the machine precision, or on the size of the mesh according to which numbers are considered as distinct.

2. Fundamental intervals and the sign algorithms

The overall purpose of this paper is an understanding of the complexity of the comparison algorithm and its cognates, in the context of continued fraction representations. The probabilistic model adopted is the uniform model over the set of legal inputs and the complexity measure of the sign algorithms is the number L of iterations performed (that is, the value of the index k of the first discrepant digit). Knuth **[28]** reports for his version of the basic HAKMEM algorithm that "*in most cases, a quick decision is reached*" while Avnaim et al. [**1**] estimate from simulations that "*the average number of iterations is [about] 1.5*".

Following good precepts, we start with an experimental approach. Simulating 10^6 executions of the BCF and CCF sign algorithms over uniform pairs of real $(0,1)$ numbers provides empirical estimates on the probability distributions of costs as well as on the average-case complexity of the algorithms. Here is a tabulation of what we find.

	Basic CF-sign	Centred CF-sign
$\Pr(L=1)$	0.710050	0.918003
$\Pr(L=2)$	0.241275	0.075710
$\Pr(L=3)$	0.038339	0.000422
$\Pr(L=4)$	0.008424	0.000035
$\Pr(L=5)$	0.001608	0.000005
$E(L)$	1.351612	1.088791

It is apparent that the probabilities of the main loop being executed k times decay roughly geometrically with k —somewhat like 5^{-k} for the basic algorithms and 13^{-k} for the centred algorithm. (The centred algorithm is based on a better approximation scheme that converges faster; in practice, the advantage may however be offset by a costlier internal loop.) In accordance with the geometric decay, the expectations are finite. For 10^6 simulations, counting on a statistical error of about 10^{-3}, we anticipate the values $E(\overline{L}) = 1.352 \pm 0.001$ and $E(\widehat{L}) = 1.089 \pm 0.001$. These heuristic observations will receive a sound mathematical basis in the forthcoming sections: see for instance the table of Figure 6 for a final summary. We also note that alternative data and complexity models can be studied by suitable adaptations of the methods presented here; examples include discrete data [**14**], nonuniform input distributions [**42**], and bit-complexity models (work in progress).

Fundamental intervals. Let us first examine the sign algorithm in the context of a numbering system arising from a general expanding map $(\mathcal{I}, U, m)$. Borrowing terminology from the standard metric theory of continued fraction, we call any $h(\mathcal{I})$, the transform by h of the interval $\mathcal{I}$, a *fundamental interval*, and the depth of h is called the *rank* of the fundamental interval. Thus, the fundamental interval contains all real numbers whose expansion starts with the k-tuple $(\mathbf{m})$ indexing h $(h = h_{\mathbf{m}})$.

We adopt here the uniform probability model (Lebesgue measure) on $\mathcal{I}$. The probability that a number of $\mathcal{I}$ belongs to the fundamental interval relative to the branch h is called the *fundamental measure* associated to h and is denoted by u_h. Without loss of generality, we restrict the interval to be of the type $\mathcal{I} = [0, \alpha]$ for some $\alpha > 0$. The fundamental measure is then from its definition

$$u_h = \frac{1}{\alpha}\,|h(0) - h(\alpha)|.$$

For analysing the 2-dimensional sign algorithm, the square $\mathcal{I} \times \mathcal{I}$ is by assumption endowed with the uniform product measure. The probabilistic event $[L \geq k+1]$ is formed with all real pairs $(x, y) \in \mathcal{I} \times \mathcal{I}$ which have expansions that coincide till depth k. Then the two real numbers x and y belong to the same fundamental interval $h(\mathcal{I})$ of rank k, and the pair (x, y) belongs to the square $h(\mathcal{I}) \times h(\mathcal{I})$. Such

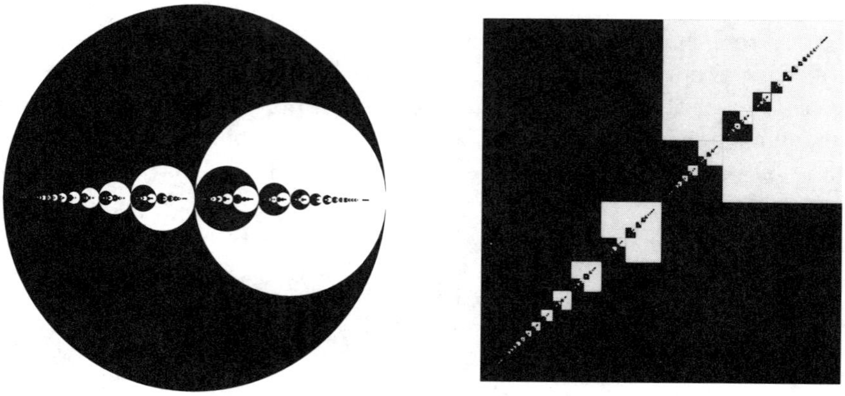

FIGURE 2. Basic continued fractions: the fundamental intervals represented by their supporting discs (left) and the fundamental squares (right). The events $[L = k]$ for ranks $k = 0\,..\,5$ are alternatively represented by dark and light colours.

a square is called a *fundamental square* of rank k. The event $[L \geq k+1]$ is thus equal to the disjoint union of all the fundamental squares of rank k,

$$[L \geq k+1] = \bigcup_{|h|=k} h(\mathcal{I}) \times h(\mathcal{I}). \tag{5}$$

Since fundamental squares are disjoint the probability that the Sign algorithm performs at least $k+1$ iterations is a sum that involves the measures of fundamental squares u_h^2, and

$$\Pr(L \geq k+1) = \sum_{|h|=k} u_h^2 = \frac{1}{\alpha^2} \sum_{|h|=k} |h(0) - h(\alpha)|^2. \tag{6}$$

The average number of iterations is then obtained by the usual summation of probabilities,

$$E[L] = \sum_h u_h^2 = \frac{1}{\alpha^2} \sum_h |h(0) - h(\alpha)|^2. \tag{7}$$

More generally, we define the *moment sums*

$$\rho^{(\ell)} := \sum_h u_h^\ell = \frac{1}{\alpha^\ell} \sum_h |h(0) - h(\alpha)|^\ell$$

which, for integer values of ℓ, are central in the the n-sorting algorithm to be discussed in Section 3.

Continued fraction algorithms. From the expression obtained in (5), the average-case analysis of any expanding-map version of the sign-algorithm is dependent upon an evaluation or at least a numerical estimation of the simplest of all moment sums, namely $\rho^{(2)}$. We now specialize the discussion to the basic and centred continued fraction algorithms. This depends upon the characterization of all the LFT's of $\overline{\mathcal{H}}$ and $\widehat{\mathcal{H}}$ provided by elementary number theory and a theorem of Hurwitz.

— The set $\overline{\mathcal{H}}$ of all possible LFT's used by the BCF–algorithm is

$$\overline{\mathcal{H}} := \left\{ \frac{az+b}{cz+d} \mid (a,b,c,d) \in \mathbb{N}^4,\ |ad-bc| = 1,\ c \le d,\ a \le c,\ b \le d \right\}.$$

— The set $\widehat{\mathcal{H}}$ of all possible LFT's used by the CCF–algorithm involves the golden ratio $\phi = (1+\sqrt{5})/2$ (Hurwitz [**25**]):

$$\widehat{\mathcal{H}} := \left\{ \frac{az+b}{cz+d} \mid (b,d) \in \mathbb{N}^2,\ (a,c) \in \mathbb{Z}^2,\ ac \ge 0, |ad-bc| = 1,\ \frac{-d}{\phi^2} < c < \frac{d}{\phi},\ |a| \le \frac{|c|}{2},\ b \le \frac{d}{2} \right\}.$$

It is observed that the numbers (a,b,c,d) are such that the pairs (a,c) and (b,d) are built with coprime integers, since the Bézout condition $ad-bc = \pm 1$ is satisfied. Thanks to this characterization, the expected cost of the sign algorithms falls as a ripe fruit.

THEOREM 1. *The expected cost of the basic and centred sign algorithms are expressible as sums over lattice points in* $\mathbb{N}^2$,

$$E(\overline{L}) = \overline{\rho}^{(2)}, \qquad E(\widehat{L}) = \widehat{\rho}^{(2)},$$

where, generally, the moment sums of index ℓ *satisfy*

$$(8)\qquad \begin{aligned} \overline{\rho}^{(\ell)} &= 1 + \frac{1}{2^\ell} + \frac{2}{\zeta(2\ell)} \sum_{d<c<2d} \frac{1}{c^\ell d^\ell}, \\ \widehat{\rho}^{(\ell)} &= \frac{2^\ell}{\zeta(2\ell)} \sum_{d\phi<c<d\phi^2} \frac{1}{c^\ell d^\ell} \qquad (\phi = (1+\sqrt{5})/2). \end{aligned}$$

PROOF. Consider first the basic case. Each pair (c,d) that satisfies $\gcd(c,d) = 1, d \ge 2$ and $0 < c < d$ is associated to two LFT's of $\overline{\mathcal{H}}$, since a rational number interior to $(0,1)$ admits at the same time a proper and an improper continued fraction expansion (the last digits is ≥ 2, resp. $= 1$). Then, taking into account boundary cases, namely pairs $(0,1)$ and $(1,1)$, we find

$$\overline{\rho}^{(\ell)} = 1 + \frac{1}{2^\ell} + 2 \sum \frac{1}{d^\ell (c+d)^\ell},$$

where the sum is taken over all the pairs (c,d) satisfying $\gcd(c,d) = 1, d \ge 2$ and $0 < c < d$. The general term in the last sum is homogeneous of degree 2ℓ, so that the gcd condition is eliminated provided one divides the sum by $\zeta(2\ell)$.

Consider next the centred case. Each pair (c,d) that satisfies $\gcd(c,d) = 1, d \ge 1$ and $-d/\phi^2 < c < d/\phi$ is associated to a unique LFT of $\widehat{\mathcal{H}}$. Then, we find

$$\widehat{\rho}^{(\ell)} = 2^\ell \sum \frac{1}{d^\ell (2d+c)^\ell},$$

where the sum is taken over all the pairs (c,d) satisfying $\gcd(c,d) = 1, d \ge 1$ and $-d/\phi^2 < c < d/\phi$. In this case, the integer $f := 2d + c$ satisfies $d\phi < f < d\phi^2$. The gcd condition is again eliminated by means of a factor $1/\zeta(2\ell)$. □

The moment sums of the basic and the centred algorithm look superficially similar. The summation condition is in each case determined geometrically as a sum over *lattice cones* defined by the two conditions,

$$d < c < 2d \qquad \text{and} \qquad \lceil d\phi \rceil \le c \le \lfloor d\phi^2 \rfloor.$$

However, the corresponding sums $\overline{\rho}^{(\ell)}$ and $\widehat{\rho}^{(\ell)}$ belong to different classes of mathematical constants. The basic ones, $\overline{\rho}^{(\ell)}$ live in a world of identities, integral representations, and special functions. Indeed, as we shall see the mean $\overline{\rho}^{(2)}$ is reducible to $\zeta(3)$ and a polylogarithm. More generally, $\rho^{(\ell)}$ is a member of the class of Euler-Zagier sums, currently a subject of active interest. In contrast, the sums $\widehat{\rho}^{(\ell)}$, starting with $\widehat{\rho}^{(2)}$, have a "nonanalytic character" due to the summation condition $\lceil d\phi \rceil \leq c \leq \lfloor d\phi^2 \rfloor$, with ϕ being irrational, and there is little hope of reducing them to some standard closed form. We shall see later that they can at least be subjected to appropriate series acceleration methods leading to efficient evaluation to high precision.

3. The sorting algorithm

The sign problem leads naturally to the more general question of testing distinctness of n real numbers, or, analogously, to the problem of sorting n numbers. For numbering systems like the ones considered here, it is the idea of *radix sorting* [**26**] that is natural. The principle is to determine the leading digit of each of the quantities $x_1, \ldots, x_n$ to be sorted, group them according to their leading digits, and then sort recursively the subgroups based on the following leading digits. Alternatively, one may view the algorithm as building successive refinements of the coarsest partition until the finest partition is obtained. These refinements are themselves best described as a tree called the *digital tree* or *trie*, which is one of the major data structures of computer science [**26, 31**]. The analysis developed here then answers the following question: How many digits in total must be determined in order to distinguish (and sort) n real numbers? In symbolic dynamical terms, it describes the way trajectories of n random points under the "shift" U evolve by sharing some common digits before branching off from each other.

Digital trees. We now formulate the sorting algorithm in its more convenient graphic tree version. To any finite set $X := \{x_1, x_2, \ldots, x_n\}$ of n real numbers of $\mathcal{I}$, one associates a *digital tree*, trie(X), defined by the two recursive rules:

(R_1) If $X = \{x\}$ has cardinality equal to 1, then trie(X) consists of a single *leaf node* that contains x.

(R_2) If X has cardinality at least 2, then trie(X) is an *internal node* represented generically by 'o' to which are attached r subtrees, where $r := \operatorname{card} m(X)$ is the number of different head digits in X. Let $b_1, b_2, \cdots, b_r$ be the different head symbols that appear in X; trie(X) is defined by

$$\text{trie}(X) = \langle o, \text{trie}(X_1), \text{trie}(X_2), \ldots, \text{trie}(X_r) \rangle,$$

where

$$X_i = \{U(x) \mid m(x) = b_i,\, x \in X\}.$$

Such a tree structure underlies classical radix sorting methods. The tree can be built by following the recursive rules (R_1), (R_2) and a suitable ordering of subtrees gives rise to an algorithm that sorts any set X either lexicographically or under the natural order on the real numbers.

Analysis of the sorting algorithm. According to standard terminology, the *level* of a node in a digital tree is the number of edges that connect it to the root. The *path length* of the tree is the sum of the levels of all leaves. The path length of trie(X) is thus equal to the total number of digits that need to be examined in

$\phi - 1$	=	/**1, 1, 1**, *1, 1, 1,...*/
γ	=	/**1, 1, 2**, *1, 2, 1,...*/
$\exp(1) - 2$	=	/**1, 2, 1**, *1, 4, 1,...*/
$\log 2$	=	/**1, 2, 3**, *1, 6, 3,...*/
$\{\exp(\pi\sqrt{163})\}$	=	/**1, 1333462407511**, *1, 8, 1, 1, ...*/
$2^{1/3} - 1$	=	/**3**, *1, 5, 1, 1, , 4,...*/
$\pi - 3$	=	/**7**, *15, 1, 292, 1, 1,...*/

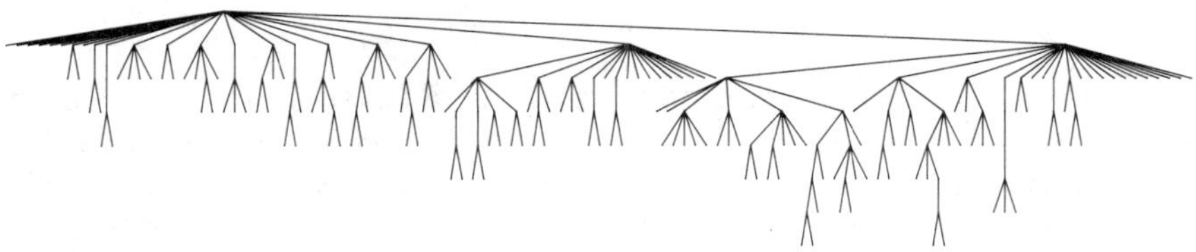

FIGURE 3. (Top) The fractional parts of 7 real numbers: the digits in bold represent the minimal prefix set of 16 digits (the sorting cost, equivalently path length) needed to distinguish the numbers from each other. (Bottom) A randomly drawn digital tree of size 100 built from basic continued fractions has path length 1323, height 16, and number of internal nodes equal to 222.

order to distinguish all elements of X and radix-sort X (see Figure 3) so that we adopt it as our definition of the *sorting cost.*

First, we discuss the case of a general numbering system. Let $P(n)$ be the expectation of the path length of trie(X) when X is drawn from $\mathcal{I}^n$ with the uniform probability measure. We claim the equality:

$$P(n) = \sum_{h\in\mathcal{H}} nu_h \left[1 - (1-u_h)^{n-1}\right]. \tag{9}$$

Indeed, let L_i be the length of the branch whose leaf contains the designated element x_i. The event $[L_i > k]$ means that there exists a fundamental interval of depth k that contains x_i and at least another element x_j. Thus, we have

$$\Pr(L_i > k) = \sum_{|h|=k} u_h \left[1 - (1-u_h)^{n-1}\right].$$

Summing over all possible values of index i and integer k then yields equation (9).

Next, a straight binomial expansion provides an expression for $P(n)$ that reduces to a linear combination of the "moment sums":

$$P(n) = n\sum_{\ell=1}^{n-1}(-1)^{\ell+1}\binom{n-1}{\ell}\rho^{(\ell+1)}. \tag{10}$$

In other words, $P(n)$ is nothing but an $(n-1)$st order difference of the moment sums.

The discussion specializes immediately to tries built on continued fractions and hence to the corresponding sorting algorithms BCF-sort and CCF-sort.

THEOREM 2. *The expectations of the number of digit inspections performed by the sorting algorithms BCF-sort (basic continued fractions) and CCF-sort (centred* *continued fractions) applied to n real numbers are given by*

(11)
$$\overline{P}(n) = n\sum_{\ell=1}^{n-1}(-1)^{\ell-1}\binom{n-1}{\ell}\overline{\rho}^{(\ell+1)}, \qquad \widehat{P}(n) = n\sum_{\ell=1}^{n-1}(-1)^{\ell-1}\binom{n-1}{\ell}\widehat{\rho}^{(\ell+1)}.$$

There, $\overline{\rho}^{(\ell)}$ and $\widehat{\rho}^{(\ell)}$ are the moment sums of Eq. (8).

The theorem predicts

$$P(2) = 2\,\rho^{(2)}, \quad P(3) = 6\,\rho^{(2)} - 3\,\rho^{(3)}, \quad P(4) = 12\,\rho^{(2)} - 12\,\rho^{(3)} + 4\,\rho^{(4)}.$$

The form $P(2) = 2\rho^{(2)}$ is in accordance with the analysis of the sign algorithm: in this special case, there are two leaves at the same level, and this level coincides with the number of iterations of the sign algorithm. Figure 3 (bottom) displays a random trie of size 100 based on the BCF system and kindly communicated to us by Julien Clément from his memoir [**11**].

We shall see in the last section that the asymptotic behaviour of path length (when n tends to infinity) involves the location of nontrivial zeros of the zeta function and hence is related to the Riemann hypothesis.

4. Multiple zeta values and the basic sign algorithm

The moment sums associated to the basic continued fraction system are double sums of simple rational functions and it is of interest to determine to what extent they relate to other classical constants of analysis. The adequate framework is that of *multiple zeta values* (MZV's), also known as Euler–Zagier sums, that we take here under the form

$$(12) \qquad \zeta^{\varepsilon_1,\dots,\varepsilon_r}(s_1,\dots,s_r) := \sum_{n_1>\dots>n_r} \frac{\varepsilon_1^{n_1}\cdots\varepsilon_r^{n_r}}{n_1^{s_1}\cdots n_r^{s_r}}, \qquad \varepsilon_j \in \{-1,+1\}.$$

The term "values" stresses the fact that instances at positive integer arguments are considered. The quantity $w = s_1 + \dots + s_r$ is called the weight and r is called the rank or the multiplicity of the sum. The simplest sums are those of rank 1,

$$\zeta^+(s) \equiv \zeta(s) = \sum_{n=1}^{\infty}\frac{1}{n^s}, \qquad \zeta^-(s) = \sum_{n=1}^{\infty}\frac{(-1)^n}{n^s} = (2^{1-s}-1)\zeta(s),$$

that is, the Riemann zeta function and its alternating variant. Especially relevant to the present discussion are the double zeta values

$$\begin{aligned}
\zeta^{++}(s,t) \equiv \zeta(s,t) &= \sum_{n=1}^{\infty}\sum_{q=1}^{n-1}\frac{1}{n^s q^t} = \sum_{n=1}^{\infty}\frac{H_{n-1}^{(t)}}{n^s}\\
\zeta^{-+}(s,t)) &= \sum_{n=1}^{\infty}\sum_{q=1}^{n-1}\frac{(-1)^n}{n^s q^t} = \sum_{n=1}^{\infty}(-1)^n\frac{H_{n-1}^{(t)}}{n^s},
\end{aligned}$$

with $H_n^{(t)} = \sum_{j\le n} j^{-t}$ the generalized harmonic number of order t.

The origin of these sums lies in a letter of Goldbach to Euler dated December 24, 1742 (see Berndt's book [**5**, Ch. 9] and references therein), with evaluations like

$$\zeta(2,1) = \sum_{n=1}^{\infty}\frac{H_{n-1}}{n^2} = \zeta(3).$$

published in 1775 by Euler and further developed by Nielsen a century later [**35**]. Interest was rekindled recently when Zagier in his address [**51**] demonstrated connections with many "deep" areas[1] of pure mathematics, at a time when, independently, Borwein and collaborators [**3, 6**] were discovering many surprising identities like the spectacular septic sum

$$\sum_{n=1}^{\infty} \frac{(H_{n-1})^4}{n^3} = \frac{185}{8}\zeta(7) - \frac{43}{2}\zeta(3)\zeta(4) + 5\zeta(2)\zeta(5),$$

that we leave as a challenge to the reader. (Hint: see for instance [**17**] for an elementary approach.) A partial evaluation is any reduction of a multiple zeta value to a polynomial form in multiple zeta values of smaller rank; a complete evaluation means a reduction to a polynomial form in zeta values $\zeta(2), \zeta(3), \ldots,$ and $\zeta^-(1) = -\log 2$. What is known regarding positive and alternating double zeta values is summarized in Fig. 4.

It is easy to reduce the moment sums to multiple zeta values. Start with

$$(13) \qquad \sum_{d<c<2d} \frac{1}{c^\ell d^\ell} = \sum_{c<2d} \frac{1}{c^\ell d^\ell} - \sum_{c} \frac{1}{c^{2\ell}} - \sum_{c<d} \frac{1}{c^\ell d^\ell}.$$

The last sum in (13) is a non-alternating zeta value $\zeta(\ell, \ell)$ that is of the "central" type and that must reduce in accordance with what is summarized in Fig 4. Indeed, distributing the sums in an expansion of the product $\zeta(s) \times \zeta(t)$ yields the "shuffle relations":

(14)

$$\zeta(s)\zeta(t) = \zeta(s,t) + \zeta(s+t) + \zeta(t,s), \quad \text{implying} \quad \zeta(\ell,\ell) = \frac{1}{2}\left(\zeta(\ell)^2 - \zeta(2\ell)\right).$$

The first sum in (13) is none other than the even part of an alternating zeta, namely,

$$(15) \qquad \sum_{c<2d} \frac{1}{c^\ell d^\ell} = 2^{\ell-1} \sum_{c<e} \frac{1+(-1)^e}{c^\ell e^\ell} = 2^{\ell-1}\left(\zeta^{++}(\ell,\ell) + \zeta^{-+}(\ell,\ell)\right).$$

Thus, by (13), (14), and (15), we find

$$(16) \qquad \rho^{(\ell)} = 2^{-\ell} - 2^{\ell-1} + \left(2^{\ell-1} - 1\right)\frac{\zeta(\ell)^2}{\zeta(2\ell)} + \frac{2^\ell}{\zeta(2\ell)}\zeta^{-+}(\ell,\ell).$$

This is as far as one can go for general values of ℓ, since a reduction of $\zeta^{-+}(\ell,\ell)$ is not there.

The particular case of the mean cost of the BCF-sign algorithm, $\ell = 2$, benefits of specific results due to Sitaramachandrarao [**40**] and De Doelder [**15**], themselves motivated by identities of Ramanujan [**5**]. First, Sitaramachandrarao relates $\zeta^{-+}(3,1)$ and $\zeta^{-+}(2,2)$ by

$$(17) \qquad 32\zeta^{-+}(3,1) + 16\zeta^{-+}(2,2) + 5\zeta(4) = 0.$$

This nontrivial identity is Eq. (5.4) of [**40**]. The next step involves a special instance of the polylogarithm function [**30**] that is classically defined by

$$\mathrm{Li}_m(z) = \frac{z}{1^m} + \frac{z^2}{2^m} + \frac{z^3}{3^m} + \cdots.$$

[1]Multiple zetas and polylogarithms are nowadays found relevant to a large body of science including knot invariants, Feynman diagrams, and even the theory of perverse sheaves.

I. Nonalternating zeta values: defined by $\varepsilon_j = +1$ in (12): $\zeta(a,b) = \sum_{m>n} \frac{1}{m^a n^b}$.

(N_0) All central double zetas are reducible: $\zeta(\ell,\ell) = \frac{1}{2}\zeta(\ell)^2 - \frac{1}{2}\zeta(2\ell)$ (Goldbach, 1742).

(N_1) All double zetas $\zeta(p,q)$ of odd weight $w = p+q$ reduce to zeta values (Borwein et al. [**6**]):

$$2\zeta(p,q) = -\left((-1)^q\binom{p+q}{p}+1\right)\zeta(p+q) + (1-(-1)^q)\zeta(p)\zeta(q)$$
$$+ 2(-1)^q \sum_j \left[\binom{2j-2}{p-1} + \binom{2j-2}{q-1}\right]\zeta(2j-1)\zeta(p+q-2j+1).$$

(N_2) All double zetas $\zeta(p,1)$ reduce to zeta values (Euler):

$$\zeta(p,1) = \frac{p}{2}\zeta(p+1) - \frac{1}{2}\sum_{k=1}^{p-2}\zeta(k+1)\zeta(p-k).$$

(N_3) The $\mathbb{Q}$-vector space of all nonalternating MZV's of weight w has a dimension d_w which, according to Zagier's conjecture, should satisfy $d_2 = d_3 = 1$, $d_w = d_{w-2} + d_{w-3}$. The upper-bound has been verified till order 9 in [**17**] and till order 12 by Hoang-Petitot [**24**] whose model conjecturally provides all identities. Consequently, all nonalternating MZV's (including double zetas) of "exceptional weights" $\{2,3,4,5,6,7,9\}$ reduce to zeta values.

II. Alternating zeta values: double ZV's defined by $\zeta^{-+}(a,b) = \sum_{m>n} \frac{(-1)^n}{m^a n^b}$.

(A_1) All alternating double zetas $\zeta^{-+}(p,q)$ of odd weight $w = p+q$ reduce to zeta values (Flajolet, Salvy [**17**, Thm. 7.2]):

$$2\zeta^{-+}(p,q) = (1-(-1)^q)\zeta^{-}(p)\zeta(q) - \zeta^{-}(p+q) + 2(-1)^q\sum_k$$
$$\left[\binom{p+q-2k-1}{p-1}\zeta^{-}(2k)\zeta^{-}(p+q-2k) + \binom{p+q-2k-1}{q-1}\zeta^{-}(2k)\zeta(p+q-2k)\right].$$

(A_2) All double zetas $\zeta^{-+}(1,q)$ reduce to zeta values (Sitaramachandrarao [**40**]):

$$2\zeta^{-+}(1,q) = -q\zeta(q+1) - 2\zeta^{-}(q+1) + \sum_{k=1}^{q}\zeta^{-}(k)\zeta^{-}(q+1-k).$$

(A_3) For weight 2, $\zeta^{-+}(1,1) = -\frac{1}{2}(\log 2)^2$ (Euler; see [**20**]). For weight 4, a special reduction is known for $\zeta^{-+}(2,2)$ in terms of zeta values augmented with $\mathrm{Li}_4(1/2)$ (Sitaramachandrarao [**40**], De Doelder [**15**]). For weight 6, each of $\zeta^{-+}(2,4), \zeta^{-+}(3,3), \zeta^{-+}(4,2), \zeta^{-+}(5,1)$ is expressible in term of any other one and zeta values. No such direct reductions seem to exist at weight 8 or higher. General relations are discussed in Borwein et al. [**6**].

FIGURE 4. A summary of known properties of double zeta values of type ζ^{++} and ζ^{-+}.

A relation between $\zeta^{-+}(3,1)$ and $\mathrm{Li}_4(1/2)$ is mentioned in [**40**] (see the formula preceding Eq. (1.2)). and proved by De Doelder in [**15**, p. 128]. Starting from the integral representation

$$\zeta^{-+}(3,1) = -\frac{1}{2}\int_0^1 \log t \log^2(1+t)\,\frac{dt}{t},$$

De Doelder deduces

$$\zeta^{-+}(3,1) = \frac{57}{64}\zeta(4) + \frac{7}{8}\zeta(3)\log(2) - \frac{\pi^2}{24}(\log 2)^2 + \frac{1}{24}(\log 2)^4 + \mathrm{Li}_4(\frac{1}{2}), \tag{18}$$

which eventually relates the constant $\rho^{(2)}$ to the tetralogarithm.

We can state:

THEOREM 3. *The mean number $\overline{\rho}^{(2)}$ of comparisons in BCF–Sign can be expressed in terms of double zeta values as*

$$\begin{aligned}\overline{\rho}^{(2)} &= \frac{3}{4} + \frac{360}{\pi^4}\zeta^{-+}(2,2) &\equiv \frac{3}{4} + \frac{360}{\pi^4}\sum_{d=1}^{\infty}\frac{(-1)^d}{d^2}\sum_{c=1}^{d-1}\frac{1}{c^2}\\ &= 2 - \frac{720}{\pi^4}\zeta^{-+}(3,1) &\equiv 2 - \frac{720}{\pi^4}\sum_{d=1}^{\infty}\frac{(-1)^d}{d^3}\sum_{c=1}^{d-1}\frac{1}{c},\end{aligned} \tag{19}$$

or with $\zeta(3)$ and the tetralogarithm $\mathrm{Li}_4(\frac{1}{2})$,

$$\overline{\rho}^{(2)} = -\frac{60}{\pi^4}\left(24\,\mathrm{Li}_4(\frac{1}{2}) - \pi^2(\log 2)^2 + 21\zeta(3)\log 2 + (\log 2)^4\right) + 17. \tag{20}$$

Thus, $\overline{\rho}^{(2)}$ lies in the class **P** *of polynomial time computable constants and*

$$\overline{\rho}^{(2)} = 1.35113\,15744\,91659\,00179\,38680\,05256\,46466\,84404\,78970\,85087 \pm 10^{-50}.$$

In the case of values $\ell \geq 3$, the quantities $\overline{\rho}^{(\ell)}$ that intervene in the analysis of the sorting algorithm do not seem to reduce to known classical constants or to other simple sums like polylogarithmic values. The expressions of (16) show that everything depends on the status of the central zeta values $\zeta^{-+}(\ell,\ell)$. Since results appear scattered in the literature, we have compiled a summary table of what is known regarding the class of alternating double zetas of type ζ^{-+} in Figure 4. It is seen from this table that $\rho^{(3)}$ is expressible in terms of either one of the quantities $\zeta^{-+}(3,3)$, $\zeta^{-+}(2,4)$, $\zeta^{-+}(4,2)$, $\zeta^{-+}(5,1)$,

$$\begin{aligned}\overline{\rho}^{(3)} &= -\frac{31}{8} + \frac{2835}{\pi^6}\zeta(3)^2 + \frac{7560}{\pi^6}\zeta^{-+}(3,3) \quad = \cdots\\ &= 18 - \frac{48195}{4\pi^6}\zeta(3)^2 + \frac{45360}{\pi^6}\zeta^{-+}(5,1),\end{aligned}$$

so that 1000 terms of the last series $\zeta^{-+}(5,1)$ suffice to determine

$$\overline{\rho}^{(3)} = 1.13662\,64940 \pm 10^{-10}, \qquad \text{and} \qquad \overline{P}(3) = 4.69690\,99648 \pm 10^{-10},$$

where the latter quantity is the mean cost of 3–sorting. However, simple polylogarithmic reductions cease to exist (probably) as soon $\ell \geq 3$ and alternative methods of evaluation must be sought.

Numerical calculation of $\overline{\rho}^{(\ell)}$. Regarding the numerical computation of the quantities $\overline{\rho}^{(\ell)}$ to high precision, Jon Borwein provides on his pages[2] efficient implementations that are visibly based on algorithms of *low* complexity (probably the

[2] URL: `www.cecm.sfu.ca/projects/EZFace/`

Euler-Maclaurin summation formula). More pragmatically, the `Pari-Gp` system includes an amazing command, `sumalt`, that is based on the general-purpose convergence acceleration process of Cohen-Villegas-Zagier [**13**] and gives very rapidly estimations of ζ^{-+} that appear to be in full agreement with the ones provided by Borwein's program.

Here is one *a priori* "reason" for the **P**-character of multiple zeta values. It is closely related to investigations of the Chudnovsky brothers [**10**] and holds for a much wider class of constants. Following Zeilberger [**52**], define a function $f(z)$ analytic at the origin to be *holonomic* if its satisfies a linear differential equation with coefficients in $\mathbb{Q}(z)$. A constant that is the value $f(z_0)$ of a holonomic function at an algebraic point z_0 where $f(z)$ is regular (i.e., analytic, holomorphic) will be called a *regular holonomic constant.* Typical holonomic constants of the regular type are π, $\log 2$ or the polylogarithmic value $\mathrm{Li}_4(1/2)$. Taylor coefficients of a holonomic function at any point satisfy linear recurrences with polynomial coefficients so that they can be determined fast and the function itself can be evaluated at high speed at any regular point that is an algebraic number (possibly using relays and analytic continuation). Consequently, all regular holonomic constants are in **P**.

The simplest type of singularity for a linear ordinary differential equation is a Fuchsian singularity (also known as singularity of the first kind or "regular" singularity [**50**]). At such a point, singular expansions that mildly generalize standard power series expansions exist and are again computable fast by means of recurrences. Define a *singular holonomic constant* as the value of a holonomic function $f(z)$ at a point z_0 that is a singular point of the Fuchsian type for a defining equation. Then, previous arguments extend, and any singular holonomic constant is also in **P**; see [**10**]. A typical singular holonomic constant is $\zeta(3) = \mathrm{Li}_3(1)$.

Any multiple zeta value is the value at ± 1 of a generating function that is holonomic and has radius of convergence 1. For instance, we have

$$\zeta^{-+}(a,b) = f(-1) \quad \text{where} \quad f(z) := \sum_{n=1}^{\infty} \frac{z^n}{n^a} H_{n-1}^b$$

satisfies the linear differential equation

$$\frac{d}{dz}\left(\frac{1-z}{z}\right)\left(z\frac{d}{dz}\right)^b \left(\frac{1-z}{z}\right)\left(z\frac{d}{dz}\right)^a f(z) = 0.$$

Therefore, any MZV is a holonomic constant. Membership of all MZV's in **P** thus follows from general principles.

5. "Nonanalytic" lattice sums and the centred sign algorithm

In a beautiful paper [**9**] titled "Strange Series and High Precision Fraud", Jon and Peter Borwein present intriguing evaluations of various types of series. One of the highlights of that paper is the fact that the near-equality

(21) $$\sum_{n=1}^{\infty} \frac{\left\lfloor n e^{\pi\sqrt{163/9}} \right\rfloor}{2^n} \doteq 1280640$$

is correct to at least half a billion digits, the difference between the two members of (21) being less that $10^{-500{,}000{,}000}$. Sums like (21) have a "nonanalytical" character due to the integer part function combined with irrational quantities, here present in the numerator. The moment sums of the centred continued fraction system are of a similar nature, with terms determined by integer-part conditions appearing in the denominator. We present here ways of estimating them that build upon methods of [**9**] and make use of variations around lattice generating function, like

$$L_\theta(z) = \sum_{m\geq 1} z^{\lfloor m\theta \rfloor}.$$

The Borwein & Borwein constant (21) obtained by differentiation from some $L_\theta(z)$ are seen to be dual to the centred moment sums of (8) that result from iterated integration.

Examination of the basic and centred moment sums of (8) shows that these quantities are sums over *lattice cones*, where the lattice cone $\mathcal{C}(\beta,\gamma)$ is defined by

$$\mathcal{C}(\beta,\gamma) := \{(m,n) \in \mathbb{N}^2 \mid m\beta < n < m\gamma\}.$$

The basic continued fraction case corresponds to a cone $\mathcal{C}(1,2)$ (or equivalently $\mathcal{C}(\frac{1}{2},1)$) and the centred case corresponds to a cone $\mathcal{C}(\phi,\phi^2)$ (or equivalently $\mathcal{C}(\phi^{-2},\phi^{-1})$) . In the latter case, what is involved in the lattice condition inherently involves the integer part function since, for an irrational θ, one has

$$m < n\theta \qquad \text{if and only if} \qquad m \leq \lfloor n\theta \rfloor.$$

It is a perhaps surprising fact that, despite their nonanalytical character, the centred moment sums $\widehat{\rho}^{(\ell)}$ can be computed in polynomial time.

THEOREM 4. *Let $\theta < 1$ be an irrational number with sequence of continued fraction approximants $\{p_n/q_n\}_{n=0}^{\infty}$. One has for any integer $\ell \geq 1$*

(22)

$$\widehat{\rho}^{(\ell)}(\theta) := \sum_{(p,q)\in\mathcal{C}(0,\theta)} \frac{1}{p^\ell q^\ell} = \sum_{n=0}^{\infty} \frac{(-1)^n}{(p_n+p_{n+1})^\ell (q_n+q_{n+1})^\ell} J_\ell\left(\frac{p_n}{p_n+p_{n+1}}, \frac{q_n}{q_n+q_{n+1}}\right)$$

where $J_s(\beta,\gamma)$ is defined for $0<\beta,\gamma<1$ and $\Re(s)>1$ as

$$J_s(\beta,\gamma) := \frac{1}{\Gamma(s)^2}\int_0^1\int_0^1 \frac{|\log x|^{s-1}\,|\log y|^{s-1}}{(1-x^\beta y^\gamma)\,(1-x^{1-\beta}y^{1-\gamma})}\,dx\,dy. \tag{23}$$

If θ is polynomial time computable, then $\widehat{\rho}^{(\ell)}(\theta)$ is polynomial time computable for each integer value of ℓ. In particular, the average number of iterations of the CCF sign algorithm is in **P** *and it satisfies*

$$\widehat{\rho}^{(2)} \equiv \widehat{\rho}^{(2)}\left(\frac{1}{\phi}\right) - \widehat{\rho}^{(2)}\left(\frac{1}{\phi^2}\right) = 1.08922\,14740\,95380 \pm 10^{-15}.$$

The theorem follows from two types of developments: on the one hand, a simple combinatorial decomposition of lattice cones, on the other hand, a classical formula for iterated integrals.

First, we state a lemma drawn from the works of Borwein, Borwein, and Mahler, see [**9**], of which we propose a direct proof.

LEMMA 1. *Let $\theta < 1$ be an irrational number with convergents $\{p_n/q_n\}_{n=0}^{\infty}$. The generating function of the lattice cone $\mathcal{C}(0,\theta)$ is given by*

$$\sum_{(m,n)\in\mathcal{C}(0,\theta)} x^m y^n = \sum_{k=0}^{\infty} (-1)^k \frac{x^{q_k+q_{k+1}} y^{p_k+p_{k+1}}}{(1-x^{q_k}y^{p_k})(1-x^{q_{k+1}}y^{p_{k+1}})}. \tag{24}$$

PROOF. The sequence of convergents $\{p_n/q_n\}$ is such that the odd subsequence $\{p_{2n+1}/q_{2n+1}\}$ is decreasing, and the even sequence $\{p_{2n}/q_{2n}\}$ is increasing starting from $p_0/q_0 = 0$. Both have limit θ. Let $A+B$ and $A-B$ denote $A \cup B$ and $A \setminus B$. Then, for any irrational θ, the following decomposition holds

$$((0,\theta)) = ((\frac{p_0}{q_0}, \frac{p_1}{q_1})) - ((\frac{p_1}{q_1}, \frac{p_2}{q_2})) + ((\frac{p_2}{q_2}, \frac{p_3}{q_3})) - \cdots = \sum (-1)^n ((\frac{p_n}{q_n}, \frac{p_{n+1}}{q_{n+1}}))$$

between various sets of the form $((\beta,\gamma))$ defined as the intersection of the open interval (β,γ) with strictly positive rational of $\mathbb{Q}^+$. A decomposition of the same type for integer open cones follows,

(25)
$$\mathcal{C}(0,\theta) = \mathcal{C}(\frac{p_0}{q_0}, \frac{p_1}{q_1}) - \mathcal{C}(\frac{p_1}{q_1}, \frac{p_2}{q_2}) + \mathcal{C}(\frac{p_2}{q_2}, \frac{p_3}{q_3}) - \cdots = \sum (-1)^n \mathcal{C}(\frac{p_n}{q_n}, \frac{p_{n+1}}{q_{n+1}}).$$

Now, the open interval $(p_n/q_n, p_{n+1}/q_{n+1})$ is the image of interval $]0,\infty[$ by the linear fractional transformation h defined as $h(x) := (p_n x + p_{n+1})/(q_n x + q_{n+1})$. Since this LFT is of determinant ± 1, this proves the equality

$$\mathcal{C}(\frac{p_n}{q_n}, \frac{p_{n+1}}{q_{n+1}}). = \mathcal{D}\begin{bmatrix} p_n & p_{n+1} \\ q_n & q_{n+1} \end{bmatrix},$$

where $\mathcal{D}$ is defined by

$$\mathcal{D}\begin{bmatrix} a & b \\ c & d \end{bmatrix} = \{\ (\lambda a + \mu b, \lambda c + \mu d)\ \mid\ \lambda,\mu \in \mathbb{N},\ \lambda,\mu \geq 1\ \}.$$

The generating function relative to $\mathcal{D}\begin{bmatrix} a & b \\ c & d \end{bmatrix}$ is seen to be

$$\sum_{\lambda,\mu=1}^{\infty} x^{\lambda a+\mu b}\, y^{\lambda c+\mu d} = \frac{x^a\, y^c}{1-x^a\, y^c}\, \frac{x^b\, y^d}{1-x^b\, y^d},$$

and this identity combined with the decomposition (25) yields (24). □

Next, a well-known formula for *iterated integration* relates power series and Dirichlet series:

(26)
$$\text{If}\quad \phi(x,y) = \sum_{p,q=1}^{\infty} \phi_{p,q} x^p y^q,$$
$$\text{then}\quad \sum_{p,q=1}^{\infty} \frac{\phi_{p,q}}{p^s q^s} = \frac{1}{\Gamma(s)^2} \int_0^\infty \int_0^\infty \phi(e^{-u}, e^{-v}) u^{s-1} v^{s-1}\, du\, dv.$$

(Proof: Expand and integrate termwise!). When applied to (24), the integration formula (26) yields the two equations (22) and (23), upon the change of variables $x = e^{-u}, y = e^{-v}$ and proper rescaling.

Consider now integer values ℓ for the parameter s. First, we remark that

$$J_\ell(\beta,\gamma) \leq \frac{1}{\beta(1-\beta)\gamma(1-\gamma)} H_\ell$$
$$\text{with}\qquad H_\ell := \frac{1}{\Gamma(\ell)^2} \int_0^1 \int_0^1 \frac{|\log(1-x)|^{\ell-1}\, |\log(1-y)|^{\ell-1}}{(x+y)^2}\, dx\, dy.$$

Furthermore, H_ℓ is $O(1)$ for $\ell \to \infty$, and the convergents p_n, q_n are at least equal to the n-th Fibonacci number F_n so that the general term of the series in (22) is $O\left(\phi^{-2\ell n}\right)$. This provides the geometrically convergent representation (22) for

$\widehat{\rho}^{(\ell)}(\theta)$. Full polynomial-time computability is formally established from there by observing that the integrands have easily computable series expansions that can be integrated termwise.

A mock Zeta function. We conclude this section with a mention of a few curious sums, simpler than $\widehat{\rho}^{(2)}$, that can be shown to lie in the class **P** by similar devices. For θ an arbitrary positive real ≥ 1, define the "mock zeta function" by

$$\zeta_\theta(s) = \sum_{n=1}^{+\infty} \frac{1}{\lfloor n\theta \rfloor^s}, \qquad \Re(s) > 1. \tag{27}$$

The basic generating function associated to the mock Zeta function, is closely related to the generating function of the cone $\mathcal{C}(0, 1/\theta)$. Using the relation (24), one finds easily for $\theta > 1$,

$$\sum_{m\geq 1} z^{\lfloor m\theta \rfloor} = \frac{z^{p_0}}{1-z^{p_0}} - \frac{1-z}{z} \sum_{n=0}^{\infty} (-1)^n \frac{z^{p_n}}{1-z^{p_n}} \frac{z^{p_{n+1}}}{1-z^{p_{n+1}}},$$

a relation that is due to Mahler. Then, via the formula for iterated integration in one variable, the mock zeta function is seen to admit the geometrically convergent representation

$$\zeta_\theta(s) = \frac{\zeta(s)}{\lfloor\theta\rfloor^s} - \frac{1}{\Gamma(s)} \sum_{n=0}^{\infty} \frac{(-1)^n}{p_n p_{n+1}^s} \int_0^\infty \Omega(t, p_{n+1}, \frac{p_n}{p_{n+1}})\, t^{s-2}\, dt, \tag{28}$$

where

$$\Omega(t, p, \lambda) = \frac{e^{t/p} - 1}{t/p} \frac{t\, e^{-t}}{1-e^{-t}} \frac{\lambda t\, e^{-\lambda t}}{1-e^{-\lambda t}}. \tag{29}$$

This is the main tool for computing in polynomial time values of the mock zeta function at the integers. For instance, we obtain "quickly"

$$\zeta_\phi(2) := \sum_{n=1}^{\infty} \frac{1}{\lfloor n\phi \rfloor^2} = 1.29106\ 03681\ 14387\ 48950\ 47876 \pm 10^{-25}, \qquad \phi = \frac{1+\sqrt{5}}{2}.$$

6. Transfer operators and sign algorithms

In 1800, Gauss conjectured that repeated applications of the continued fraction transformation $\{1/x\}$ to a uniform real number over the interval $(0, 1)$ gives rise to a nonuniform number whose probability density is approximately

$$\overline{\psi}(x) = \frac{1}{\log 2} \frac{1}{1+x}. \tag{30}$$

Such a property, once supplemented with uniform convergence, says quite a bit about the stochastic behaviour of continued fraction representations of real numbers. For instance, the probability of observing a digit equal to d is in the asymptotic limit

$$\int_{1/(d+1)}^{1/d} \overline{\psi}(x)\, dx = \log_2 \frac{(d+1)^2}{d(d+2)}.$$

The situation is now well understood thanks to the works of Kuzmin, Lévy, and Wirsing, who defined the *density transformer* associated to the continued fraction map

$$\overline{\mathcal{G}}[f](x) = \sum_{m=1}^{\infty} \frac{1}{(m+x)^2} f\left(\frac{1}{m+x}\right). \tag{31}$$

The characteristic property is that $\overline{\mathcal{G}}[f]$ describes the density of the transform of a random variable with density f by the continued fraction map $\overline{U}$. Gauss's assertion is then synonymous to

$$\overline{\mathcal{G}}^k[\mathbf{1}] \to \overline{\psi} \qquad \text{as} \qquad k \to +\infty, \tag{32}$$

since $\mathbf{1}$ is the density function of a uniform variate. The point of view has now been changed and problems concerning the metric theory of continued fractions are rephrased as questions regarding the functional-analytic properties of an (infinite dimensional) operator over some suitable function space. Spectral properties and eigenfunctions are obviously important objects in this perspective.

In the case of an arbitrary expanding map, with U the shift and $\mathcal{H}_1$ the set of inverse branches of depth 1, Ruelle introduced the *transfer operator* that depends on a (real or complex) parameter s,

$$\mathcal{G}_s[f](x) = \sum_{h \in \mathcal{H}_1} |h'(x)|^s \, f \circ h(x), \tag{33}$$

and gives back the density transformer when $s = 1$. The interest of the extension is the increased descriptive power afforded by the extra parameter s. Here, we shall make use of the case $s = 2$ (the sign algorithm) and the case of a complex s (the sorting algorithm in Section 7).

For a general (analytic) expanding map, we assume from now on two conditions that supplement the axioms (E_1) and (E_2) of Section 1:

- (E_3) There exists a complex domain $\mathcal{V}$ that contains $\mathcal{I}$ where each inverse branch $h \in \mathcal{H}_1$ is analytic and satisfies $|h'(z)| \geq \delta > 1$.
- (E_4) The derivatives are summable in the sense that $\sum_{h \in \mathcal{H}_1} |h'(z)|^\eta < \infty$ for some fixed $\eta < 1$ and all z in $\mathcal{V}$.

Quite clearly, the transfer operator operates on space $A_\infty(\mathcal{V})$ formed with all functions f that are holomorphic in $\mathcal{V}$ and continuous on the closed domain $\overline{\mathcal{V}}$. Endowed with the sup-norm, $A_\infty(\mathcal{V})$ is a Banach space. The operator $\mathbf{G}_s$ is bounded for $\Re(s) > \eta$ and is also compact. (The argument uses the "selection principle" granted by Vitali's theorem and properties of normal families of functions; see [**23**, Ch. 15].) It follows by one of the most basic theorems of functional analysis [**37**] that the transfer operator has a discrete spectrum: the collection of its eigenvalues $\{\lambda_j(s)\}_{j=1}^{\infty}$ has no accumulation point, except 0. (More is known since Grothendieck's theory of nuclear spaces applies at full strength and the eigenvalues converge fast to zero in such a way that $\sum |\lambda_j(s)|^\varepsilon$ is summable for all $\varepsilon > 0$.) In addition, for a positive real value $s = \sigma$, the operator is in a strong sense positive; general Perron-Fronbenius properties [**29**] then imply that the dominant eigenvalue $\lambda_1(\sigma)$ (the one of largest modulus) is unique and simple, so that

$$\lambda_1(\sigma) > |\lambda_2(\sigma)| \geq |\lambda_3(\sigma)| \geq \cdots \geq |\lambda_j(\sigma)| \to 0.$$

As a pleasant outcome of these properties, one can essentially treat the operator $\mathcal{G}_\sigma$ as though it were simply a finite nonnegative matrix.

The properties specific to the transfer operator $\overline{\mathcal{G}}_s$ of the continued fraction map have been worked out by Dieter Mayer and by Babenko in a series of papers [**2, 32,**

33, 34]. In particular, these authors discovered the existence of a hidden Hilbert space structure resulting in the additional property that the $\overline{\lambda}_j(\sigma)$ are real for real σ. Building on their results, Hensley [**21**] was able to establish that the Euclidean gcd algorithm has an asymptotically Gaussian behaviour. In a different direction, the authors have demonstrated in [**14, 19, 43, 45, 47**] that the transfer operators can be put to use in order to analyse a variety of continued fraction algorithms.

The secant operators. For a general dynamical system, the transfer operator does not describe the moment sums $\sum_h u_h^\ell$, but a modified form due to Vallée [**44**] does. The *secant operator* $\mathbf{G}_s$ is defined for complex numbers s satisfying $\Re(s) > \eta$ as

$$\mathbf{G}_s[F](x, y) = \sum_{|h|=1} \left| \frac{h(x) - h(y)}{x - y} \right|^s F(h(x), h(y)).$$

By the chain rule for secants, the iterate of order k is expressible in terms of all inverse branches of depth k

$$\mathbf{G}_s^k[F](x, y) = \sum_{|h|=k} \left| \frac{h(x) - h(y)}{x - y} \right|^s F(h(x), h(y)).$$

In particular, this identity specialized at $x = 0$ and $y = \alpha$ provides an alternative expression for the probability distribution of the cost of the sign algorithm given in (6):

$$\Pr(L \geq k + 1) = \mathbf{G}_2^k[\mathbf{1}](0, \alpha).$$

The secant operator extends Ruelle's transfer operator of (33) in the sense that "the diagonal of the secant is the tangent":

$$\mathbf{G}_s[F](x, x) = \mathcal{G}_s[f](x) \qquad \text{where} \qquad f(x) = F(x, x).$$

Here, asymptotic properties are needed, and they involve dominant spectral properties of the operator $\mathbf{G}_s$ that are closely related to the corresponding ones for $\mathcal{G}_s$. The operator $\mathbf{G}_s$ acts on the space $B_\infty(\mathcal{V})$ formed with all functions f that are holomorphic in the product $\mathcal{V} \times \mathcal{V}$ and continuous on the closed domain $\bar{\mathcal{V}} \times \bar{\mathcal{V}}$. Endowed with the sup-norm, $B_\infty(\mathcal{V})$ is a Banach space and the operator $\mathbf{G}_s$ is bounded, and therefore compact (again by classical properties of bounded sequences of analytic functions). Its spectrum is thus discrete, with only an accumulation point at 0.

The spectrum of the secant operator $\mathbf{G}_s$ is completely determined in [**44**]. In particular, when $s = \sigma$ is real, the dominant eigenvalue $\mu_1(\sigma)$ of $\mathbf{G}_s$ is the same as the dominant eigenvalue $\lambda_1(\sigma)$ of $\mathcal{G}_s$ while the subdominant eigenvalue $\mu_2(\sigma)$ satisfies $|\mu_2(\sigma)| < \mu_1(\sigma)$.

Such dominant spectral properties provide an asymptotic form for the iterates of Ruelle-Mayer operators where the k-th iterate of a positive function will have a regime dictated by $\lambda_1(\sigma)^k$ with an error term of the order of $\mu_2(\sigma)^k$. Specializing the discussion to moment sums, we thus find for the sign algorithm and an arbitrary analytic expanding map

$$\Pr(L \geq k + 1) \equiv \sum_{|h|=k} u_h^2 \equiv \mathbf{G}_2^k\ [\mathbf{1}](0, \alpha) = C\lambda_1(2)^k + O(\mu_2(2)^k).$$

Thus, for an arbitrary analytic expanding map, the sign algorithm exhibits an exponential decay of probabilities and the rate of decay is the dominant eigenvalue of the standard transfer operator $\mathcal{G}_2$.

In the particular case of the basic continued fraction algorithm, we find, by methods discussed later in this section, a value $\overline{\lambda}_1(2) \approx 0.19945$ with at least 25 decimal places, and $\overline{\mu}_2(2) \approx -0.07573$; see (34) below. (The constant $\overline{\lambda}_1(2)$ is described under the name of "Vallée's constant" in Finch's repertoire of mathematical constants [**16**].) In the case of the centred algorithm, we estimate $\widehat{\lambda}_1(2) \approx 0.07738$; see (35). These values explain quite well the empirical observations on the geometric decay of probabilities presented at the beginning of Section 3.

THEOREM 5. *The probability that the BCF–Sign Algorithm performs at least $k+1$ iterations satisfies*

$$\Pr(\overline{L} \geq k+1) = \overline{C}\,\overline{\lambda}_1(2)^k + O(\overline{\mu}_2(2)^k),$$

where $\overline{\lambda}_1(2) \approx 0.19945$ and $\overline{\mu}_2(2) \approx 0.07573$ are given by (34), $\overline{C} \approx 1.3$.

The probability that the CCF–Sign Algorithm performs at least $k+1$ iterations satisfies

$$\Pr(\widehat{L} \geq k+1) = \widehat{C}\,\widehat{\lambda}_1(2)^k + O(\widehat{\mu}_2(2)^k),$$

where $\widehat{\lambda}_1(2) \approx 0.07738$ is given by (35), $|\widehat{\mu}_2(2)| < \widehat{\lambda}_1(2)$, and $\widehat{C} > 0$.

Numerical estimates of the spectrum. We now describe a piece of experimental mathematics aimed at providing *inter alia* estimates on dominant and subdominant eigenvalues of the transfer operator $\mathcal{G}_s$ in the particular case of the continued fraction algorithms. Let $\Pi_{m,a} f(z)$ be the operation (a "projection") that selects the terms till order $m-1$ inclusive in the Taylor expansion of the (analytic) function $f(z)$ at the point a. We introduce the truncated operators

$$\Pi_{m,a} \circ \mathcal{G}_s \circ \Pi_{m,a},$$

that is to say, we examine the effect of $\mathcal{G}_s$ on initial segments of the Taylor expansion of functions. Such truncated operators are finite-dimensional. In the case of the continued fraction maps, their eigenvalues appear to convey quite a lot of information on the spectrum of the full operator $\mathcal{G}_s$ while being accessible to numerical analysis.

The transform of a monomial x^m by a continued fraction transfer operator involves the Hurwitz zeta function that is classically defined by

$$\zeta(s,w) = \sum_{m=0}^{\infty} \frac{1}{(m+w)^s},$$

and is such that its Taylor expansion at a point a is expressible in terms of the collection of values $\zeta(s+j,a)$. Thus, viewed as acting on series expansions at $x=a$, the operator $\mathcal{G}_s$ is expressed by an infinite matrix $M = (M_{i,j})$, where $M_{i,j}$ is the coefficient of $(x-a)^i$ in $\mathcal{G}_s[(x-a)^j]$. For instance, the BCF operator with $a=0$ corresponds to the infinite matrix $(\overline{M})_{i,j}$ with

$$\overline{M}_{i,j} = (-1)^i \binom{2s+i+j-1}{i} \zeta(2s+i+j),$$

and the infinite matrix for $s = 2$ starts as

$$\begin{bmatrix} \zeta(4) & \zeta(5) & \zeta(6) & \zeta(7) \\ -4\,\zeta(5) & -5\,\zeta(6) & -6\,\zeta(7) & -7\,\zeta(8) \\ 10\,\zeta(6) & 15\,\zeta(7) & 21\,\zeta(8) & 28\,\zeta(9) \\ -20\,\zeta(7) & -35\,\zeta(8) & -56\,\zeta(9) & -84\,\zeta(10) \end{bmatrix}$$

We now restrict attention to the basic continued fraction operator, and, guided by numerical experiments, adopt the value $a = 1/2$ that gives faster convergence. Using computer algebra, we have determined the eigenvalues of the truncated matrices $T^{[m]}$ for many values of $m \leq 100$ and numerical accuracy up to 200 digits. Examination of all the eigenvalues of these truncations reveals that most of them stabilize to a fixed set of definite values, with the occasional occurrence of some ghost values that eventually disappear as m increases. It is then natural to conjecture that the stable limit values do yield the complete spectrum of $\overline{\mathcal{G}}_s$.

We have developed a campaign of experiments based on the `Pari-Gp` and Maple systems (see **[14]** for an earlier account) in order to determine the spectrum of $\overline{\mathcal{G}}_1$ and $\overline{\mathcal{G}}_2$. It is based on the truncation principle and on the following three additional points.

— Examination of the first few values of the spectrum led us to conjecture that $\overline{\lambda}_j(\sigma)/\overline{\lambda}_{j+1}(\sigma)$ tends to $-\phi^{2\sigma}$. A heuristic reason is that the operator $\overline{\mathcal{G}}_s$ is in a sense (e.g., for large $s = \sigma$) "dominated" by its first term, which is a composition operator

$$\overline{\mathcal{C}}_s[f](x) = \frac{1}{(1+x)^s} f(\frac{1}{1+x}),$$

whose eigenfunctions and spectrum are explicit **[39]**. The surprise is that the approximation of the spectrum holds both for small values of σ ($\sigma = 1, 2$) *and* for large eigenvalue indices. This observation serves as a convergence accelerator as it enables one to filter out ghost eigenvalues, thereby making it possible to access smaller eigenvalues much faster.

— The traces of the continued fraction operators are computable in polynomial time, as shown in **[14]** by means of a multivariate version of Vardi's "zeta summation trick" **[48]**. Filtered eigenvalues can then be confirmed by comparing with the trace $\operatorname{Tr} \overline{\mathcal{G}}_s^2$ and this check is reliable since cancellations cannot occur for $s = \sigma$ as the spectrum is real, by virtue of the hidden Hilbert space structure **[2, 32]**.

— For real $s = \sigma$, the estimate of the dominant eigenvalue $\overline{\lambda}_1(\sigma)$ at least can be *certified* thanks to a technique of "test functions" detailed in **[14]**.

In this way, we have obtained highly convincing values for the first 37 eigenvalues of the basic operator $\overline{\mathcal{G}}_2$ with an accuracy almost certainly better than 10^{-25}. Here is a listing of our first estimates:

$$\begin{aligned} \overline{\lambda}_1(2) &\doteq +0.19945\,88183\,43767\,26019\,18456 \\ \overline{\lambda}_2(2) &\doteq -0.07573\,95140\,84360\,60892\,78089 \\ \overline{\lambda}_3(2) &\doteq +0.02856\,64037\,69818\,52783\,00174 \\ \overline{\lambda}_4(2) &\doteq -0.01077\,74165\,76612\,69829\,31408. \end{aligned} \tag{34}$$

The same process has also given us values of the spectrum when $s = 1$, including an estimate to more than 30 digits of accuracy of Wirsing's constant $\overline{\lambda}_2(1)$ (that dictates the speed on convergence to the stationary regime in (32)),

$$\overline{\lambda}_2(1) = -0.30366\,30028\,98732\,65859\,74481\,21901 \pm 10^{-30}.$$

This has made it possible to correct some spurious values in Knuth's classic account of the subject [**27**]. Another application of detailed spectral estimates is the determination of Hensley's constant $H = \overline{\lambda}_1''(1)$ that expresses the variance of Euclid's algorithm, where we obtain:

$$H = 9.08037\,31646 \cdots .$$

The spectrum of the centred operator $\widehat{\mathcal{G}}_2$ is at the moment less well understood. By the truncation method, we have found the dominant eigenvalue

$$\widehat{\lambda}_1(2) = +0.07738\,53773\,83629\,09062\,03319 \pm 10^{-25}, \tag{35}$$

which can be certified by the already mentioned method of test functions. Some of the subdominant eigenvalues probably include values near

$$-0.00375\,30752\,61163\,03856, \; -0.00010\,51799, \; +0.00000\,51256, \tag{36}$$

but the spectrum seems to obey a more complicated pattern than in the basic case.

The truncation technique being versatile enough, has also enabled Hensley [**22**] and Vallée [**46**] to determine numerically the Hausdorff dimension of various Cantor sets for continued fraction expansions.

7. Riemann hypothesis and sorting algorithms

We conclude this paper by examining the cost $\overline{P}(n)$ of sorting n uniform real numbers by means of the basic continued fraction algorithm. What is needed is a characterization of the asymptotic dependency of $\overline{P}(n)$ on n. The starting point is the exact expression of $P(n)$ provided by Theorem 2 as a difference of the moment sums $\rho^{(\ell)}$.

We make use here of a well-known fact of complex analysis: *Asymptotic properties of a number sequence are strongly related to singularities of an analytic function that extrapolates the number sequence.* In the case of finite differences, this principle is vindicated by the method of Nörlund-Rice integrals [**18, 36**]. Let f_n be a number sequence and $\phi(s)$ a holomorphic function that extrapolates the sequence f_n in the sense that $f_n = \phi(n)$. Then, the equality

$$R_n := \frac{1}{2i\pi}\int_L \phi(s)\,\frac{n!}{s(s-1)(s-2)\cdots(s-n)}\,ds = \sum_{k=1}^{n}\binom{n}{k}(-1)^{n-k} f_k, \tag{37}$$

holds, by virtue of the residue theorem, provided L is a simple contour that encircles the points $1, 2, \ldots, n$ while avoiding the singularities of $\phi(s)$. What happens next is dependent upon $\phi(s)$ admitting a meromorphic continuation to the complex plane and being of moderate growth in right half-planes. Then, a shift of the contour till $\Re(s) = -T$ yields

$$R_n = -\sum_s \operatorname{Res}\left(\phi(s)\frac{n!}{s(s-1)\ldots(s-n)}\right), \tag{38}$$

where the sum is extended to all poles s of $\phi(s)$ in $\Re(s) > -T$. A simple pole of the integrand at some nonintegral point $s = s_0$ will for instance contribute a quantity

$$r_0\frac{\Gamma(n+1)\Gamma(-s_0)}{\Gamma(n+1-s_0)} = r_0\Gamma(-s_0)n^{s_0}\left(1 + O(\frac{1}{n})\right). \tag{39}$$

Poles farther to the right induce "dominant" contributions, multiple poles introduce logarithmic terms (see [**18**] for details), and nonreal poles $s_0 = \sigma_0 + it_0$ lead to periodic fluctuations since

$$n^{s_0} = n^{\sigma_0 + it_0} = n^{\sigma_0} e^{it_0 \log n}.$$

This situation specializes to the case of the sorting cost as follows. We start from the integral representation induced by (37),

$$\frac{\overline{P}(n+1)}{n+1} = \frac{1}{2i\pi} \int_{1/2-i\infty}^{1/2+i\infty} \rho^{(s+1)} \frac{n!}{s(s-1)\cdots(s-n)}\, ds, \tag{40}$$

assuming that the moment sums of complex index $\rho^{(s+1)}$ exist. According to the principles expressed above, we thus need to determine the nature of the singularities of the integrand to the left of $\Re(s) = 1/2$, and more specifically when $\Re(s)$ lies in $(-\frac{3}{2}, \frac{1}{2})$.

Moment sums of complex index. The expressions stated earlier in (11) for $\rho^{(\ell)}$ with ℓ an integer clearly lift to complex values of s. In particular, we have for $\Re(s) > 1$

$$\rho^{(s)} = 2^{-s} - 2^{1-s} + \left(2^{s-1} - 1\right) \frac{\zeta(s)^2}{\zeta(2\,s)} + 2^s \frac{\zeta^{-+}(s,s)}{\zeta(2\,s)} \tag{41}$$

where ζ^{-+} is the alternating double zeta function of Section 4. The singularities arising from the zeta terms are apparent: there is a simple pole at $s = 1$ due to $\zeta(s)$ and there are poles at the zeros of the term $\zeta(2s)$ present in the denominator. So everything eventually rests on the nature of $\zeta^{-+}(s,s)$. For this function, Euler-Maclaurin summation implies that $\zeta^{-+}(s,s)$, like its simpler counterpart the alternating zeta function $\zeta^{-}(s)$, is an entire function. Furthermore, the identity

$$\zeta^{-+}(1,1) = -\frac{1}{2}(\log 2)^2 \tag{42}$$

is elementary and known since Euler's time; see [**20**] and Figure 4.

Thus, $\rho^{(s)}$ is meromorphic for all s. The expansion of $\rho^{(s)}$ at $s = 1$ is then found by a simple calculation:

$$\rho^{(s)} = K_0 \frac{1}{s-1} + K + O(s-1), \tag{43}$$

with

$$K_0 = \frac{6 \log 2}{\pi^2}, \qquad K = 12\,\frac{\gamma \log 2}{\pi^2} + 9\,\frac{(\log 2)^2}{\pi^2} - 72\,\frac{\zeta'(2) \log 2}{\pi^4} - \frac{1}{2}.$$

The first order constant K_0 coincides with the inverse of Lévy's entropic constant, that is, the entropy of the continued fraction map. The second order constant turns out to be a variant of Porter's constant P defined classically by the constant term in the average number of steps of Euclid's algorithm: we find by identification with the values listed in the reference site for constants [**16**] the equality

$$P = 2K - 2K_0 + \frac{1}{2}. \tag{44}$$

The other singularities of $\rho(s)$ in the critical strip $0 < \Re(s) < 1$ are those of the denominator $\zeta(2s)$. Each of them is at $1/2$ times a nontrivial zero of the Riemann zeta function. If such a zero is at $s_0 = \sigma + it$ and is simple, then, the contribution has an order given by (39). The details of contour surgery are easily supplied and

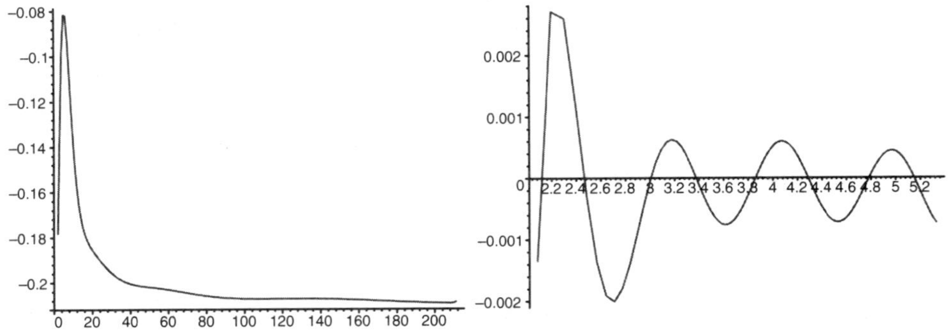

FIGURE 5. A plot (left) of $\overline{P}(n) - (K_0 n \log n + K_1 n)$ for $n = 2\,..\,200$ suggests a smooth convergence to a constant. A plot (bottom) of the empirically determined

$$P(n) - \left(K_0 n \log n + K_1 n - 0.21035 + \frac{0.365}{n} + \frac{2.02}{n^2} + \frac{13.5}{n^3}\right)$$

against $\log n$, for $n = 8\,..\,200$, reveals the first nontrivial zeros of $\zeta(s)$. In fact, both graphs represent functions that will eventually oscillate unboundedly between $-\infty$ and $+\infty$.

---

they are very similar to the ones used in Mellin transform asymptotics. See [**18**] or the treatment of sums of Gram, Ramanujan, Hardy-Littlewood, Riesz in analytic number theory [**41**, Ch. 14].

We then obtain:

THEOREM 6. *The expected cost of sorting n uniform real numbers given by their basic continued fraction representations satisfies*

$$\overline{P}(n) = K_0 n \log n + K_1 n + Q(n) + K_2 + o(1),$$

for constants $K_0, K_1, K_2 \in \mathbb{R}$, where K_0 is Lévy's entropic constant and K_1 is a Porter-like constant:

$$K_0 = \frac{6 \log 2}{\pi^2}, \qquad K_1 = 18\frac{\gamma \log 2}{\pi^2} + 9\frac{(\log 2)^2}{\pi^2} - 72\frac{\log 2 \zeta'(2)}{\pi^4} - \frac{1}{2}.$$

The function $Q(u)$ is an oscillating function with mean value 0 that satisfies

$$Q(n) = O(u^{\delta/2}),$$

where δ is any number such that

$$\delta > \sup\left\{\Re(s) \;\middle|\; \zeta(s) = 0\right\}.$$

The overall shape of this result is given in [**12**] where the derivation was based on a technique of "Dirichlet depoissonization" and Mellin transforms. We add here an explicit determination of K_1 and a perhaps more straightforward approach to the asymptotic analysis via Nörlund-Rice integrals.

Comparing numerically $P(n)$ and the initial terms of its asymptotic expansion as given by Theorem 6 leads to interesting observations. Using Cohen-Villegas-Zagier acceleration of summation provided by `Pari-Gp` as discussed in Section 4, we have determined the exact values of $P(n)$ for $n = 2\,..\,220$ with at least 16 digits of accuracy, which was achieved by determining the $\{\rho^{(\ell)}\}_{\ell=2}^{220}$ to about 100 digits of accuracy. First, in the given range, there is apparently a smooth convergence of $\overline{P}(n) - K_0 n \log n - K_1 n$ to a constant (reflecting K_2) that is near -0.2; see Figure 5 (left). By successive experimental adjustments—the process looks like restoring a blurred photograph in a crime fiction!—we are led to considering

$$B(n) = K_0 n \log n + K_1 n - 0.21035 + \frac{0.365}{n} + \frac{2.02}{n^2} + \frac{13.5}{n^3}. \tag{45}$$

For n in the interval $[8\,..\,220]$, the difference $\overline{P}(n) - B(n)$ then shows four slightly damped oscillations of minute amplitude that are between -0.002 and $+0.002$ and have a period slightly less than 1. However, we know that this difference must eventually tend to $\pm\infty$ with a regime of $O(n^{1/4})$ at least! The oscillations that are displayed in Fig. 5 (right) are reflexes of the nontrivial zeta zeros. In effect, corresponding to the first two nontrivial zeros at $z_0 = \frac{1}{2} \pm 14.13472i$, we have $|\Gamma(-z_0)| = 2.32 \cdot 10^{-5}$, so that the corresponding fluctuation has amplitude of the order of $10^{-4} n^{1/4}$ with an oscillating term similar to $\cos(2\pi \frac{\log n}{0.89})$. These facts match the numerical data quite well. In addition, due to the fast decay of the gamma function towards $\pm i\infty$, we estimate that any failure of the Riemann hypothesis could only be detected on $\overline{P}(n)$ for values of n well beyond the extraordinary bound of $10^{1,000,000,000}$ (see [**12**] for a discussion).

Constants and operators. The operator theory of Section 6 nicely completes the picture as regards the irruption of Lévy's entropic constant and of a Porter-like constant in Theorem 6. What happens is that the quantity $\rho^{(s)}$, which is the Dirichlet series of fundamental intervals, admits a representation as a quasi-inverse,

$$\rho^{(s)} = (I - \overline{\mathbf{G}}_s)^{-1}[\mathbf{1}](0,1) = (I - \overline{\mathcal{G}}_s)^{-1}\,[\frac{1}{(1+x)^s}](0), \tag{46}$$

so that its pole at $s = 1$ must involve the derivative $\overline{\lambda}'(1)$ of the dominant eigenvalue function $s \to \overline{\lambda}(s)$ at $s = 1$ as its first order constant, while $\overline{\lambda}'(1)$ is otherwise known to equal the entropy of the continued fraction map as appears in K_0. The quantity K is the second order constant in (43) and, under the operator framework (46), it must be related to derivatives of $\overline{\lambda}(s)$ and projectors at $s = 1$. The situation then parallels that of Euclid's algorithm where Porter's constant is also a second order constant at $s = 1$ in the expansion of another expression

$$\zeta(2s)(I - \overline{\mathcal{G}}_s)^{-1}\,[\frac{1}{(1+x)^{2s}}](0), \tag{47}$$

so that the relation between K and P is perhaps not totally unexpected.

We summarize in Fig. 6 the major constants encountered in the paper.

Open problems. We conclude this paper with a few open problems that we feel of interest.

(P_1) Why does Cohen-Villegas-Zagier acceleration work so well with the moment sums of the basic continued fraction system? It can apparently even be used to provide analytic continuation beyond the convergence region of double zeta values.

(P_2) Which of the alternating double zetas relate to other constants of analysis? What is the exact nature of relations with incomplete beta integrals, Nielsen-Ramanujan constants [**16**], and other polylogarithmic constants of [**7**]?

(P_3) Prove a version of the conjecture (Section 6) regarding the spectrum of the BCF operator $\overline{\mathcal{G}}_\sigma$.

Cost of BCF-sign	$\overline{\rho}^{(2)}$	1.35113 15744	$\leftrightarrow$ double zeta $\zeta^{-+}(2,2)$
Cost of CCF-sign	$\widehat{\rho}^{(2)}$	1.08922 14740	Lattice sum
Prob. decay of BCF-sign	$\overline{\lambda}_1(2)$	0.19945 88183	Dominant eigenvalue of $\overline{\mathcal{G}}_2$
Prob. decay of CCF-sign	$\widehat{\lambda}_1(2)$	0.07738 53773	Dominant eigenvalue of $\widehat{\mathcal{G}}_2$
Sorting, 1st order (BCF)	K_0	0.42138 29566	$\leftrightarrow$ Lévy's entropic constant
Sorting, 2nd order (BCF)	K_1	1.14815 08398	$\leftrightarrow$ Porter's constant

FIGURE 6. The major constants encountered in the analysis of the sign and sorting algorithms.

(P_4) Why is the operator truncation method so effective (and obviously sound) for the BCF operator?

(P_5) Investigate theoretically and numerically the spectrum of the CCF operator $\widehat{\mathcal{G}}_\sigma$.

(P_6) Explain the occurrence of Porter's constant in the analysis of sorting. In other words, is there some direct relation between sorting and the GCD algorithm, as possibly suggested by (46) and (47)?

(P_7) Analyse the limit distribution of characteristic parameters of the continued fraction fraction sorting algorithm. Candidates are path length, depth of nodes, and number of nodes; see [**12**] for what is currently known (e.g., height). In view of what exists for binary tries (the Jacquet-Régnier theorems [**31**]), we expect Gaussian laws for these parameters.

References

[1] F. Avnaim, J.-D. Boissonnat, O. Devillers, F. P. Preparata, and M. Yvinec, *Evaluating signs of determinants using single-precision arithmetic*, Algorithmica **17** (1997), no. 2, 111–132.

[2] K. I. Babenko, *On a problem of Gauss*, Soviet Mathematical Doklady **19** (1978), no. 1, 136–140.

[3] David H. Bailey, Jonathan M. Borwein, and Roland Girgensohn, *Experimental evaluation of Euler sums*, Experimental Mathematics **3** (1994), no. 1, 17–30.

[4] M. Beeler, R. W. Gosper, and R. Schroeppel, *HAKMEM*, Memorandum 239, M.I.T., Artificial Intelligence Laboratory, February 1972, Available on the WorldWide Web at `http://www.inwap.com/pdp10/hbaker/hakmem/hakmem.html`.

[5] Bruce C. Berndt, *Ramanujan's notebooks, part I*, Springer Verlag, 1985.

[6] David Borwein, Jonathan Borwein, and Roland Girgensohn, *Explicit evaluation of Euler sums*, Proceedings of the Edinburgh Mathematical Society **38** (1995), 277–294.

[7] Jonathan Borwein, David Bradley, David Broadhurst, and Petr Lisonek, *Special values of multiple polylogarithms*, Transactions of the American Mathematical Society (1998), To appear.

[8] Jonathan M. Borwein and Peter B. Borwein, *Pi and the AGM*, John Wiley, 1987.

[9] ———, *Strange series and high precision fraud*, American Mathematical Monthly **99** (1992), no. 7, 622–640.

[10] David V. Chudnovsky and Gregory V. Chudnovsky, *Computer algebra in the service of mathematical physics and number theory*, Computers in mathematics (Stanford, CA, 1986) (New York), Dekker, New York, 1990, pp. 109–232.

[11] Julien Clément, *Analyse en moyenne des tries basés sur la numération en fraction continue*, Mémoire de DEA, 1996.

[12] Julien Clément, Philippe Flajolet, and Brigitte Vallée, *Dynamical sources in information theory: A general analysis of trie structures*, Research Report 3645, Institut National de Recherche en Informatique et en Automatique, March 1999, 61 pages. To appear in *Algorithmica*, 1999.

[13] Henri Cohen, Fernando Rodriguez Villegas, and Don Zagier, *Convergence acceleration of alternating series*, Preprint, 1999, To appear in *Experimental Mathematics.*

[14] Hervé Daudé, Philippe Flajolet, and Brigitte Vallée, *An average-case analysis of the Gaussian algorithm for lattice reduction*, Combinatorics, Probability and Computing **6** (1997), no. 4, 397–433.

[15] P. J. De Doelder, *On some series containing* $\psi(x) - \psi(y)$ *and* $(\psi(x) - \psi(y))^2$ *for certain values of* x *and* y, Journal of Computational and Applied Mathematics **37** (1991), 125–141.

[16] Steven Finch, *Favorite mathematical constants*, Available under World Wide Web at `http: //www.mathsoft.com/asolve/constant/constant.html`, 1995.

[17] Philippe Flajolet and Bruno Salvy, *Euler sums and contour integral representations*, Experimental Mathematics **7** (1998), no. 1, 15–35.

[18] Philippe Flajolet and Robert Sedgewick, *Mellin transforms and asymptotics: finite differences and Rice's integrals*, Theoretical Computer Science **144** (1995), no. 1–2, 101–124.

[19] Philippe Flajolet and Brigitte Vallée, *Continued fraction algorithms, functional operators, and structure constants*, Theoretical Computer Science **194** (1998), no. 1–2, 1–34.

[20] Eldon R. Hansen, *A table of series and products*, Prentice-Hall, 1975.

[21] Doug Hensley, *The number of steps in the Euclidean algorithm*, Journal of Number Theory **49** (1994), no. 2, 142–182.

[22] ———, *A polynomial time algorithm for the Hausdorff dimension of continued fraction Cantor sets*, Journal of Number Theory **58** (1996), no. 1, 9–45.

[23] E. Hille, *Analytic function theory*, Blaisdell Publishing Company, Waltham, 1962, 2 Volumes.

[24] Hoang Ngoc Minh and Michel Petitot, *Lyndon words, polylogarithms, and the Riemann zeta function*, Discrete Mathematics (1997), To be published.

[25] A. Hurwitz, *Über eine besondere Art der Kettenbruch–Entwicklung reeller Grössen*, Advances in Mathematics **12** (1889), 367–405.

[26] Donald E. Knuth, *The art of computer programming*, vol. 3: Sorting and Searching, Addison-Wesley, 1973.

[27] ———, *The art of computer programming*, 2nd ed., vol. 2: Seminumerical Algorithms, Addison-Wesley, 1981.

[28] ———, *Metafont the program*, Addison Wesley, Reading, Massachusetts, 1986, Volume D of *Computers and Typesetting.*

[29] M. Krasnoselskii, *Positive solutions of operator equations*, P. Noordhoff, Groningen, 1964.

[30] L. Lewin (ed.), *Structural properties of polylogarithms*, American Mathematical Society, 1991.

[31] Hosam Mahmoud, *Evolution of random search trees*, John Wiley, New York, 1992.

[32] D. Mayer and G. Roepstorff, *On the relaxation time of Gauss's continued fraction map. I. The Hilbert space approach*, Journal of Statistical Physics **47** (1987), no. 1/2, 149–171.

[33] ———, *On the relaxation time of Gauss's continued fraction map. II. The Banach space approach (transfer operator approach)*, Journal of Statistical Physics **50** (1988), no. 1/2, 331–344.

[34] Dieter H. Mayer, *On a* ζ *function related to the continued fraction transformation*, Bulletin de la Société Mathématique de France **104** (1976), 195–203.

[35] Niels Nielsen, *Die Gammafunktion*, Chelsea Publishing Company, New York, 1965, Reprinted from *Handbuch der Theorie der Gammafunktion* (1906) and *Theorie der Integrallogarithmus und verwandter Transzendenten* (1906).

[36] Niels Erik Nörlund, *Vorlesungen über Differenzenrechnung*, Chelsea Publishing Company, New York, 1954.

[37] Frigyes Riesz and Béla Sz.-Nagy, *Functional analysis*, Dover, 1990, A reprint of the 1955 edition, Ungar Publishing Company.

[38] David Ruelle, *Dynamical zeta functions for piecewise monotone maps of the interval*, CRM Monograph Series, vol. 4, American Mathematical Society, Providence, 1994.

[39] Joel H. Shapiro, *Composition operators and classical function theory*, Springer-Verlag, New York, 1993.

[40] R. Sitaramachandrarao, *A formula of S. Ramanujan*, Journal of Number Theory **25** (1987), 1–19.

[41] E. C. Titchmarsh and D. R. Heath-Brown, *The theory of the Riemann zeta-function*, second ed., Oxford Science Publications, 1986.

[42] Brigitte Vallée, *Algorithms for computing signs of* 2×2 *determinants: dynamics and average-case analysis*, , Lecture Notes in Computer Science, no. 1136, 1997, Proceedings of the Fifth European Symposium on Algorithms, Graz, September 1997., pp. 486–499.
[43] ______, *Opérateurs de Ruelle-Mayer généralisés et analyse en moyenne des algorithmes d'Euclide et de Gauss*, Acta Arithmetica **81** (1997), no. 2, 101–144.
[44] ______, *Dynamical sources in information theory: Fundamental intervals and word prefixes*, Les cahiers du GREYC, Université de Caen, 1998, 32p. To appear in *Algorithmica*, 1999.
[45] ______, *Dynamics of the binary Euclidean algorithm: Functional analysis and operators*, Algorithmica **22** (1998), no. 4, 660–685.
[46] ______, *Dynamique des fractions continues à contraintes périodiques*, Journal of Number Theory **72** (1998), no. 2, 183–235.
[47] ______, *A unifying framework for the analysis of a class of Euclidean algorithms*, Preprint, 2000, To appear in *Proceedings of LATIN'2000, Lecture Notes in Computer Science*, in press.
[48] Ilan Vardi, *Computational recreations in Mathematica*, Addison Wesley, 1991.
[49] Jean Vuillemin, *Exact real computer arithmetic with continued fractions*, IEEE Transactions on Computers **39** (1990), no. 8, 1087–1105.
[50] W. Wasow, *Asymptotic expansions for ordinary differential equations*, Dover, 1987, A reprint of the John Wiley edition, 1965.
[51] Don Zagier, *Values of zeta functions and their applications*, Proceedings of the First European Congress of Mathematics, Paris (A. Joseph *et al.*, ed.), vol. II, Birkhäuser Verlag, 1994, (Progress in Mathematics, volume 120.), pp. 497–512.
[52] Doron Zeilberger, *A holonomic approach to special functions identities*, Journal of Computational and Applied Mathematics **32** (1990), 321–368.

INRIA, ROCQUENCOURT, F-78150 LE CHESNAY (FRANCE)
E-mail address: `Philippe.Flajolet@inria.fr`

DÉPARTEMENT D'INFORMATIQUE, UNIVERSITÉ DE CAEN, F-14032 CAEN (FRANCE)
E-mail address: `Brigitte.Vallee@info.unicaen.fr`

Canadian Mathematical Society
Conference Proceedings
Volume **27**, 2000

A class of dynamical systems with a convex Lyapunov function

Manal Gabour, Simeon Reich and Alexander J. Zaslavski

ABSTRACT. We consider a metric space of sequences of mappings acting on a closed bounded convex subset K of a Banach space which share a common convex Lyapunov function f. We introduce the concept of normality and show that a generic element taken from this space is normal. The sequence of values of the Lyapunov function f along any (unrestricted) trajectory of such an element tends to the infimum of f on K.

1. Introduction

Assume that $(X, ||\cdot||)$ is a Banach space with norm $||\cdot||$, $K \subset X$ is a nonempty bounded closed convex subset of X, and $f : K \to R^1$ is a convex uniformly continuous function. Set

$$\inf(f) = \inf\{f(x) : x \in K\}.$$

Observe that this infimum is finite because K is bounded and f is uniformly continuous. We consider the topological subspace $K \subset X$ with the relative topology. Denote by $\mathfrak{A}$ the set of all self-mappings $A : K \to K$ such that

$$f(Ax) \le f(x) \text{ for all } x \in K \tag{1.1}$$

and by $\mathfrak{A}_c$ the set of all continuous mappings $A \in \mathfrak{A}$. In [10, Section 4] we constructed many mappings belonging to $\mathfrak{A}_c$.

For the set $\mathfrak{A}$ we define a metric $\rho : \mathfrak{A} \times \mathfrak{A} \to R^1$ by

$$\rho(A, B) = \sup\{||Ax - Bx|| : x \in K\},\ A, B \in \mathfrak{A}. \tag{1.2}$$

1991 *Mathematics Subject Classification.* 49M45, 90C25.

Key words and phrases. Banach space, complete metric space, convex function, generic property, infinite product.

Clearly the metric space $\mathfrak{A}$ is complete and $\mathfrak{A}_c$ is a closed subset of $\mathfrak{A}$. In the sequel we will consider the metric space $(\mathfrak{A}_c, \rho)$. Denote by $\mathfrak{M}$ the set of all sequences $\{A_t\}_{t=1}^{\infty} \subset \mathfrak{A}$ and by $\mathfrak{M}_c$ the set of all sequences $\{A_t\}_{t=1}^{\infty} \subset \mathfrak{A}_c$. Members $\{A_t\}_{t=1}^{\infty}$, $\{B_t\}_{t=1}^{\infty}$ and $\{C_t\}_{t=1}^{\infty}$ of $\mathfrak{M}$ will occasionally be denoted by boldface $\mathbf{A}$, $\mathbf{B}$ and $\mathbf{C}$, respectively. For the set $\mathfrak{M}$ we will consider two uniformities and the topologies induced by them. The first uniformity is determined by the following base:

(1.3)
$$E_w(N, \epsilon) = \{(\{A_t\}_{t=1}^{\infty}, \{B_t\}_{t=1}^{\infty}) \in \mathfrak{M} \times \mathfrak{M} : \rho(A_t, B_t) \le \epsilon, \ t = 1, \dots N\},$$

where N is a natural number and $\epsilon > 0$. Clearly the uniform space $\mathfrak{M}$ with this uniformity is metrizable (by a metric $\rho_w : \mathfrak{M} \times \mathfrak{M} \to R^1$) and complete (see [8]). We equip the set $\mathfrak{M}$ with the topology induced by this uniformity. This topology will be called weak and denoted by τ_w. Clearly $\mathfrak{M}_c$ is a closed subest of $\mathfrak{M}$ with the weak topology.

The second uniformity is determined by the following base:

$$E_s(\epsilon) = \{(\{A_t\}_{t=1}^{\infty}, \{B_t\}_{t=1}^{\infty}) \in \mathfrak{M} \times \mathfrak{M} : \rho(A_t, B_t) \le \epsilon, \ t \ge 1\}, \tag{1.4}$$

where $\epsilon > 0$. Clearly this uniformity is metrizable (by a metric $\rho_s : \mathfrak{M} \times \mathfrak{M} \to R^1$) and complete (see [8]). Denote by τ_s the topology induced by this uniformity in $\mathfrak{M}$. Since τ_s is clearly stronger than τ_w, it will be called strong. We consider the topological subspace $\mathfrak{M}_c \subset \mathfrak{M}$ with the relative weak and strong topologies.

From the point of view of the theory of dynamical systems each element of $\mathfrak{M}$ describes a nonstationary dynamical system with a Lyapunov function f. Also, some optimization procedures in Banach spaces can be represented by elements of $\mathfrak{M}$ (see examples in [10, Section 4]). In [10] we showed that for a generic sequence taken from the space $\mathfrak{M}_c$ the sequence of values of the Lyapunov function f along any trajectory tends to the infimum of f. For recent studies of the minimization of convex functionals on abstract spaces see, for example, [1], [2] and [7].

In [10], instead of considering a certain convergence property for a single sequence of operators, we investigated it for the space $\mathfrak{M}_c$ of all such sequences, and showed that this property holds for most of them. This approach has already been successfully applied in global analysis and the theory of dynamical systems ([4], [5] and [9]), as well as in optimization theory and the calculus of variations (see [2], [3], [6], [7], [10], [11] and [12]).

A mapping $A \in \mathfrak{A}$ is called normal if given $\epsilon > 0$ there is $\delta(\epsilon) > 0$ such that for each $x \in K$ satisfying $f(x) \ge \inf(f) + \epsilon$, the inequality

$$f(Ax) \le f(x) - \delta(\epsilon)$$

is true.

A sequence $\{A_t\}_{t=1}^{\infty} \in \mathfrak{M}$ is called normal if given $\epsilon > 0$ there is $\delta(\epsilon) > 0$ such that for each $x \in K$ satisfying $f(x) \ge \inf(f) + \epsilon$ and each integer $t \ge 1$, the inequality

$$f(A_t x) \le f(x) - \delta(\epsilon)$$

holds.

In this paper we show that a generic element taken from the spaces $\mathfrak{A}$, $\mathfrak{A}_c$, $\mathfrak{M}$ and $\mathfrak{M}_c$ is normal. This is important because it turns out that the sequence of values of the Lyapunov function f along any (unrestricted) trajectory of such an element tends to the infimum of f on K.

For $\alpha \in (0,1)$, $\mathbf{A} = \{A_t\}_{t=1}^{\infty}$ and $\mathbf{B} = \{B_t\}_{t=1}^{\infty} \in \mathfrak{M}$ define $\alpha\mathbf{A} + (1-\alpha)\mathbf{B} = \{\alpha A_t + (1-\alpha)B_t\}_{t=1}^{\infty} \in \mathfrak{M}$.

We can easily prove the following fact.

PROPOSITION 1.1. *Let $\alpha \in (0,1)$, $\mathbf{A}, \mathbf{B} \in \mathfrak{M}$ and let $\mathbf{A}$ be normal. Then $\alpha\mathbf{A} + (1-\alpha)\mathbf{B}$ is also normal.*

We now state the main results of our paper. Their proofs will be given in Sections 4-6.

THEOREM 1.1. *Let $\mathbf{A} = \{A_t\}_{t=1}^{\infty} \in \mathfrak{M}$ be normal and let $\epsilon > 0$. Then there exists a neighborhood U of $\mathbf{A}$ in $\mathfrak{M}$ with the strong topology and a natural number N such that for each $\mathbf{C} = \{C_t\}_{t=1}^{\infty} \in U$, each $x \in K$ and each $r : \{1, 2, \dots\} \to \{1, 2, \dots\}$,*

$$f(C_{r(N)} \dots C_{r(1)} x) \le \inf(f) + \epsilon.$$

THEOREM 1.2. *Let $\mathbf{A} = \{A_t\}_{t=1}^{\infty} \in \mathfrak{M}$ be normal and let $\epsilon > 0$. Then there exists a neighborhood U of $\mathbf{A}$ in $\mathfrak{M}$ with the weak topology and a natural number N such that for each $\mathbf{C} = \{C_t\}_{t=1}^{\infty} \in U$ and each $x \in K$,*

$$f(C_N \dots C_1 x) \le \inf(f) + \epsilon.$$

THEOREM 1.3. *There exists a set $\mathcal{F} \subset \mathfrak{M}$ which is a countable intersection of open everywhere dense sets in $\mathfrak{M}$ with the strong topology and a set $\mathcal{F}_c \subset \mathcal{F} \cap \mathfrak{M}_c$ which is a countable intersection of open everywhere dense sets in $\mathfrak{M}_c$ with the strong topology such that each $\mathbf{A} \in \mathcal{F}$ is normal.*

THEOREM 1.4. *There exists a set $\mathcal{F} \subset \mathfrak{A}$ which is a countable intersection of open everywhere dense sets in $\mathfrak{A}$ and a set $\mathcal{F}_c \subset \mathcal{F} \cap \mathfrak{A}_c$ which is a countable intersection of open everywhere dense sets in $\mathfrak{A}_c$ such that each $\mathbf{A} \in \mathcal{F}$ is normal.*

2. Existence of a normal $A \in \mathfrak{A}_c$

If there is $x_{min} \in K$ for which $f(x_{min}) = \inf(f)$, then we can set $A(x) = x_{min}$ for all $x \in K$ and this A is normal. Therefore in order to show the existence of a normal $A \in \mathfrak{A}_c$ we may assume that

$$\{x \in K : f(x) = \inf(f)\} = \emptyset. \tag{2.1}$$

In [10] a normal $A \in \mathfrak{A}_c$ was constructed by using a continuous partition of unity. J. M. Borwein suggested that its existence should follow from Michael's selection theorem.

PROPOSITION 2.1. *There exists a normal* $A_* \in \mathfrak{A}_c$.

PROOF. We may assume that (2.1) is true. Define a set-valued map $a: K \to 2^K$ as follows: For each $x \in K$ denote by $a(x)$ the closure (in the norm topology of X) of the set

$$\{y \in K : f(y) < 2^{-1}(f(x) + \inf(f))\}. \tag{2.2}$$

Clearly for each $x \in K$ the set $a(x)$ is nonempty, closed and convex. We will show that a is lower semicontinuous.

Let $x_0 \in K$, $y_0 \in a(x_0)$ and let $\epsilon > 0$. To prove that a is lower semicontinuous we need to show that there is $\delta > 0$ such that for each $x \in K$ satisfying $||x - x_0|| < \delta$,

$$a(x) \cap \{y \in K : ||y - y_0|| < \epsilon\} \neq \emptyset.$$

By the definition of $a(x_0)$ there exists $y_1 \in K$ such that

$$f(y_1) < 2^{-1}(f(x_0) + \inf(f)) \text{ and } ||y_1 - y_0|| < \epsilon/2.$$

Since the function f is continuous, there is $\delta > 0$ such that for each $x \in K$ satisfying $||x - x_0|| < \delta$,

$$f(y_1) < 2^{-1}(f(x) + \inf(f)).$$

Hence $y_1 \in a(x)$ by definition. Therefore a is indeed lower semicontinuous. By Michael's selection theorem there exists a continuous mapping $A_* : K \to K$ such that $A_*x \in a(x)$ for all $x \in K$. It follows from the definition of a (see (2.2)) that for each $x \in K$,

$$f(A_*x) \leq 2^{-1}(f(x) + \inf(f)).$$

This implies that A_* is normal. This completes the proof of Proposition 2.1.

3. Preliminary results

By Proposition 2.1, there exists a normal mapping $A_* \in \mathfrak{A}_c$. For each $\{A_t\}_{t=1}^{\infty} \in \mathfrak{M}$ and each $\gamma \in (0,1)$ we define a sequence of mappings $\mathbf{A}^\gamma = \{A_t^\gamma\}_{t=1}^{\infty} \in \mathfrak{M}$ by

$$A_t^\gamma x = (1-\gamma)A_t x + \gamma A_* x, \quad x \in K,\ t = 1, 2, \dots \tag{3.1}$$

Clearly for each $\mathbf{A} = \{A_t\}_{t=1}^{\infty} \in \mathfrak{M}_c$ and each $\gamma \in (0,1)$, $\mathbf{A}^\gamma \in \mathfrak{M}_c$. By (3.1) and Proposition 1.1, $\mathbf{A}^\gamma$ is normal for each $\mathbf{A} \in \mathfrak{M}$ and each $\gamma \in (0,1)$. It is obvious that for each $\mathbf{A} \in \mathfrak{M}$,

$$\mathbf{A}^\gamma \to \mathbf{A} \text{ as } \gamma \to 0^+ \text{ in the strong topology.} \tag{3.2}$$

LEMMA 3.1. *Let* $\mathbf{A} = \{A_t\}_{t=1}^{\infty} \in \mathfrak{M}$ *be normal and let* $\epsilon > 0$. *Then there exist a neighborhood* U *of* $\mathbf{A}$ *in* $\mathfrak{M}$ *with the strong topology and a number* $\delta > 0$ *such that for each* $\mathbf{B} = \{B_t\}_{t=1}^{\infty} \in U$, *each* $x \in K$ *satisfying*

$$f(x) \geq \inf(f) + \epsilon \tag{3.3}$$

and each integer $t \geq 1$,

$$f(B_t x) \leq f(x) - \delta.$$

PROOF. Since $\mathbf{A}$ is normal there is $\delta_0 > 0$ such that for each integer $t \geq 1$ and each $x \in K$ satisfying (3.3),

$$f(A_t x) \leq f(x) - \delta_0. \tag{3.4}$$

Since f is uniformly continuous, there is $\delta \in (0, 4^{-1}\delta_0)$ such that

$$|f(y) - f(z)| \leq 4^{-1}\delta_0 \tag{3.5}$$

for each $y, z \in K$ satisfying $||y - z|| \leq 2\delta$. Set

$$U = \{\mathbf{B} \in \mathfrak{M} : \ (\mathbf{A}, \mathbf{B}) \in E_s(\delta)\} \tag{3.6}$$

(see (1.4)).

Assume that $\mathbf{B} = \{B_t\}_{t=1}^{\infty} \in U$, let $t \geq 1$ be an integer and let $x \in K$ satisfy (3.3). By (3.3) and the definition of δ_0, (3.4) is true. The definitions of δ and U (see (3.5) and (3.6)) imply that

$$||A_t x - B_t x|| \leq \delta \text{ and } |f(A_t x) - f(B_t x)| \leq \delta_0/4.$$

Combined with (3.4) this implies that

$$f(B_t x) \leq f(x) + 4^{-1}\delta_0 - \delta_0 \leq f(x) - \delta.$$

This completes the proof of the lemma.

4. Proof of Theorem 1.1

We use the definitions and notations of Section 3.

PROOF OF THEOREM 1.1. Assume that $\mathbf{A} = \{A_t\}_{t=1}^{\infty} \in \mathfrak{M}$ is normal and let $\epsilon > 0$. By Lemma 3.1 there exist a neighborhood U of $\mathbf{A}$ in $\mathfrak{M}$ with the strong topology and a number $\delta > 0$ such that the following property holds:

(Pi) For each $\{B_t\}_{t=1}^{\infty} \in U$, each integer $t \geq 1$ and each $x \in K$ satisfying (3.3), the inequality

$$f(B_t x) \leq f(x) - \delta \tag{4.1}$$

holds. Choose a natural number $N \geq 4$ such that

$$\delta N > 2(\epsilon + 1) + 2\sup\{|f(z)| : \ z \in K\}. \tag{4.2}$$

Assume that

$$\mathbf{C} = \{C_t\}_{t=1}^{\infty} \in U,\ x \in K \text{ and } r:\ \{1,2,\dots\} \to \{1,2,\dots\}. \tag{4.3}$$

We will show that

$$f(C_{r(N)} \dots C_{r(1)}x) \le \inf(f) + \epsilon. \tag{4.4}$$

Assume the contrary. Then, since $f(C_t y) \le f(y)$ for all $t \ge 1$ and all $y \in K$, we have

$$f(x) > \inf(f) + \epsilon,\ f(C_{r(n)} \dots C_{r(1)}x) > \inf(f) + \epsilon,\ n = 1, \dots N. \tag{4.5}$$

It follows from (4.5), (4.3) and property (Pi) that

$$f(C_{r(1)}x) \le f(x) - \delta,\ f(C_{r(n+1)}C_{r(n)} \dots C_{r(1)}x) \le f(C_{r(n)} \dots C_{r(1)}x) - \delta,$$
$$n = 1, \dots N-1.$$

This implies that

$$f(C_{r(N)} \dots C_{r(1)}x) \le f(x) - N\delta \le -2 - \sup\{|f(z)| :\ z \in K\},$$

a contradiction. Therefore (4.4) is valid and Theorem 1.1 is proved.

5. Proof of Theorem 1.2

Assume that $\mathbf{A} = \{A_t\}_{t=1}^{\infty} \in \mathfrak{M}$ is normal and $\epsilon > 0$. Since $\mathbf{A}$ is normal there is $\delta \in (0,1)$ such that for each integer $t \ge 1$ and each $x \in K$ satisfying

$$f(x) \ge \inf(f) + \epsilon, \tag{5.1}$$

the following inequality is valid:

$$f(A_t x) \le f(x) - \delta. \tag{5.2}$$

Choose a natural number $N > 4$ for which

$$N > 4\delta^{-1} + 4\delta^{-1}\sup\{|f(z)| :\ z \in K\}. \tag{5.3}$$

Since f is uniformly continuous there is $\Delta \in (0, 4^{-1}\delta)$ such that

$$|f(z) - f(y)| \le 8^{-1}\delta \tag{5.4}$$

for each $y, z \in K$ satisfying $||z - y|| \le 4\Delta$. Set

$$U = \{\mathbf{B} \in \mathfrak{M} :\ (\mathbf{A}, \mathbf{B}) \in E_w(N, \Delta)\} \tag{5.5}$$

(see (1.3)).

Assume that

$$\mathbf{C} = \{C_t\}_{t=1}^{\infty} \in U \text{ and } x \in K. \tag{5.6}$$

We will show that

$$f(C_N \dots C_1 x) \le \inf(f) + \epsilon. \tag{5.7}$$

Assume the contrary. Then, since $f(C_t y) \le f(y)$ for all $t \ge 1$ and all $y \in K$, we have

$$f(x) > \inf(f) + \epsilon,\ f(C_n \dots C_1 x) > \inf(f) + \epsilon,\ n = 1, \dots N. \tag{5.8}$$

Define $C_0 : K \to K$ by $C_0 x = x$ for all $x \in K$. Let $t \in \{0, \dots N-1\}$. It follows from (5.8) and the definition of δ (see (5.1) and (5.2)) that

$$f(A_{t+1} C_t \dots C_0 x) \le f(C_t \dots C_0 x) - \delta. \tag{5.9}$$

The definition of U (see (5.5)) and (5.6) imply that

$$||A_{t+1} C_t \dots C_0 x - C_{t+1} C_t \dots C_0 x|| \le \Delta.$$

By this inequality and the definition of Δ (see (5.4)),

$$|f(A_{t+1} C_t \dots C_0 x) - f(C_{t+1} C_t \dots C_0 x)| \le 8^{-1}\delta.$$

Combined with (5.9) this implies that

$$f(C_{t+1} C_t \dots C_0 x) \le f(C_t \dots C_0 x) - 2^{-1}\delta.$$

Since this inequality is true for all $t \in \{0, \dots N-1\}$ we conclude that

$$f(C_N \dots C_1 x) \le f(x) - 2^{-1} N\delta.$$

Together with (5.3) this implies that

$$-\sup\{|f(z)| :\ z \in K\} \le \sup\{|f(z)| :\ z \in K\} - 2^{-1}\delta N$$

$$\le -2 - \sup\{|f(z)| :\ z \in K\},$$

a contradiction. Therefore (5.7) holds and Theorem 1.2 is proved.

6. Proof of Theorem 1.3

Once again we use the notations and definitions of Section 3.

Let $\mathbf{A} \in \mathfrak{M}$, $\gamma \in (0,1)$ and let $i \ge 1$ be an integer. Consider the sequence $\mathbf{A}^\gamma \in \mathfrak{M}$ defined by (3.1). By Proposition 1.1, $\mathbf{A}^\gamma$ is normal. By Lemma 3.1 there exists an open neighborhood $U(\mathbf{A}, \gamma, i)$ of $\mathbf{A}^\gamma$ in $\mathfrak{M}$ with the strong topology and a number $\delta(\mathbf{A}, \gamma, i) > 0$ such that the following property holds:

(Pii) For each $\mathbf{B} = \{B_t\}_{t=1}^\infty \in U(\mathbf{A}, \gamma, i)$, each integer $t \ge 1$ and each $x \in K$ satisfying $f(x) \ge \inf(f) + 2^{-i}$,

$$f(B_t x) \le f(x) - \delta(\mathbf{A}, \gamma, i).$$

Define

$$\mathcal{F} = \cap_{i=1}^\infty \cup \{U(\mathbf{A}, \gamma, i) :\ \mathbf{A} \in \mathfrak{M},\ \gamma \in (0,1)\} \tag{6.1}$$

and

$$\mathcal{F}_c = [\cap_{i=1}^\infty \cup \{U(\mathbf{A}, \gamma, i) :\ \mathbf{A} \in \mathfrak{M}_c,\ \gamma \in (0,1)\}] \cap \mathfrak{M}_c.$$

Clearly $\mathcal{F}_c \subset \mathcal{F}$, $\mathcal{F}$ is a countable intersection of open everywhere dense sets in $\mathfrak{M}$ with the strong topology and $\mathcal{F}_c$ is a countable intersection of open everywhere dense sets in $\mathfrak{M}_c$ with the strong topology.

Assume that $\mathbf{B} = \{B_t\}_{t=1}^{\infty} \in \mathcal{F}$. We will show that $\mathbf{B}$ is normal.

Let $\epsilon > 0$. Choose an integer $i \geq 1$ such that

$$2^{-i} < \epsilon/8. \tag{6.2}$$

By (6.1) there exist $\mathbf{A} \in \mathfrak{M}$ and $\gamma \in (0,1)$ such that

$$\mathbf{B} \in U(\mathbf{A}, \gamma, i). \tag{6.3}$$

Let $t \geq 1$ be an integer, $x \in K$, and $f(x) \geq \inf(f) + \epsilon$. Then by (6.2), (6.3) and property (Pii),

$$f(B_t x) \leq f(x) - \delta(\mathbf{A}, \gamma, i).$$

Thus $\mathbf{B}$ is normal and Theorem 1.3 is proved.

The proof of Theorem 1.4 is analogous to that of Theorem 1.3.

Acknowledgments. The second author was partially supported by the Israel Science Foundation founded by the Israel Academy of Sciences and Humanities, by the Fund for the Promotion of Research at the Technion, and by the Technion VPR Fund. The authors are grateful to Professor Jonathan M. Borwein for his fruitful suggestions.

References

1. Alber Y.I., Iusem A.N. and Solodov M.V., *Minimization of nonsmooth convex functionals in Banach spaces*, Journal of Convex Analysis **4** (1997), 235-255.
2. Beer J. and Lucchetti R., *Convex optimization and the epi-distance topology*, Trans. Amer. Math. Soc. **327** (1991), 795 - 813.
3. Borwein J.M. and Preiss D., *A smooth variational principle with applications to subdifferentiability and to differentiability of convex functions*, Trans. Amer. Math. Soc. **303** (1987), 517 - 527.
4. De Blasi F.S. and Myjak J., *Sur la convergence des approximations successives pour les contractions non linéaires dans un espace de Banach*, C.R. Acad. Sc. Paris **283** (1976), 185-187.
5. De Blasi F.S. and Myjak J., *Generic flows generated by continuous vector fields in Banach spaces*, Adv. in Math. **50** (1983), 266-280.
6. Donchev A.L. and Zolezzi T., *Well-Posed Optimization Problems*, Lecture Notes in Math. **1543** (1993), Springer, Berlin-Heidelberg-New York.
7. Ioffe A.D. and Zaslavski A.J., *Variational principles and well-posedness in optimization and calculus of variations*, SIAM Journal on Control and Optimization, in press.
8. Kelley J.L., *General Topology*, D. Van Nostrand, New York, 1955.
9. Reich S. and Zaslavski A.J., *Convergence of generic infinite products of nonexpansive and uniformly continuous operators*, Nonlinear Analysis: Theory, Methods and Applications **36** (1999), 1049-1065.
10. Reich S. and Zaslavski A.J., *On the minimization of convex functionals*, Calculus of Variations and Differential Equations, Chapman & Hall/CRC Research Notes in Mathematics Series, CRC Press, Boca Raton, FL **410** (1999), 200-209.
11. Zaslavski A.J., *Optimal programs on infinite horizon, 1 and 2*, SIAM Journal on Control and Optimization **33** (1995), 1643-1686.

12. Zaslavski A.J., *Existence of solutions of optimal control problems without convexity assumptions*, Nonlinear Analysis: Theory, Methods and Applications, in press.

DEPARTMENT OF MATHEMATICS, THE TECHNION-ISRAEL INSTITUTE OF TECHNOLOGY,
32000 HAIFA, ISRAEL
E-mail address: gmanal, sreich, ajzasl @tx.technion.ac.il

Canadian Mathematical Society
Conference Proceedings
Volume **27**, 2000

On a Convergence of Lower Semicontinuous Functions linked with the Graph Convergence of their Subdifferentials

Michel Geoffroy and Marc Lassonde

To Jonathan Borwein, on the occasion of his award of a Dhc

ABSTRACT. We introduce and discuss a new notion of epigraphical convergence for lower semicontinuous functions, which we call *ball-affine convergence*. This convergence subsumes the slice convergence when the functions are convex as well as various other convergence notions used for studying the stability of subdifferentials of nonconvex functions. We show that the ball-affine convergence of lower semicontinuous functions is linked with the convergence of their local infima and with the graph convergence of their subdifferentials in any space suitable for them. The latter result covers and extends several convergence theorems concerning convex, smooth or arbitrary lower semicontinuous functions, partly because our considerations do not depend on any specific subdifferential.

1. Introduction

Throughout the paper, X is a real Banach space and X^* is its topological dual with $\langle \cdot, \cdot \rangle$ being the duality pairing on $X \times X^*$. We denote by $\mathrm{LSC}(X)$ the space of all proper lower semicontinuous functions from X into $\mathbb{R} \cup \{\infty\}$ and by $\Gamma(X)$ the subspace of $\mathrm{LSC}(X)$ consisting of convex functions. For f in $\mathrm{LSC}(X)$, we let $\operatorname{dom} f = \{x \in X \mid f(x) < \infty\}$, and, given a subdifferential operator ∂, we let $\partial f = \{(x, x^*) \in X \times X^* \mid x^* \in \partial f(x)\}$ and $\widetilde{\partial} f = \{(x, f(x), x^*) \in X \times \mathbb{R} \times X^* \mid (x, x^*) \in \partial f\}$. The purpose of this paper is to study the behavior of the sequence $(\widetilde{\partial} f_n)$ as (f_n) converges to f in some sense.

The first major contribution to this stability problem goes back to Attouch, [**1**] and [**2**, Theorem 3.66], who showed that, in a reflexive Banach space X, the Mosco convergence of a sequence of functions in $\Gamma(X)$ is equivalent to the graph convergence of their subdifferentials plus a so-called normalization condition. This theorem was subsequently generalized by Attouch and Beer [**3**] to any Banach space, replacing Mosco convergence by slice convergence. For a simple proof of this result, see Combari-Thibault [**15**]. For partial extensions of Attouch's theorem to the class

1991 *Mathematics Subject Classification.* Primary 49J52, Secondary 58C20.
Key words and phrases. Lower semicontinuous function, slice convergence, set convergence, subdifferential, variational principles, fuzzy principles.

of primal-lower-nice functions on a Hilbert space, see Levy-Poliquin-Thibault [**25**]. For variants of Attouch's theorem involving a stronger notion of convergence, see Attouch-Ndoutoume-Théra [**4**] and Beer-Théra [**10**].

As is well known, in the equivalence theorem of Attouch, the implication concerning the integration of subdifferentials cannot be generalized outside the convex or smooth setting. On the other hand, the converse implication can be expressed in a slightly sharper and more flexible form. Precisely, Beer, [**8**, Theorem 4.11] or [**9**, Theorem 8.3.7], showed that, in any Banach space, the slice convergence of a sequence (f_n) to f in $\Gamma(X)$ entails the inclusion

$$\widetilde{\partial} f \subset \operatorname{Li} \widetilde{\partial} f_n,$$

where $\operatorname{Li} \widetilde{\partial} f_n$ stands for the Painlevé-Kuratowski lower limit of the sequence $(\widetilde{\partial} f_n)$. Results in this spirit also exist outside the convex setting, e.g.,

– Aubin and Frankowska [**5**, Theorem 7.6.1] consider Fréchet differentiable functions on a Banach space, using the uniform convergence on bounded sets and the Fréchet subdifferential;

– Deville [**16**] deals with proper lower semicontinuous functions on a Banach space with C^1 Lipschitz bump functions, using an extension of the slice convergence on $\mathrm{LSC}(X)$ and the Fréchet subdifferential;

– Penot [**27**] studies proper lower semicontinuous functions on an Asplund space, using a somewhat technical concept of convergence on $\mathrm{LSC}(X)$ and the Fréchet subdifferential.

There is no link between the above results, since the classes of functions, the underlying spaces and the modes of convergence are all different. We observe however that the Fréchet subdifferential is involved in all the results; indeed, as we shall see, this subdifferential plays a crucial role. It is the purpose of this work to devise a general theorem that covers these results and extends them to arbitrary subdifferentials.

To carry out this objective, we are led to define a new notion of convergence on $\mathrm{LSC}(X)$. This convergence is derived from a topology on the space $\mathrm{LSC}(X)$ which we call the *ball-affine* topology, written τ_{ba}. We show that the τ_{ba}-convergence of a sequence of functions can be characterized in terms of the lower convergence of their local infima. We next define the weaker notion of local ball-affine convergence which is sufficient when dealing with the stability of subdifferentials. We show that this local τ_{ba}-convergence subsumes all the convergences considered in the above-mentioned results. These considerations are the object of Section 2.

The next section provides a description of the class of Banach spaces for which the convergence theorem holds. Roughly speaking, these Banach spaces should satisfy a Brøndsted-Rockafellar type property with respect to the subdifferential involved. This property, denoted $(\mathrm{BR})_\partial$, is *a priori* weaker than the smoothness properties considered so far in nonsmooth analysis, e.g., ∂-smoothness of an equivalent norm [**11, 7, 24**], existence of β-smooth Lipschitz bump functions [**16, 11, 29**], ∂-trustworthiness [**20, 21**]. Let us mention that Hilbert spaces satisfy $(\mathrm{BR})_\partial$ with respect to the proximal subdifferential, Asplund spaces with respect to the Fréchet subdifferential, any Banach space with respect to the subdifferential of Clarke, of Michel-Penot, or of Ioffe. Variational principles and fuzzy first-order rules are the main ingredients of this section.

The last section is devoted to the proof of our convergence theorem and to some of its consequences. One feature of this theorem is that it involves both an arbitrary subdifferential for the functions of the sequence and a Fréchet-like subdifferential for the limit function. Its precise statement reads thus: *Let ∂ be any subdifferential and let X be a Banach space such that* $(\mathrm{BR})_\partial$ *is satisfied. Then, the local τ_{ba}-convergence of a sequence (f_n) to f in* $\mathrm{LSC}(X)$ *entails the inclusion*

$$\widetilde{\Delta} f \subset \operatorname{Li} \widetilde{\partial} f_n,$$

where Δf contains $\partial^F f$, the Fréchet subdifferential of f. The results of Beer, Aubin-Frankowska, Deville and Penot mentioned above are among its main consequences.

2. The ball-affine convergence on LSC(X)

2.1. Notations. The set of all nonempty closed and bounded convex subsets of X is denoted by $CB(X)$; the gap between two subsets A, B of X is given by

$$D(A, B) := \inf\{\, \|a - b\| \mid a \in A,\ b \in B\,\};$$

the open ε-neighborhood of A is written

$$O_\varepsilon(A) := \{\, x \in X \mid d(x, A) < \varepsilon\,\}$$

whereas the closed ε-ball centered at point x is written $B_\varepsilon(x)$; the indicator of A is the function $\delta_A : X \to \mathbb{R} \cup \{\infty\}$ defined by

$$\delta_A(x) := \begin{cases} 0 & \text{if } x \in A \\ \infty & \text{otherwise.} \end{cases}$$

For $f : X \to \mathbb{R} \cup \{\infty\}$, we denote by $\operatorname{graph} f := \{\, (x, \alpha) \mid f(x) = \alpha\,\}$ the graph of f, by $\operatorname{epi} f := \{\, (x, \alpha) \mid f(x) \leq \alpha\,\}$ the epigraph of f, and by $\operatorname{hypo} f := \{\, (x, \alpha) \mid x \in \operatorname{dom} f,\ f(x) \geq \alpha\,\}$ the hypograph of f.

For B a closed ball in X and $\varphi : X \to \mathbb{R}$ a continuous affine function, we let $\varphi_B := \varphi + \delta_B$ and denote by $AB(X)$ the space of all such functions, that is,

$$AB(X) := \{\, \varphi + \delta_B \mid \varphi : X \to \mathbb{R} \text{ continuous affine, } B \text{ closed ball of } X\,\}.$$

We endow the space $X \times \mathbb{R}$ with the norm $\|(x, t)\| := \max\{\|x\|, |t|\}$ and we denote by d the associated metric.

2.2. The ball-affine topology. Associated with the space $AB(X)$, we define a topology on $\mathrm{LSC}(X)$ as follows:

DEFINITION 2.1. The *ball-affine topology* τ_{ba} on $\mathrm{LSC}(X)$ has as a sub-base all sets of the form

$$\begin{aligned} U(\varphi_B, \varepsilon) &:= \{f \in \mathrm{LSC}(X) \mid D(\operatorname{epi} f, \operatorname{graph} \varphi_B) < \varepsilon\}, \\ L(\varphi_B) &:= \{f \in \mathrm{LSC}(X) \mid D(\operatorname{epi} f, \operatorname{graph} \varphi_B) > 0\}, \end{aligned}$$

where $\varphi_B \in AB(X)$ and $\varepsilon > 0$. The topology generated by all sets of the form $U(\varphi_B, \varepsilon)$ (respectively $L(\varphi_B)$) is denoted by τ_{ba}^+ (respectively by τ_{ba}^-).

The following lemma is elementary but useful:

LEMMA 2.2. *Let $f, g : X \to \mathbb{R} \cup \{\infty\}$ be proper functions on X. Then,*

$$D(\operatorname{epi} f, \operatorname{hypo} g) = D(\operatorname{graph} f, \operatorname{hypo} g) = D(\operatorname{epi} f, \operatorname{graph} g).$$

PROOF. Let $\varepsilon > 0$, and pick $(x, \alpha) \in \operatorname{epi} f$ and $(y, \beta) \in \operatorname{hypo} g$ such that

$$d((x,\alpha),(y,\beta)) \leq D(\operatorname{epi} f, \operatorname{hypo} g) + \varepsilon.$$

Since $\alpha \geq f(x)$, we have $(y, \beta - \alpha + f(x)) \in \operatorname{hypo} g$. Therefore,

$$\begin{aligned} d((x,\alpha),(y,\beta)) &= \max\{\|x-y\|, |\alpha-\beta|\} \\ &= d((x,f(x)),(y,\beta-\alpha+f(x))) \\ &\geq D(\operatorname{graph} f, \operatorname{hypo} g), \end{aligned}$$

and so, $D(\operatorname{graph} f, \operatorname{hypo} g) \leq D(\operatorname{epi} f, \operatorname{hypo} g) + \varepsilon$. Since $\varepsilon > 0$ was arbitrary, we conclude that $D(\operatorname{graph} f, \operatorname{hypo} g) \leq D(\operatorname{epi} f, \operatorname{hypo} g)$. The converse inequality being obviously true, the first equality is proved.

The second equality, $D(\operatorname{epi} f, \operatorname{hypo} g) = D(\operatorname{epi} f, \operatorname{graph} g)$, can be proved exactly in the same way. □

PROPOSITION 2.3. *The topology τ_{ba}^+ agrees with the topology generated by all sets of the form*

$$(V\times]-\infty,\alpha[)^- := \{\, f \in \operatorname{LSC}(X) \mid \operatorname{epi} f \cap (V\times]-\infty,\alpha[) \neq \emptyset \,\},$$

where V is open in X and α belongs to $\mathbb{R}$.

PROOF. First, we show that the topology generated by all sets of the form $(V\times]-\infty,\alpha[)^-$ contains τ_{ba}^+. By Lemma 2.2,

$$U(\varphi_B, \varepsilon) = \{\, f \in \operatorname{LSC}(X) \mid D(\operatorname{epi} f, \operatorname{hypo} \varphi_B) < \varepsilon \,\},$$

that is,

$$U(\varphi_B, \varepsilon) = \{\, f \in \operatorname{LSC}(X) \mid \operatorname{epi} f \cap O_\varepsilon(\operatorname{hypo} \varphi_B) \neq \emptyset \,\}.$$

But

$$O_\varepsilon(\operatorname{hypo} \varphi_B) = \bigcup_{(y,\beta)\in \operatorname{hypo} \varphi_B} O_\varepsilon(y)\times]-\infty,\beta+\varepsilon[,$$

hence,

$$U(\varphi_B, \varepsilon) = \bigcup_{(y,\beta)\in \operatorname{hypo} \varphi_B} \{\, f \in \operatorname{LSC}(X) \mid \operatorname{epi} f \cap (O_\varepsilon(y)\times]-\infty,\beta+\varepsilon[) \neq \emptyset \,\},$$

proving that $U(\varphi_B, \varepsilon)$ belongs to the topology generated by the sets $(V\times]-\infty,\alpha[)^-$.

Conversely, we show that τ_{ba}^+ contains all sets of the form $(V\times]-\infty,\alpha[)^-$. For each v in V there is $\varepsilon_v > 0$ such that $O_{\varepsilon_v}(v) \subset V$, so

$$V\times]-\infty,\alpha[= \bigcup_{v\in V} O_{\varepsilon_v}(v)\times]-\infty,\alpha[.$$

But

$$O_{\varepsilon_v}(v)\times]-\infty,\alpha[= O_{\varepsilon_v}(\{v\}\times]-\infty,\alpha-\varepsilon_v[\,) = O_{\varepsilon_v}(\operatorname{hypo} \varphi_{\{v\}}),$$

where $\varphi_{\{v\}} := \alpha - \varepsilon_v + \delta_{\{v\}} \in AB(X)$. Consequently,

$$\begin{aligned} (V\times]-\infty,\alpha[)^- &= \bigcup_{v\in V} \{\, f \in \operatorname{LSC}(X) \mid \operatorname{epi} f \cap O_{\varepsilon_v}(\operatorname{hypo} \varphi_{\{v\}}) \neq \emptyset \,\} \\ &= \bigcup_{v\in V} \{\, f \in \operatorname{LSC}(X) \mid D(\operatorname{epi} f, \operatorname{graph} \varphi_{\{v\}}) < \varepsilon_v \,\}, \end{aligned}$$

proving that $(V\times]-\infty,\alpha[)^-$ belongs to τ_{ba}^+. □

Thanks to Proposition 2.3, we can compare the ball-affine topology with the slice topology on $\Gamma(X)$. We first recall the definition of the slice topology in this space.

DEFINITION 2.4. (Beer [**8, 9**]). A sub-base for the *slice topology* τ_s on $\Gamma(X)$ consists of all sets of the form

$$\begin{aligned}(V\times]-\infty,\alpha[)^- &:= \{\, f\in\Gamma(X) \mid \operatorname{epi} f\cap (V\times]-\infty,\alpha[) \neq \emptyset \,\},\\ (B^c)^{++} &:= \{\, f\in\Gamma(X) \mid D(\operatorname{epi} f, B) > 0 \,\},\end{aligned}$$

where V is open in X, $\alpha\in\mathbb{R}$ and $B\in CB(X\times\mathbb{R})$.

PROPOSITION 2.5. *The topologies τ_{ba} and τ_s agree on $\Gamma(X)$.*

PROOF. It follows from Proposition 2.3 that the topology τ_{ba}^+ agrees with the one generated by all sets of the form $(V\times]-\infty,\alpha[)^-$. Moreover, since the graph of any element of $AB(X)$ belongs to $CB(X\times\mathbb{R})$, the topology τ_{ba}^- is contained in the topology generated by all sets of the form $(B^c)^{++}$ on $\Gamma(X)$. The converse is also true since, according to Beer, [**8**, Theorem 3.5] or [**9**, Proposition 8.1.2], the topology τ_s can be recovered if we just let the subsets B in Definition 2.4 run over the slices of $X\times\mathbb{R}$, where a slice S is given by

$$S := \{\, (x,\alpha)\in X\times\mathbb{R} \mid \|x\|\le\lambda,\ \alpha = \langle x^*, x\rangle + \eta \,\},$$

that is, $S = \operatorname{graph}\varphi_B$, where $\varphi_B := \langle x^*, .\,\rangle + \eta + \delta_{B_\lambda(0)} \in AB(X)$. □

2.3. Infima of functions and the ball-affine convergence. The definition of the ball-affine topology forces the following definition for the ball-affine convergence of a sequence in $\mathrm{LSC}(X)$:

DEFINITION 2.6. A sequence (f_n) in $\mathrm{LSC}(X)$ τ_{ba}*-converges* to $f\in\mathrm{LSC}(X)$, written $f = \tau_{ba} - \lim f_n$, if and only if for all $\varphi_B\in AB(X)$ the following holds

$$(\mathrm{A})\begin{cases}\text{(i)} & \limsup_n D(\operatorname{epi} f_n, \operatorname{graph}\varphi_B) \le D(\operatorname{epi} f, \operatorname{graph}\varphi_B);\\ \text{(ii)} & D(\operatorname{epi} f, \operatorname{graph}\varphi_B) > 0 \Rightarrow D(\operatorname{epi} f_n, \operatorname{graph}\varphi_B) > 0, \text{ for large } n.\end{cases}$$

The main theorem of this section provides a characterization of the τ_{ba}-convergence of functions in terms of the lower convergence of their local infima. The proof is based on the following enlargement lemma:

LEMMA 2.7. *Let $x_0\in X$, $x^*\in X^*, \eta\in\mathbb{R}$ and $\lambda>0$. For $\delta\ge 0$, set*

$$S_\delta = \{(x,\alpha)\in X\times\mathbb{R} \mid \|x - x_0\| \le \lambda + \delta,\ \alpha = \langle x^*, x\rangle + \eta + \delta\}.$$

(a) *For $f\in\mathrm{LSC}(X)$, $\varepsilon>0$, $\gamma\in]0,\varepsilon[$, and $\delta = \dfrac{\varepsilon-\gamma}{1+\|x^*\|}$, the following holds:*

$$D(\operatorname{epi} f, S_0) > \varepsilon \Rightarrow D(\operatorname{epi} f, S_\delta) \ge \gamma.$$

(b) *For $f\in\mathrm{LSC}(X)$, $\delta>0$, and $\varepsilon = \dfrac{\delta}{1+\|x^*\|}$, the following holds:*

$$\operatorname{epi} f\cap S_\delta = \emptyset \Rightarrow D(\operatorname{epi} f, S_0) \ge \varepsilon.$$

PROOF. (a) Let $(x,\alpha)\in S_\delta$, i.e., $\|x - x_0\|\le\lambda+\delta$ and $\alpha = \langle x^*, x\rangle + \eta + \delta$, and pick (y,β) in $\operatorname{epi} f$. We must show that

$$d((x,\alpha),(y,\beta)) \ge \gamma. \tag{1}$$

If $\|y - x\|\ge\gamma$, we are done. If $\|y-x\|<\gamma$, we consider two cases:

(i) $x \in B_\lambda(x_0)$. Then $(x, \alpha - \delta) \in S_0$, hence $d((x, \alpha - \delta), (y, \beta)) > \varepsilon$. Since $\|y - x\| < \gamma < \varepsilon$, necessarily $\beta - \alpha + \delta > \varepsilon$, i.e., $\beta - \alpha > \varepsilon - \delta \geq \gamma$. This gives (1).

(ii) $x \notin B_\lambda(x_0)$. Let $x_\lambda = x_0 + t(x - x_0)$, where $t = \lambda/\|x - x_0\|$, and let $\alpha_\lambda = \langle x^*, x_\lambda \rangle + \eta$. Then $x_\lambda \in B_\lambda(x_0)$, hence $(x_\lambda, \alpha_\lambda) \in S_0$ and consequently $d((x_\lambda, \alpha_\lambda), (y, \beta)) > \varepsilon$. Since

$$\|x - x_\lambda\| = \|(1 - t)(x - x_0)\| = \|x - x_0\| - \lambda \leq \delta,$$

we have $\|y - x_\lambda\| \leq \|y - x\| + \|x - x_\lambda\| < \gamma + \delta \leq \varepsilon$, so

$$\begin{aligned}\beta &> \alpha_\lambda + \varepsilon \\ &= \langle x^*, x_\lambda - x \rangle + \langle x^*, x \rangle + \eta + \varepsilon \\ &\geq -\|x^*\|\delta + \alpha - \delta + \varepsilon = \gamma + \alpha,\end{aligned}$$

from which we derive that $\beta - \alpha \geq \gamma$, proving that (1) also holds in this case.

(b) Let $(x, \alpha) \in S_0$, i.e., $\|x - x_0\| \leq \lambda$ and $\alpha = \langle x^*, x \rangle + \eta$, and pick (y, β) in $\operatorname{epi} f$. As above, we must show that

$$d((x, \alpha), (y, \beta)) \geq \varepsilon. \tag{2}$$

If $\|y - x\| \geq \varepsilon$, this is obvious. If $\|y - x\| < \varepsilon$, then $\|y - x_0\| \leq \lambda + \varepsilon \leq \lambda + \delta$. Because $\operatorname{epi} f \cap S_\delta = \emptyset$, we necessarily have

$$\begin{aligned}\beta &> \langle x^*, y \rangle + \eta + \delta \\ &= \langle x^*, y - x \rangle + \langle x^*, x \rangle + \eta + \delta \\ &\geq -\|x^*\|\varepsilon + \alpha + \delta \\ &= \alpha + \varepsilon.\end{aligned}$$

It follows that $\beta - \alpha > \varepsilon$, hence (2) holds. □

In the following characterization of τ_{ba}-convergence we use the concept of uniform infimum as introduced in [**7, 24**]. We recall that the *uniform infimum* of a function f on a ball $B_\lambda(x_0)$ is defined by

$$r_{B_\lambda(x_0)}(f) := \sup_{\delta > 0} \inf_{B_{\lambda+\delta}(x_0)} f.$$

It is shown in [**7**] that, if $f \in \mathrm{LSC}(X)$ is convex and finite at x_0, or uniformly continuous near x_0, or inf-compact near x_0, then $r_{B_\lambda(x_0)}(f) = \inf_{B_\lambda(x_0)} f$ (see [**24**] for extensions of these properties to families of functions).

THEOREM 2.8. *Let (f_n) be a sequence in* $\mathrm{LSC}(X)$ *and let $f \in \mathrm{LSC}(X)$. The following are equivalent:*

(A) $f = \tau_{ba} - \lim f_n$;

(B) $\begin{cases} \text{(i) } \forall x_0 \in \operatorname{dom} f,\ \exists (x_n) \subset X:\ x_n \to x_0,\ f_n(x_n) \to f(x_0); \\ \text{(ii) } \forall x_0 \in X,\ \forall \lambda \geq 0,\ \forall x^* \in X^*:\ r_{B_\lambda(x_0)}(f - x^*) \leq \liminf_n r_{B_\lambda(x_0)}(f_n - x^*). \end{cases}$

PROOF. **(A) ⇒ (B) (i)**. Let x_0 in $\operatorname{dom} f$, and consider $\varphi_B := f(x_0) + \delta_{\{x_0\}} \in AB(X)$. Since $\operatorname{graph} \varphi_B = \{(x_0, f(x_0))\}$, we have $D(\operatorname{epi} f, \operatorname{graph} \varphi_B) = 0$, so, by (A) (i),

$$\limsup_n D(\operatorname{epi} f_n, \operatorname{graph} \varphi_B) = 0.$$

Hence, there exists a sequence (x_n, α_n) in $\operatorname{epi} f_n$ such that $x_n \to x_0$ and $\alpha_n \to f(x_0)$. Therefore,

(3) $$\limsup_n f_n(x_n) \leq \limsup_n \alpha_n \leq f(x_0).$$

Now, consider $\varphi'_B = f(x_0) - \varepsilon + \delta_{\{x_0\}} \in AB(X)$ where $\varepsilon > 0$ is arbitrary. Since $\operatorname{graph} \varphi'_B = \{(x_0, f(x_0) - \varepsilon)\}$ and $\operatorname{epi} f$ is closed, we have $D(\operatorname{epi} f, \operatorname{graph} \varphi'_B) > 0$. It follows from Lemma 2.7 (a) that $D(\operatorname{epi} f, \operatorname{graph} \varphi'_{B_\delta}) > 0$ for some $\delta > 0$, where $B_\delta = B_\delta(x_0)$ and $\varphi'_{B_\delta} = f(x_0) - \varepsilon + \delta_{B_\delta}$. By (A) (ii), there exists an integer N_1 such that for all $n \geq N_1$ it holds

(4) $$D(\operatorname{epi} f_n, \operatorname{graph} \varphi'_{B_\delta}) > 0.$$

Furthermore, $x_n \to x_0$, so there exists $N_2 \geq N_1$ such that $x_n \in B_\delta$ for all $n \geq N_2$. Hence $(x_n, f(x_0) - \varepsilon)$ belongs to $\operatorname{graph} \varphi'_{B_\delta}$ for all $n \geq N_2$, and, consequently, by (4), does not belong to $\operatorname{epi} f_n$. It follows that

$$f_n(x_n) > f(x_0) - \varepsilon$$

for all $n \geq N_2$. Since ε is arbitrary, we infer that

(5) $$\liminf_n f_n(x_n) \geq f(x_0).$$

Combining (3) and (5) we deduce that (B) (i) holds, which completes the proof of (A) $\Rightarrow$ (B) (i).

(A) (ii) $\Rightarrow$ (B) (ii). Let $\eta < r_{B_\lambda(x_0)}(f - x^*)$, and take $\gamma > 0$ such that $\eta + \gamma < r_{B_\lambda(x_0)}(f - x^*)$. By the very definition of $r_{B_\lambda(x_0)}(f - x^*)$, there exists some $\varepsilon > 0$ such that

(6) $$\eta + \gamma < \inf_{B_{\lambda+2\varepsilon}(x_0)} (f - x^*),$$

and we can furthermore assume that $\varepsilon \leq \gamma/(1 + \|x^*\|)$.

Let $\varphi_B = \langle x^*, . \rangle + \eta + \delta_B \in AB(X)$, where $B = B_{\lambda+\varepsilon}(x_0)$. Pick (x, α) in $\operatorname{graph} \varphi_B$ and (y, β) in $\operatorname{epi} f$. We claim that $d((x, \alpha), (y, \beta)) \geq \varepsilon$. Indeed, in case $\|y - x\| \geq \varepsilon$, this is obvious. Otherwise, $\|y - x\| < \varepsilon$, hence $\|y - x_0\| \leq \lambda + 2\varepsilon$, so, in view of (6),

$$\begin{aligned} \beta \geq f(y) &\geq \eta + \gamma + \langle x^*, y \rangle \\ &= \eta + \gamma + \langle x^*, x \rangle + \langle x^*, y - x \rangle \\ &\geq \alpha + \gamma - \|x^*\| \varepsilon \\ &\geq \alpha + \varepsilon, \end{aligned}$$

whence $d((x, \alpha), (y, \beta)) \geq \beta - \alpha \geq \varepsilon$, proving the claim.

Therefore, $D(\operatorname{epi} f, \operatorname{graph} \varphi_B) \geq \varepsilon > 0$, so, by (A) (ii), $D(\operatorname{epi} f_n, \operatorname{graph} \varphi_B) > 0$ for large n. This clearly implies that $f_n(x) > \langle x^*, x \rangle + \eta$ for every x in $B_{\lambda+\varepsilon}(x_0)$ and large n, hence

$$\liminf_n r_{B_\lambda(x_0)}(f_n - x^*) \geq \liminf_n \inf_{B_{\lambda+\varepsilon}(x_0)} (f_n - x^*) \geq \eta.$$

The proof of (A)(ii) $\Rightarrow$ (B)(ii) is thus complete.

(B) (i) $\Rightarrow$ (A) (i). Fix x_0 in $\operatorname{dom} f$ and, using (B) (i), let $(x_n) \subset X$ such that $x_n \to x_0$ and $f_n(x_n) \to f(x_0)$. We have

$$\limsup_n d((x_n, f_n(x_n)), \operatorname{hypo} \varphi_B) \geq \limsup_n D(\operatorname{epi} f_n, \operatorname{hypo} \varphi_B),$$

so, from the continuity of $(x, y) \mapsto d((x, y), S)$, we infer that

$$d((x_0, f(x_0)), \operatorname{hypo} \varphi_B) \geq \limsup_n D(\operatorname{epi} f_n, \operatorname{hypo} \varphi_B).$$

This inequality being valid for all $(x_0, f(x_0))$ in $\operatorname{graph} f$, we conclude that

$$D(\operatorname{graph} f, \operatorname{hypo} \varphi_B) \geq \limsup_n D(\operatorname{epi} f_n, \operatorname{hypo} \varphi_B),$$

and it remains to invoke Lemma 2.2 to get (A) (i).

(B) (ii) $\Rightarrow$ (A) (ii). Assume that $D(\operatorname{epi} f, \operatorname{graph} \varphi_{B_\lambda}) = \varepsilon > 0$, where $B_\lambda = B_\lambda(x_0)$ and $\varphi_{B_\lambda} = \langle x^*, . \rangle + \eta + \delta_{B_\lambda}$. By Lemma 2.7 (a), there exists $\delta > 0$ such that

$$D(\operatorname{epi} f, \operatorname{graph} \varphi_{B_{\lambda+2\delta}}) = \gamma > 0.$$

Let $x \in B_{\lambda+2\delta} \cap \operatorname{dom} f$ and $\alpha = \langle x^*, x \rangle + \eta$. Since $(x, \alpha) \in \operatorname{graph} \varphi_{B_{\lambda+2\delta}}$, we have $(x, \alpha) \notin \operatorname{epi} f$, that is, $f(x) - \alpha > 0$. Therefore,

$$f(x) - \alpha = d((x, f(x)), (x, \alpha)) \geq D(\operatorname{epi} f, \operatorname{graph} \varphi_{B_{\lambda+2\delta}}) = \gamma,$$

so $f(x) - \langle x^*, x \rangle \geq \eta + \gamma$ for every $x \in B_{\lambda+2\delta}$. Thus,

$$r_{B_{\lambda+\delta}}(f - x^*) \geq \inf_{B_{\lambda+2\delta}} (f - x^*) > \eta.$$

Now, applying (B) (ii), we find that $\inf_{B_{\lambda+\delta}}(f_n - x^*) > \eta$ for large n, hence $\operatorname{epi} f_n \cap \operatorname{graph} \varphi_{B_{\lambda+\delta}} = \emptyset$ for large n. Finally, applying Lemma 2.7 (b), we derive that

$$D(\operatorname{epi} f_n, \operatorname{graph} \varphi_{B_\lambda}) > 0$$

for large n, as required. □

2.4. The local ball-affine convergence. The object of this subsection is to compare the ball-affine convergence with other convergences of functions linked with the graph convergence of their subdifferentials. It turns out that to handle this problem it is sufficient that the sequence of functions converges locally rather than globally, where the local ball-affine convergence is defined as follows:

DEFINITION 2.9. A sequence (f_n) in $\mathrm{LSC}(X)$ is said to be *locally τ_{ba}-convergent* to $f \in \mathrm{LSC}(X)$, written $f = \mathrm{loc}{-}\tau_{ba} - \lim f_n$, if and only if for every $x \in \operatorname{dom} f$ there exists $\bar\lambda > 0$ such that for all $B = B_\lambda(x)$ with $0 \leq \lambda < \bar\lambda$ and all $\varphi_B = \varphi + \delta_B$ with φ affine continuous the following holds

$$(\mathrm{A\_loc}) \begin{cases} \text{(i)} & \limsup_n D(\operatorname{epi} f_n, \operatorname{graph} \varphi_B) \leq D(\operatorname{epi} f, \operatorname{graph} \varphi_B); \\ \text{(ii)} & D(\operatorname{epi} f, \operatorname{graph} \varphi_B) > 0 \Rightarrow D(\operatorname{epi} f_n, \operatorname{graph} \varphi_B) > 0, \text{ for large } n. \end{cases}$$

It is clear that τ_{ba}-convergence implies local τ_{ba}-convergence. Obvious modifications in the proof of Theorem 2.8 yields the following characterization of local τ_{ba}-convergence:

THEOREM 2.10. *Let (f_n) be a sequence in* $\mathrm{LSC}(X)$ *and let* $f \in \mathrm{LSC}(X)$. *The following are equivalent:*

(A_loc) $f = \mathrm{loc}{-}\tau_{ba} - \lim f_n$;

$$(\text{B\_loc})\begin{cases} \text{(i)} & \forall x_0 \in \operatorname{dom} f,\ \exists (x_n) \subset X:\ x_n \to x_0,\ f_n(x_n) \to f(x_0); \\ \text{(ii)} & \forall x_0 \in \operatorname{dom} f,\ \exists \bar{\lambda} > 0,\ \forall \lambda \in [0, \bar{\lambda}[,\ \ \forall x^* \in X^*: \\ & \qquad r_{B_\lambda(x_0)}(f - x^*) \leq \liminf_n r_{B_\lambda(x_0)}(f_n - x^*). \end{cases}$$

REMARK 2.11. The convergence hypothesis made in Deville [**16**, Lemma 2.7] can be formulated as follows: for any $x_0 \in \operatorname{dom} f$ it is assumed that

(i) There exists $(x_n) \subset X$ such that $x_n \to x_0$ and $f_n(x_n) \to f(x_0)$;

(ii) There exists $\varepsilon_0 > 0$ such that for all $B = B_\varepsilon(x_0)$ with $\varepsilon < \varepsilon_0$ and all $\varphi_B = \varphi + \delta_B$ with φ affine continuous, if $D(\operatorname{epi} f, \operatorname{hypo} \varphi_B) > 0$ then $D(\operatorname{epi} f_n, \operatorname{hypo} \varphi_B) > 0$ for n large enough.

In view of Lemma 2.2, (ii) above amounts to (A_loc) (ii), hence to (B_loc) (ii) as the proof of Theorem 2.8 shows. Thus, the above convergence hypothesis is actually equivalent to the local ball-affine convergence.

Before proceeding with the comparison of local ball-affine convergence with other convergence notions, we need to recall some definitions. Let (f_n) be a sequence in $\operatorname{LSC}(X)$ and let $f \in \operatorname{LSC}(X)$. Then:

• The sequence (f_n) is said to be *epi-convergent* to f, written $f = \textit{epi} - \lim f_n$, if

$$\operatorname{Ls} \operatorname{epi} f_n \subset \operatorname{epi} f \subset \operatorname{Li} \operatorname{epi} f_n,$$

where $\operatorname{Ls} A_n$ and $\operatorname{Li} A_n$ denote, respectively, the Painlevé-Kuratowski upper and lower limits of the sequence of sets (A_n). We recall that

$$\operatorname{Li} A_n = \{x \in X \mid \exists x_n \in A_n, \quad x_n \to x\},$$

and

$$\operatorname{Ls} A_n = \{x \in X \mid \exists x_{n_k} \in A_{n_k}, \quad x_{n_k} \to x\}.$$

• The sequence (f_n) is said to be *locally uniformly convergent* to f on X, written $f = \operatorname{LU} - \lim f_n$, if for every $x \in \operatorname{dom} f$ there exists $\bar{\lambda} > 0$ such that (f_n) converges uniformly to f on the ball $B_{\bar{\lambda}}(x)$.

• According to Deville [**16**, Definition 2.3], the sequence (f_n) is said to be *slice convergent* to f, written $f = \tau_s - \lim f_n$, if for every φ_B, restriction of a continuous affine function to a closed ball of X, one has

$$D(\operatorname{epi} f_n, \operatorname{hypo} \varphi_B) \to D(\operatorname{epi} f, \operatorname{hypo} \varphi_B).$$

• According to Penot [**27**, Definition 2.3], a sequence (f_n) is said to be *locally pseudo-uniformly lower convergent* to f, written $f = \operatorname{lpul} - \lim f_n$, if for every $x_0 \in \operatorname{dom} f$ there is $\mu > f(x_0)$ and $\rho > 0$ such that for each $\alpha > 0$ there exists $N \in \mathbb{N}$ such that for any $n \geq N$ and $x \in B_\rho(x_0) \cap [f_n < \mu]$ there exists $x' \in B_\alpha(x)$ for which

$$f_n(x) \geq f(x') - \alpha.$$

For a comprehensive account on convergences of sets and functions, the reader is referred to the monographs by Attouch [**2**] and by Beer [**9**], and to the survey paper of Sonntag-Zălinescu [**28**].

PROPOSITION 2.12. *Let (f_n) be a sequence in* $\operatorname{LSC}(X)$ *and let* $f \in \operatorname{LSC}(X)$. *Then,* $f = \operatorname{loc-}\tau_{ba} - \lim f_n$ *in each of the following cases:*

(a) $f = \operatorname{LU} - \lim f_n$;

(b) $f = \tau_s - \lim f_n$ *in the sense of Deville;*

(c) $f = \operatorname{epi} - \lim f_n$ *and* $f = \operatorname{lpul} - \lim f_n$;

(d) (f_n) *and* f *are convex and* $f = \tau_s - \lim f_n$ *for the topology* τ_s *of Definition 2.4.*

PROOF. (a) Assume that $f = \mathrm{LU} - \lim f_n$ and fix $x_0 \in \operatorname{dom} f$. Then there exists $\bar{\lambda} > 0$ such that for every $\varepsilon > 0$

$$\sup_{x \in B_{\bar{\lambda}}(x_0)} |f(x) - f_n(x)| < \varepsilon$$

for large n. Let $\lambda \in [0, \bar{\lambda}[$ and $x^* \in X^*$. It is easily seen that

$$r_{B_\lambda(x_0)}(f - x^*) < r_{B_\lambda(x_0)}(f_n - x^*) + \varepsilon$$

for large n, so

$$r_{B_\lambda}(x_0)(f - x^*) \leq \liminf_n r_{B_\lambda(x_0)}(f_n - x^*)$$

proving that (B_loc) (ii) holds. Moreover, (B_loc) (i) is clearly satisfied by the stationary sequence $x_n := x_0$.

(b) In view of Lemma 2.2, it is evident that τ_s-convergence in the sense of Deville implies ball-affine convergence, hence local τ_{ba}-convergence.

(c) Notice that $f = \mathrm{epi} - \lim f_n$ immediately yields (B_loc) (i). We claim that $f = \mathrm{lpul} - \lim f_n$ implies (B_loc) (ii). Indeed, observe first that $f = \mathrm{lpul} - \lim f_n$ implies that $f - x^* = \mathrm{lpul} - \lim(f_n - x^*)$ for every $x^* \in X^*$, so it suffices to prove (B_loc) (ii) for $x^* = 0$. Then, fix $x_0 \in \operatorname{dom} f$ and consider $\mu > f(x_0)$ and $\bar{\lambda} := \rho$ as given by the definition of lpul-convergence. Let $\lambda \in [0, \bar{\lambda}[$ and $\eta < r_{B_\lambda(x_0)}(f)$. Take $\varepsilon > 0$ such that $\lambda + 2\varepsilon < \rho$ and

$$\eta + \varepsilon < \inf_{B_{\lambda+2\varepsilon}(x_0)} f.$$

To prove the claim it suffices to show that

$$\eta \leq \liminf_n \inf_{B_{\lambda+\varepsilon}(x_0)} f_n.$$

By definition of lpul-convergence, there exists $N \in \mathbb{N}$ such that for any $n \geq N$ and $x \in B_{\lambda+\varepsilon}(x_0) \cap \{f_n < \mu\}$ there exists $x' \in B_\varepsilon(x)$ for which $f_n(x) \geq f(x') - \varepsilon$.

For $y \in B_{\lambda+\varepsilon}(x_0) \cap \{f_n \geq \mu\}$, we have

$$f_n(y) \geq \mu > f(x_0) \geq r_{B_\lambda(x_0)}(f) > \eta,$$

while for $x \in B_{\lambda+\varepsilon}(x_0) \cap \{f_n < \mu\}$ there exists $x' \in B_\varepsilon(x) \subset B_{\lambda+2\varepsilon}(x_0)$ such that

$$f_n(x) \geq f(x') - \varepsilon \geq \inf_{B_{\lambda+2\varepsilon}(x_0)} f - \varepsilon > \eta.$$

Therefore, there exists $N \in \mathbb{N}$ such that for any $n \geq N$ and $x \in B_{\lambda+\varepsilon}(x_0)$ one has $f_n(x) > \eta$, that is,

$$\eta \leq \liminf_n \inf_{B_{\lambda+\varepsilon}(x_0)} f_n,$$

as required.

(d) follows from Proposition 2.5. □

REMARK 2.13. Proposition 2.12 (d) asserts that local ball-affine convergence is weaker than slice convergence on $\Gamma(X)$. We do not know whether the converse is true.

3. Subdifferentials and associated spaces

3.1. Subdifferentials. Following [**24**] (see also [**6, 7, 21, 23**]), we call *subdifferential*, denoted by ∂, any operator which associates a subset $\partial f(x)$ of X^* to any Banach space X, any $f \in \mathrm{LSC}(X)$ and any $x \in X$, such that the following axioms are satisfied:

(**A1**) If f is convex, then $\partial f(x) = \{x^* \in X^* \mid \langle x^*, y-x\rangle + f(x) \leq f(y),\ \forall y \in X\}$;
(**A2**) If $F\colon X \times X \to \mathbb{R}\cup\{+\infty\}$ is such that $F(x,y) = f(x)+g(y)$, then $\partial F(x,y) \subset \partial f(x) \times \partial g(y)$, and equality holds whenever f or g is a constant function;
(**A3**) If $f + \varphi + \psi$ attains a finite local minimum at x_0 and φ, $\psi : X \to \mathbb{R}$ are convex continuous and ∂-differentiable at x_0, then $0 \in \partial f(x_0) + \partial\varphi(x_0) + \partial\psi(x_0)$, where g ∂-differentiable at x means that both $\partial g(x)$ and $\partial(-g)(x)$ are nonempty.

All known subdifferentials satisfy axioms (A1)–(A3). In the sequel, we shall mainly refer to the following:

– the Fenchel subdifferential;

– the elementary subdifferentials, which are naturally associated with a common concept of derivative, e.g., Gâteaux, Hadamard, Fréchet, Lipschitz, Hölder; following Ioffe [**21, 23**], we call *canonical* the subdifferentials obtained by suitably modifying the definition of the derivative, and we call *viscosity* the subdifferentials built from the supporting locally Lipschitz differentiable functions (see [**11, 21, 23, 6, 7, 24**] and the references therein);

– the more elaborate subdifferentials, like those of Clarke [**14**], Michel-Penot [**26**], Ioffe [**21**], etc.

3.2. ∂-trustworthy and ∂-regular spaces. To be able to work with a given subdifferential on a space, we need that some basic properties involving the subdifferential be available on that space. This led Ioffe to introduce the concept of trustworthiness, first for the (canonical) Hadamard and Fréchet subdifferentials [**20**], and more recently for arbitrary subdifferentials [**21**]. Roughly speaking, a Banach space is said to be ∂-trustworthy provided a so-called basic fuzzy principle involving ∂ holds in that space. Such a principle provides necessary conditions for a point to minimize a sum of functions in terms of the subdifferential of each function.

Several other principles have proved to be key tools in nonsmooth analysis, e.g., fuzzy calculus rules, extremal principles, multi-directional mean value inequalities, or first-order necessary conditions for various constrained optimization problems (see, e.g., the survey paper of Borwein-Zhu [**12**]). It has recently been discovered that most of these principles are in fact equivalent, in the sense that, if one of them holds in X^N for every natural N, then so do the other ones (see [**29, 22, 24**]). The basic fuzzy principle mentioned above belongs to the equivalence list, as well as the following special case of it:

(R1)$_\partial$ Rule for a local minimum: convex Lipschitz version. *Let f be in* LSC(X) *and let $\varphi\colon X \to \mathbb{R}$ be convex Lipschitz. Suppose that $f+\varphi$ attains a finite local minimum at x_0. Then, there exist sequences $(\bar{x}_n, \bar{x}_n^*) \subset \partial f$ and $(\bar{y}_n, \bar{y}_n^*) \subset \partial\varphi$ such that*

(i) $\bar{x}_n \to x_0$, $\bar{y}_n \to x_0$, $f(\bar{x}_n) \to f(x_0)$;
(ii) $\bar{x}_n^* + \bar{y}_n^* \to 0$.

We find it convenient to give a name to the class of spaces identified by the equivalence theorem:

DEFINITION 3.1. A Banach space X is said to be *∂-regular* (with respect to a subdifferential ∂) if the above rule (R1) holds in X^N for every natural N.

Said equivalently, a Banach space X is ∂-regular if and only if X^N is ∂-trustworthy for every natural N. Examples of ∂-regular spaces of course abound. Indeed, it can be shown (see [**24**, Theorem 4.4]) that, for any subdifferential ∂ satisfying axioms (A1)–(A3), Banach spaces with a ∂-smooth renorm in the sense of [**6, 7**] are ∂-regular, while, for any viscosity β-subdifferential ∂^β, Banach spaces with a β-smooth Lipschitz bump function are ∂^β-regular. We only quote the main examples of ∂-regular spaces associated with the most standard subdifferentials:

– Hilbert spaces are ∂^P-regular with respect to the proximal subdifferential ∂^P;

– Separable Banach spaces are ∂^H-regular with respect to the Hadamard subdifferential ∂^H;

– It follows from [**19**] that a Banach space is an Asplund space if and only if it is ∂^F-regular with respect to the canonical Fréchet subdifferential ∂^F;

– Any Banach space is ∂-regular with respect to the subdifferential of Clarke, of Michel-Penot, or of Ioffe.

3.3. Spaces with Brøndsted-Rockafellar property. The regularity properties discussed in the previous subsection are not essential for proving the convergence theorem we have in mind. What is needed is a simpler property, involving just one function, which is similar to the Brøndsted-Rockafellar theorem of convex analysis [**13**] and to the approximate minimum theorem for non-convex functions of Aussel *et al.* [**6**] (see also assumption (A) in Penot [**27**, Theorem 2.8]). This property reads as follows:

$(\mathrm{BR})_\partial$ Brøndsted-Rockafellar rule. *Let $f \in \mathrm{LSC}(X)$ and let $(x_0, x_0^*) \in X \times X^*$, $\varepsilon > 0$ and $\lambda > 0$ such that*

$$(f - x_0^*)(x_0) < \inf_{B_\lambda(x_0)} (f - x_0^*) + \varepsilon.$$

Then, there exists $(\bar{x}, \bar{x}^*) \in \partial f$ such that:

(i) $\|\bar{x} - x_0\| < \lambda, \quad |f(\bar{x}) - f(x_0)| < \varepsilon + \lambda\|x_0^*\|$;

(ii) $\|\bar{x}^* - x_0^*\| < \varepsilon/\lambda$.

The following theorem shows that $(\mathrm{BR})_\partial$ is indeed weaker than the regularity properties considered above:

THEOREM 3.2. *For any subdifferential ∂, $(\mathrm{R1})_\partial$ implies $(\mathrm{BR})_\partial$.*

PROOF. Let $\varepsilon > 0$, $\lambda > 0$ and $(x_0, x_0^*) \in X \times X^*$ such that

$$(f - x_0^*)(x_0) < \inf_{B_\lambda(x_0)} (f - x_0^*) + \varepsilon.$$

We first apply Ekeland's variational principle [**17**] to $g := f - x_0^* + \delta_{B_\lambda(x_0)}$, with $0 < \varepsilon' < \varepsilon$ such that $g(x_0) < \inf_X g + \varepsilon'$ and with λ' such that $\lambda\varepsilon'/\varepsilon < \lambda' < \lambda$. This produces a point $z \in \operatorname{dom} g \subset \operatorname{dom} f$ satisfying

$$\|z - x_0\| \le \lambda'; \tag{7}$$

$$(f - x_0^*)(z) \le (f - x_0^*)(x_0) \le (f - x_0^*)(z) + \varepsilon'; \tag{8}$$

$$f - x_0^* + (\varepsilon'/\lambda')\|\,.-z\| \quad \text{admits a local minimum at } z. \tag{9}$$

Next we apply $(\mathrm{R1})_\partial$ to f and $\varphi := (\varepsilon'/\lambda')\| \cdot - z\| - x_0^*$, with z satisfying (9). Then, for $\varepsilon'' = \min\{\lambda - \lambda', \varepsilon - \varepsilon', \varepsilon/\lambda - \varepsilon'/\lambda'\} > 0$, we get $(\bar{x}, \bar{x}^*) \in \partial f$ and $(\bar{y}, \bar{y}^*) \in \partial\varphi$ such that

$$\|\bar{x} - z\| < \varepsilon'', \tag{10}$$

$$|f(\bar{x}) - f(z)| < \varepsilon'', \tag{11}$$

$$\|\bar{x}^* + \bar{y}^*\| < \varepsilon''. \tag{12}$$

Combining the above inequalities, we obtain

$$\|\bar{x} - x_0\| < \lambda' + \varepsilon'' \leq \lambda,$$

and

$$\begin{aligned} |f(\bar{x}) - f(x_0)| &\leq |f(\bar{x}) - f(z)| + |f(z) - f(x_0)| \\ &\leq \varepsilon'' + \varepsilon' + |\langle x_0^*, z - x_0\rangle| \\ &\leq \varepsilon + \lambda\|x_0^*\|. \end{aligned}$$

On the other hand, we have $\bar{y}^* = \bar{z}^* - x_0^*$ with $\bar{z}^*$ in $(\varepsilon'/\lambda')\partial(\| \cdot - z\|)(\bar{y})$, so, using (12) and standard calculus of convex analysis, we derive that

$$\|\bar{x}^* - x_0^*\| \leq \|\bar{x}^* + \bar{y}^*\| + \|\bar{z}^*\| < \varepsilon'' + \frac{\varepsilon'}{\lambda'} \leq \frac{\varepsilon}{\lambda}.$$

Thus, $(\bar{x}, \bar{x}^*) \in \partial f$ satisfies the required inequalities of $(\mathrm{BR})_\partial$. □

REMARK 3.3. It follows from Theorem 3.2 that ∂-regular spaces and, more generally, ∂-trustworthy spaces have Property $(\mathrm{BR})_\partial$. It is not clear whether the converse is true for an arbitrary subdifferential. However, for the canonical Fréchet subdifferential ∂^F, all these spaces are the same. More precisely, as is easily seen from the characterization of Asplund spaces established in [**20, 18, 19**] and Theorem 3.2, the following statements are equivalent:

(i) X is an Asplund space;
(ii) X is ∂^F-regular;
(iii) X is ∂^F-trustworthy;
(iv) X satisfies $(\mathrm{BR})_{\partial^F}$;
(v) Any $f \in \mathrm{LSC}(X)$ is ∂^F-subdifferentiable on a dense subset of $\operatorname{dom} f$.

Before closing this section, we show how Theorem 3.2 enables to recover variants of the two above-mentioned theorems of Brøndsted-Rockafellar and Aussel-Corvellec-Lassonde.

COROLLARY 3.4. *Let X be a Banach space and let $f \in \Gamma(X)$. Suppose that $(x_0, x_0^*) \in X \times X^*$ is given such that*

$$(f - x_0^*)(x_0) < \inf_X (f - x_0^*) + \varepsilon.$$

Then, for each $\lambda > 0$ there exists $(\bar{x}, \bar{x}^*) \in \partial^{Fen} f$ such that

(i) $\|\bar{x} - x_0\| < \lambda$;
(ii) $\|\bar{x}^* - x_0^*\| < \varepsilon/\lambda$.

PROOF. Any Banach space being ∂^C-regular with respect to the Clarke subdifferential ∂^C, it follows from Theorem 3.2 that $(BR)_{\partial^C}$ holds in X. To conclude, it suffices to recall that $\partial^C f = \partial^{Fen} f$ whenever f belongs to $\Gamma(X)$. □

The statement of Brøndsted-Rockafellar's theorem is that of Corollary 3.4 with all "<" being replaced by "≤".

COROLLARY 3.5. *Let X be a Banach space with a ∂-smooth renorm and let $f \in \mathrm{LSC}(X)$. Let $A \subset X$ be a closed subset and let $\varepsilon > 0$. Suppose that $x_0 \in A$ and $\lambda > 0$ are such that*

$$B_\lambda(x_0) \subset A \quad \text{and} \quad f(x_0) < \inf_A f + \varepsilon.$$

Then, there exists $\bar{x} \in B_\lambda(x_0)$ such that

$$f(\bar{x}) < \inf_A f + 2\varepsilon \quad \text{and} \quad \partial f(\bar{x}) \cap \frac{\varepsilon}{\lambda}\mathbf{B}^* \neq \emptyset,$$

where $\mathbf{B}^*$ denotes the dual unit ball.

PROOF. Since by [**24**, Theorem 4.4] a space with a ∂-smooth renorm is ∂-regular, it follows from Theorem 3.2 that $(BR)_\partial$ holds in X. It thus remains to apply this rule with $x_0^* = 0$. □

The statement of Aussel-Corvellec-Lassonde's theorem is that of Corollary 3.4 with the conclusion being replaced by

$$f(\bar{x}) < \inf_A f + \varepsilon \quad \text{and} \quad \partial f(\bar{x}) \cap 2\frac{\varepsilon}{\lambda}\mathbf{B}^* \neq \emptyset.$$

4. Subdifferentials and the local ball-affine convergence

4.1. The main result. One feature of our convergence theorem is that it involves both an arbitrary subdifferential for the functions of the sequence and a Fréchet-like subdifferential for the limit function. The latter operator is defined as follows: for $f \in \mathrm{LSC}(X)$, we let

$$\Delta f := \{ (x, x^*) \in X \times X^* \mid \forall \varepsilon > 0,\ \exists \lambda \in]0, \varepsilon] : \\ (f - x^*)(x) < (f - x^*)(y) + \lambda\varepsilon,\ \forall y \in B_\lambda(x) \}$$

and

$$\widetilde{\Delta} f := \{(x, f(x), x^*) \in X \times \mathbb{R} \times X^* \mid (x, x^*) \in \Delta f\}.$$

This operator has the following nice properties (the easy proofs are left to the reader):

(a) $\partial^F f \subset \Delta f$, for any $f \in \mathrm{LSC}(X)$, and

(b) $\partial^{Fen} f = \Delta f$, for any $f \in \Gamma(X)$.

THEOREM 4.1. *Let ∂ be any subdifferential and let X be a Banach space such that $(\mathrm{BR})_\partial$ is satisfied. If (f_n) locally τ_{ba}-converges to f in $\mathrm{LSC}(X)$, then*

$$\widetilde{\Delta} f \subset \mathrm{Li}\, \widetilde{\partial} f_n.$$

PROOF. Let $(x, x^*) \in \Delta f$ and $\varepsilon \in]0, \bar{\lambda}[$ where $\bar{\lambda}$ is the scalar given by (B_loc) (ii). By definition of Δf, there exists $\lambda \in]0, \varepsilon]$ such that

$$(f - x^*)(x) < \inf_{B_\lambda(x)} (f - x^*) + \lambda\varepsilon. \tag{13}$$

According to (B_loc) (i) there exists a sequence $(x_n) \subset X$ such that $x_n \to x$ and $f_n(x_n) \to f(x)$, hence for n large enough it holds

$$(f_n - x^*)(x_n) < (f - x^*)(x) + \lambda\varepsilon \tag{14}$$

and

$$B_{\lambda/4}(x_n) \subset B_{\lambda/2}(x). \tag{15}$$

On the other hand, (B_loc) (ii) yields

$$r_{B_{\lambda/2}(x)}(f - x^*) \leq \liminf_n \inf_{B_{\lambda/2}(x)} (f_n - x^*). \tag{16}$$

¿From (13) and (16), we derive that

$$(f - x^*)(x) < \liminf_n \inf_{B_{\lambda/2}(x)} (f_n - x^*) + \lambda\varepsilon,$$

and combining this inequality with (14) and (15), we conclude that for large n it holds

$$(f_n - x^*)(x_n) < \inf_{B_{\lambda/4}(x_n)} (f_n - x^*) + 2\lambda\varepsilon. \tag{17}$$

Now, applying $(\mathrm{BR})_\partial$ to f_n and x_n satisfying (17), we get $(\bar{x}_n, \bar{x}_n^*)$ in ∂f_n verifying

$$\|\bar{x}_n - x_n\| < \lambda/4 < \varepsilon, \tag{18}$$

$$|f_n(\bar{x}_n) - f_n(x_n)| < 2\lambda\varepsilon + (\lambda/4)\|x^*\| < 2\varepsilon^2 + \varepsilon\|x^*\|, \tag{19}$$

$$\|\bar{x}_n^* - x^*\| < 8\varepsilon. \tag{20}$$

The rest of the proof is routine. For each $\varepsilon = 1/k < \bar{\lambda}$, denote by $(\bar{x}_{n,k}, \bar{x}_{n,k}^*)$ the sequence in ∂f_n verifying (18)–(20) for all $n \geq N_k$. Clearly, we may assume that $N_{k+1} > N_k$, so that $N_k \to \infty$. For any large $n \in \mathbb{N}$, set $(\bar{x}_n, \bar{x}_n^*) := (\bar{x}_{n,k}, \bar{x}_{n,k}^*)$, where k is such that $N_k \leq n < N_{k+1}$. It easily follows from (18)–(20) that the sequence so defined verifies

$$\begin{cases} (\bar{x}_n, f_n(\bar{x}_n), \bar{x}_n^*) \in \widetilde{\partial} f_n, \\ \bar{x}_n \to x, \\ f_n(\bar{x}_n) \to f(x), \\ \bar{x}_n^* \to x^*. \end{cases}$$

This completes the proof of the theorem. □

4.2. Applications. Theorem 4.1 yields several convergence theorems for specific subdifferentials. In this last subsection, we mention a few of them.

THEOREM 4.2. *Let X be a Hilbert space. If (f_n) locally τ_{ba}-converges to f in* LSC(X), *then*

$$\widetilde{\partial}^P f \subset \mathrm{Li}\, \widetilde{\partial}^P f_n.$$

PROOF. A Hilbert space being ∂^P-regular with respect to the proximal subdifferential ∂^P, X satisfies $(\mathrm{BR})_{\partial^P}$. Applying Theorem 4.1 with $\partial = \partial^P$, we obtain $\widetilde{\Delta} f \subset \mathrm{Li}\, \widetilde{\partial}^P f_n$. To complete the proof, it remains to observe that $\partial^P f \subset \partial^F f \subset \Delta f$. □

The next theorem provides a common generalization of results of Deville [**16**, Theorem 2.5 and Lemma 2.7] and Penot [**27**, Theorem 2.8]:

THEOREM 4.3. *Let X be an Asplund space. If (f_n) locally τ_{ba}-converges to f in* LSC(X), *then*

$$\widetilde{\partial}^F f \subset \mathrm{Li}\, \widetilde{\partial}^F f_n.$$

PROOF. We know from the previous section that an Asplund space satisfies $(\mathrm{BR})_{\partial^F}$ for the canonical Fréchet subdifferential ∂^F. So it suffices to apply Theorem 4.1 with $\partial = \partial^F$ and to remind that $\partial^F f \subset \Delta f$. □

In Deville [**16**, Theorem 2.5], the space X is supposed to have C^1 Lipschitz bump functions (which is stronger than being Asplund) and the sequence is supposed to be τ_s-convergent in the sense of [**16**, Definition 2.3], which is a notion stronger than local τ_{ba}-convergence (see Proposition 2.12). In Deville [**16**, Lemma 2.7], the space is as above and the convergence agrees with our local τ_{ba}-convergence (see Remark 2.11). In Penot [**27**, Theorem 2.8], the space is Asplund, but the sequence is supposed to be both epi- and lpul-convergent, which is stronger than being locally τ_{ba}-convergent (see Proposition 2.12).

Theorem 4.1 also recovers the result of Beer mentioned in the introduction:

THEOREM 4.4 (Beer [**8**]). *Let X be a Banach space. If (f_n) slice converges to f in $\Gamma(X)$, then*

$$\widetilde{\partial}^{Fen} f \subset \operatorname{Li} \widetilde{\partial}^{Fen} f_n.$$

PROOF. It suffices to apply Theorem 4.1 to the Clarke subdifferential and to recall that τ_{ba}-convergence and slice convergence agree in $\Gamma(X)$ and that $\partial^C g = \partial^{Fen} g = \Delta g$ for g in $\Gamma(X)$. □

For a refinement of Theorem 4.4, see Combari-Thibault [**15**, Corollary 2.7].

Our last application concerns the smooth setting:

THEOREM 4.5. *Let X be a Banach space and let (f_n) be a sequence of Gâteaux differentiable functions. If (f_n) locally τ_{ba}-converges to f in* LSC(X), *then*

$$\partial^F f \subset \operatorname{Li} \nabla^G f_n.$$

PROOF. Any Banach space satisfies $(\mathrm{BR})_{\partial^{MP}}$ where ∂^{MP} is the subdifferential of Michel-Penot (see the previous section), so we may apply Theorem 4.1 to this subdifferential. The result then follows from the fact that $\partial^{MP} g(x) = \{\nabla^G g(x)\}$ whenever g is Gâteaux differentiable at x. □

Theorem 4.5 refines a result of Aubin and Frankowska [**5**, Theorem 7.6.1] where the f_n are supposed to be either Fréchet differentiable or Gâteaux differentiable and locally Lipschitz, f is supposed to be lower bounded, and the sequence is assumed to converge uniformly on the bounded subsets of X (which is stronger than local τ_{ba}-convergence, as follows from Proposition 2.12).

References

[1] H. ATTOUCH. Convergence de fonctions convexes, des sous différentiels et semigroupes associés. *C.R. Acad. Sci. Paris Sér. I Math.*, 284:539–542, 1977.

[2] H. ATTOUCH. *Variational convergence for functions and operators.* Pitman, 1984.

[3] H. ATTOUCH and G. BEER. On the convergence of subdifferentials of convex functions. *Arch. Math.*, 60:389–400, 1993.

[4] H. ATTOUCH, J. L. NDOUTOUME, and M. THÉRA. Epigraphical convergence of functions and convergence of their derivatives in Banach spaces. *Séminaire d'Analyse Convexe–Montpellier*, 1990.

[5] J.-P. AUBIN and H. FRANKOWSKA. *Set-valued analysis.* Birkhäuser, 1990.

[6] D. AUSSEL, J.-N. CORVELLEC, and M. LASSONDE. Mean value theorem and subdifferential criteria for lower semi-continuous functions. *Trans. Amer. Math. Soc.*, 347:4147–4161, 1995.

[7] D. AUSSEL, J.-N. CORVELLEC, and M. LASSONDE. Nonsmooth constrained optimization and multidirectional mean value inequalities. *SIAM J. Optim.*, 9:690–706, 1999.

[8] G. BEER. The slice topology: a viable alternative to Mosco convergence in nonreflexive spaces. *Nonlinear Anal.*, 19(3):271–290, 1992.

[9] G. BEER. *Topologies on closed and closed convex sets.* Kluwer Academic Publishers, 1993.

[10] G. BEER and M. THÉRA. Attouch-Wets convergence and a differential operator for convex functions. *Proc. Amer. Math. Soc.*, 122:851–858, 1994.

[11] J. M. BORWEIN and Q. J. ZHU. Viscosity solutions and viscosity subderivatives in smooth Banach spaces with application to metric regularity. *SIAM J. Control Optim.*, 34:1568–1591, 1996.

[12] J. M. BORWEIN and Q. J. ZHU. A survey of subdifferential calculus with applications. *Nonlinear Anal.*, to appear.

[13] A. BRØNDSTED and R. T. ROCKAFELLAR. On the subdifferentiability of convex functions. *Proc. Amer. Math. Soc.*, 16:605–611, 1965.

[14] F.H. CLARKE. *Optimization and Nonsmooth Analysis.* Wiley-Interscience, 1983.

[15] C. COMBARI and L. THIBAULT. On the graph convergence of subdifferentials of convex functions. *Proc. Amer. Math. Soc.*, 126:2231–2240, 1998.

[16] R. DEVILLE. Stability of subdifferentials of nonconvex functions in Banach spaces. *Set-Valued Anal.*, 2:141–157, 1994.

[17] I. EKELAND. On the variational principle. *J. Math. Anal. Appl.*, 47:324–353, 1974.

[18] M. FABIAN. On classes of subdifferentiability spaces of Ioffe. *Nonlinear Anal.*, 12:63–74, 1988.

[19] M. FABIAN. Subdifferentiability and trustworthiness in the light of the new variational principle of Borwein and Preiss. *Acta Univ. Carolin. Math. Phys.*, 30:51–56, 1989.

[20] A. D. IOFFE. On subdifferentiability spaces. *Ann. New York Acad. Sci.*, 410:107–119, 1983.

[21] A. D. IOFFE. Codirectional compactness, metric regularity and subdifferential calculus. *Preprint*, 1996.

[22] A. D. IOFFE. Fuzzy principles and characterization of trustworthiness. *Set-Valued Anal.*, 6:265–276, 1998.

[23] A. D. IOFFE. Variational methods in local and global analysis. In F.H. Clarke and R.J. Stern, editors, *Nonlinear Analysis, Differential Equations and Control*, NATO Science Series, pages 447–502. Kluwer Academic Publishers, 1999.

[24] M. LASSONDE. First-order rules for nonsmooth constrained optimization. *Nonlinear Anal.*, to appear.

[25] A. LEVY, R. POLIQUIN, and L. THIBAULT. Partial extensions of Attouch's theorem with applications to proto-derivatives of subgradients mapping. *Trans. Amer. Math. Soc.*, 347:1269–1294, 1995.

[26] P. MICHEL and J.-P. PENOT. Calcul sous-différentiel pour des fonctions lipschitziennes et non-lipschitziennes. *C. R. Acad. Sci. Paris*, 298:269–272, 1985.

[27] J.-P. PENOT. On the interchange of subdifferentiation and epi-convergence. *J. Math. Anal. Appl.*, 196:676–698, 1995.

[28] Y. SONNTAG and C. ZĂLINESCU. Set convergences. An attempt of classification. *Trans. Amer. Math. Soc.*, 340:199–226, 1993.

[29] Q. J. ZHU. The equivalence of several basic theorems on subdifferentials. *Set-Valued Anal.*, 6:171–185, 1998.

DÉPARTEMENT DE MATHÉMATIQUES, UNIVERSITÉ DES ANTILLES ET DE LA GUYANE, 97159 POINTE À PITRE, GUADELOUPE, FRANCE

E-mail address: Michel.Geoffroy@univ-ag.fr

E-mail address: Marc.Lassonde@univ-ag.fr

Canadian Mathematical Society
Conference Proceedings
Volume **27**, 2000

Rotundity related to Lipschitz separation

John R. Giles and Jon D. Vanderwerff

Dédié à Jonathan Borwein
à l'occasion de la remise de son titre de Docteur Honoris Causa

ABSTRACT. Previous work by J.M. Borwein and the authors showed that a Banach space possessing desirable extensions of Lipschitz functions is characterised by a curious rotundity property of the space. Here we explore related rotundity properties: we examine a weaker form of weak uniform rotundity sufficient for such extensions and we show that related forms of rotundity characterise certain rotundity properties of the second dual.

1. Introduction

In paper [B-G-V] study was made of those Banach spaces where there are sufficient Lipschitz extensions to take different values at points of the space and it was shown that such spaces can be characterised as those satisfying a certain rotundity property.

A Banach space X is said to be *Lipschitz separated* if given a proper closed linear subspace Y of X, every 1-Lipschitz function f on Y with $\sup\{|f(x)| : x \in Y\} \leq M$ has the property that, for each $x \in X \backslash Y$,

$$\sup \tilde{f}(x) > \inf \tilde{f}(x)$$

over all $\tilde{f}$ 1-Lipschitz extensions of f with $\sup\{|\tilde{f}(x)| : x \in X\} \leq M$. A Banach space X is Lipschitz separated if and only if for every pair of sequences $\{x_n\}$, $\{y_n\}$ in $S(X)$ such that $\lim\limits_{m,n\to\infty} \|x_n + y_m\| = 2$ there exists

1991 *Mathematics Subject Classification.* 46B20, 49J52.
Key words and phrases. Lipschitz functions, Lipschitz separation, rotundity, weak and weak* uniform rotundity, uniformly Gâteaux differentiable.

no $\phi \in S(X^*)$ such that

$$\limsup_{n\to\infty} \phi(y_n) < 0 < \liminf_{n\to\infty} \phi(x_n), \qquad \text{[B-G-V, Theorem 3.2]}.$$

A Banach space X is *rotund* (R) if for $x, y \in S(X)$ and $\|x + y\| = 2$ we have $x = y$. The space X is *weakly uniformly rotund* (WUR) if for sequences $\{x_n\}$,$\{y_n\}$ in $S(X)$ and $\lim_{n\to\infty} \|x_n + y_n\| = 2$ we have $\{x_n - y_n\}$ converges weakly to 0; if X is a dual space and $\{x_n - y_n\}$ converges weak* to 0, the space is *weak* uniformly rotund* (W*UR). Clearly, if X is WUR then X is Lipschitz separated but examples were given to show that the converse does not hold in general, [B-G-V, Theorem 4.1]. However, it was shown [B-G-V, Theorem 3.3(b)] that if X is Lipschitz separated then X^{**} is R, but again examples were given to show that the converse does not hold in general, [B-G-V, Example 4.3].

Our aim in this paper is to explore the nature of the rotundity property characterising Lipschitz separation by examining closely related properties. In §2 we define a rotundity property weaker than WUR but one which still implies but is not implied by Lipschitz separation. We show that it satisfies duality properties similar to WUR and like WUR implies that the space is Asplund. In §3 we characterise rotundity properties in the second dual. Our characterisation of the second dual being R enables an illuminating comparison with the rotundity property characterising Lipschitz separation.

2. A variant of weak uniform rotundity

We say that a Banach space X is said to be *equatorially weakly uniformly rotund* (EWUR) if for every pair of sequences $\{x_n\}$,$\{y_n\}$ in $S(X)$ such that $\lim_{n\to\infty} \|x_n + y_n\| = 2$ there exists no continuous linear functional $\phi \in S(X^*)$ such that

$$\limsup \phi(y_n) < 0 < \liminf \phi(x_n).$$

If X is a dual and there exists no weak* continuous linear functional $\phi \in S(X^*)$ such that

$$\limsup \phi(y_n) < 0 < \liminf \phi(x_n)$$

we say that X is *equatorially weak* uniformly rotund* (EW*UR).

It is evident that a Banach space X being WUR is EWUR. However, the following example [G-V, Theorem 1], which builds on ideas in [S1, Example 2.9] shows that EWUR is a weaker property than WUR.

EXAMPLE 2.1. *A Banach space X with separable dual can be equivalently renormed to be EWUR and not WUR.*

SKETCH OF PROOF. Since X has separable dual it can be equivalently renormed to be WUR, [D-G-Z, Corollary II.6.9(ii)] and we consider X so renormed. Consider a closed hyperplane Y in X and $e \neq Y$ so that given $x \in X$ there exists $\lambda \in \mathbb{R}$ and $y \in Y$ such that $x = \lambda e + y$ and we define an equivalent norm $||\cdot||_S$ on X by $\|x\|_S = |\lambda| + \|y\|$. We now define an equivalent norm $|||\cdot|||$ on X by

$$|||x||| = \left(\|x\|_S^2 + \sum_{i=1}^{\infty} 2^{-i} f_i^2(y) \right)^{\frac{1}{2}}$$

where $\{f_i : i \in \mathbb{N}\}$ is a dense subset of $S(Y^*)$. The details that $|||\cdot|||$ is EWUR but not WUR can be found in [G-V].

Clearly, the space X being EWUR implies that it is Lipschitz separated, but it has been shown that the converse does not hold generally, [B-G-V, Theorem 4.1 (b)].

Using the property that a Lipschitz separated space X does not contain a subspace topologically isomorphic to ℓ_1, [B-G-V, Corollary 3.4 (b)], we have the following characterisation of EWUR for separable spaces.

THEOREM 2.2. *A separable Banach space X is EWUR if and only if given sequences $\{x_n\}$, $\{y_n\}$ in $S(X)$ where $\lim_{n\to\infty} \|x_n + y_n\| = 2$ we have either $\{\alpha x_n - \beta y_n\}$ for some $0 < \alpha \leq 1$ and $0 < \beta \leq 1$, or $\{x_n\}$ or $\{y_n\}$ has a subsequence converging weakly to* 0.

PROOF. Suppose X is EWUR but does not have the property. Then there exist sequences $\{x_n\}$, $\{y_n\}$ in $S(X)$ where $\lim_{n\to\infty} \|x_n + y_n\| = 2$ but $\{\alpha x_n - \beta y_n\}$ has no subsequences weakly convergent to 0 for any $0 < \alpha \leq 1$, $0 < \beta \leq 1$ and neither does $\{x_n\}$ nor $\{y_n\}$. Since X is separable and EWUR then $X \not\supseteq \ell_1$ and so $B(X^{**})$ is weak* sequentially compact. Now $\{\hat{x}_n\}$ has a subsequence $\{\hat{x}_{n_k}\}$ weak* convergent to some $F \in B(X^{**})\backslash\{0\}$ and $\{\hat{y}_n\}$ has a subsequence $\{\hat{y}_{n_{k_\ell}}\}$ weak* convergent to some $G \in B(X^{**})\backslash\{0\}$. Now if $G = \lambda F$ for some $0 < \lambda \leq 1$ then $\{\lambda x_{n_{k_\ell}} - y_{n_{k_\ell}}\}$ is weakly convergent to 0 contrary to hypothesis. Similarly, we may assume $F \neq \lambda G$ for $0 < \lambda \leq 1$ and so there exists $\phi \in S(X^*)$ and $\delta > 0$ such that $\phi(F) > \delta$ and $\phi(G) < -\delta$. But then there exists a subsequence $\{\hat{x}_{n_{k_{\ell_m}}}\}$ such that $\phi(x_{n_{k_{\ell_m}}}) > \delta$ and a subsequence $\{\hat{y}_{n_{k_{\ell_{m_p}}}}\}$ such that $\phi(y_{n_{k_{\ell_{m_p}}}}) < -\delta$. We conclude that

$$\limsup \phi(y_{n_{k_{\ell_{m_p}}}}) < 0 < \liminf \phi(x_{n_{k_{\ell_{m_p}}}})$$

which contradicts X being EWUR.

Conversely, suppose X is not EWUR. Then there exists a pair of sequences $\{x_n\}$, $\{y_n\}$ in $S(X)$ where $\lim_{n\to\infty} \|x_n + y_n\| = 2$ and $\phi \in S(X^*)$ such that

$$\phi(y_n) < -\delta \quad \text{and} \quad \phi(x_n) > \delta \quad \text{for all } n \in \mathbb{N}.$$

Then neither $\{x_n\}$ nor $\{y_n\}$ has a subsequence weakly convergent to 0 and given $0 < \alpha \leq 1$, $0 < \beta \leq 1$ we have

$$\phi(\alpha x_n - \beta y_n) > (\alpha + \beta)\delta \quad \text{for all } n \in \mathbb{N}.$$

So X does not have the property.

It is not difficult to see that the previous theorem extends to nonseparable spaces if we replace subsequences with subnets.

An attractive feature of the EWUR property is that it mimics the duality properties of WUR. A Banach space X has *uniformly Gâteaux differentiable* norm, (we say that X is UG), if given $\epsilon > 0$ and $z \in S(X)$ there exists a $\delta(\epsilon, z) > 0$ such that for all $x \in S(X)$,

$$\|x + tz\| + \|x - tz\| - 2\|x\| < \epsilon t \quad \text{for all } 0 < t < \delta.$$

It is well known that a Banach space X is WUR if and only if X^* is UG and that X is UG if and only if X^* is W*UR, [D-G-Z, Theorem II.6.7 p.63]. To establish similar duality relations for EWUR we introduce the corresponding weaker differentiability property.

We say that a Banach space X has norm *equatorially uniformly Gâteaux differentiable*, (we say that X is EUG), if given $z \in S(X)$ and $\epsilon > 0$, there exists a $\delta(\epsilon, z) > 0$ such that for all $x \in S(X)$,

$$\min\{\|x + tz\| - \|x\|, \|x - tz\| - \|x\|\} < \epsilon t \quad \text{for all } 0 < t < \delta.$$

It is interesting to note that this condition is equivalent to a smoothness property used by Vlasov [V, p. 403] in connection with the study of the convexity of Chebychev sets; see [G, p. 399].

LEMMA 2.3. *Consider a Banach space X with dual X^*.*
(i) *If X is EWUR then X^* is EUG.*
(ii) *If X is EUG then X^* is EW* UR.*

PROOF. (i) Suppose X^* is not EUG. Then there exists $\phi \in S(X^*)$, $r > 0$ and a sequence $\{f_n\} \subset S(X)$ and $t_n \to 0^+$ such that

$$\|f_n + t_n\phi\| - \|f_n\| > rt_n \quad \text{and} \quad \|f_n - t_n\phi\| - \|f_n\| > rt_n.$$

For each $n \in \mathbb{N}$ choose x_n, y_n in $S(X)$ such that

$$(f_n + t_n\phi)(x_n) > \|f_n + t_n\phi\| - \frac{1}{n}\, t_n \quad \text{and} \quad (f_n - t_n\phi)(y_n) > \|f_n - t_n\phi\| - \frac{1}{n}\, t_n$$

Then $f_n(x_n) \to 1$ and $f_n(y_n) \to 1$ as $n \to \infty$ and so $\lim_{n\to\infty} \|x_n + y_n\| = 2$. But also

$$t_n\phi(x_n) \geq \|f_n + t_n\phi\| - f_n(x_n) - \frac{1}{n}\, t_n \geq \|f_n + t_n\phi\| - \|f_n\| - \frac{1}{n}\, t_n > rt_n - \frac{1}{n}\, t_n$$

and then $\liminf \phi(x_n) \geq r$.

Similarly, $-t_n\phi(y_n) > rt_n - \frac{1}{n}\, t_n$ and then $\limsup \phi(y_n) \leq -r$. So X is not EWUR.

(ii) Suppose X^* is not EW*UR. Then there exist sequences $\{f_n\}$, $\{g_n\}$ in $S(X^*)$ such that $\|f_n + g_n\| \to 2$ as $n \to \infty$ and $z \in S(X), r > 0$ such that

$$g_n(z) < -r < 0 < r < f_n(z).$$

Now choose $x_n \in S(X)$ such that $(f_n + g_n)(x_n) \to 2$ as $n \to \infty$. Then $f_n(x_n) \to 1$ and $g_n(x_n) \to 1$ as $n \to \infty$. Given $t > 0$ and $0 < \epsilon < \frac{r}{2}$ there exists $\nu \in \mathbb{N}$ such that

$$\|x_n + tz\| - \|x_n\| \geq f_n(x_n + tz) - \|x_n\| > -\epsilon t + tr > \frac{tr}{2} \text{ for all } n > \nu.$$

Similarly $\|x_n - tz\| - \|x_n\| \geq \frac{tr}{2}$ for all $n > \nu$. So X is not EUG.

THEOREM 2.4. *Consider a Banach space X with dual X^*.*
(i) *X is EWUR if and only if X^* is EUG.*
(ii) *X is EUG if and only if X^* is EW* UR.*

PROOF. (i) X is EWUR $\Rightarrow$ X^* is EUG $\Rightarrow$ X^{**} is EW*UR $\Rightarrow$ X is EWUR.
(ii) X^* is EW*UR $\Rightarrow$ EUG follows along the same lines as Lemma 2.3(i). The converse implication follows from Lemma 2.3(ii).

An EUG norm is Gâteaux differentiable away from the origin. However, the duality between WUR and UG norms, along with Example 2.1 and Theorem 2.4 shows that an EUG norm need not be UG. Therefore, the EUG condition is different from a norm having a right-hand derivative that is uniform in each direction because such a norm is UG, [F-P, p. 67]. Consequently, the subdifferential mapping of an EUG norm is not necessarily norm to weak* uniformly continuous on its sphere, nevertheless the following result shows one could say that it is norm to weak* "equatorially" uniformly continuous.

THEOREM 2.5. *A Banach space X is EUG if and only if for all sequences $\{x_n\}$, $\{y_n\}$ in $S(X)$ such that $\lim_{n\to\infty} \|x_n - y_n\| = 0$ there exists no $z \in S(X)$ such that*

$$\limsup f_{y_n}(z) < 0 < \liminf f_{x_n}(z) \quad \text{where } f_{x_n} \in \partial\|x_n\| \text{ and } f_{y_n} \in \partial\|y_n\|.$$

PROOF. Now $\|x-y\| + \|f_x + f_y\| \geq 2$ for all $x, y \in S(X)$ and $f_x \in \partial\|x\|$, $f_y \in \partial\|y\|$. So if $\|x_n - y_n\| \to 0$ then $\|f_{x_n} + f_{y_n}\| \to 2$ as $n \to \infty$. By Corollary 2.4 (ii) there exists no $z \in S(X)$ such that

$$\limsup f_{y_n}(z) < 0 < \liminf f_{x_n}(z).$$

Conversely, if X is not EUG, there exists a $z \in S(X)$ and $r > 0$ and sequence $\{x_n\}$ in D_z and $t_n \to 0^+$ such that

$$\|x_n + t_n z\| - \|x_n\| > rt_n \quad \text{and} \quad \|x_n - t_n z\| - \|x_n\| > rt_n.$$

But $f_{x_n+t_nz}(t_nz) \geq \|x_n + t_nz\| - \|x_n\|$ and $f_{x_n-t_nz}(-t_nz) \geq \|x_n - t_nz\| - \|x_n\|$. So

$$\limsup f_{x_n-t_nz}(z) < 0 < \liminf f_{x_n+t_nz}(z).$$

But $\left\| \dfrac{x_n - t_nz}{\|x_n - t_nz\|} - \dfrac{x_n + t_nz}{\|x_n + t_nz\|} \right\| \to 0$ as $n \to \infty$.

It has recently been shown that a Banach space X with WUR dual is not necessarily reflexive, [H, Proposition 3]. So there is some point in determining the differentiability condition for the norm of X which corresponds to its dual being WUR and EWUR.

We say that a Banach space X has norm *very uniformly Gâteaux differentiable*, (we say that X is VUG) if given $\epsilon > 0$ and $F \in S(X^{**})$ there exists a $\delta(\epsilon, F) > 0$ such that for all $x \in S(X)$,

$$\|\hat{x} + tF\| + \|\hat{x} - tF\| - 2\|\hat{x}\| < \epsilon t \quad \text{for all } 0 < t < \delta.$$

We say that X has norm *very equatorially uniformly Gâteaux differentiable*, (we say that X is VEUG), if given $\epsilon > 0$ and $F \in S(X^{**})$ there exists a $\delta(\epsilon, F) > 0$ such that for all $x \in S(X)$,

$$\min\{\|\hat{x} + tF\| - \|\hat{x}\|, \|\hat{x} - tF\| - \|\hat{x}\|\} < \epsilon t \quad \text{for all } 0 < t < \delta.$$

THEOREM 2.6. *Consider a Banach space X with dual X^*,*

(i) *X is VUG if and only if X^* is WUR.*

(ii) *X is EVUG if and only if X^* is EWUR.*

PROOF. The proofs in one direction follow from the duality properties associated with WUR and EWUR. The proofs in the other direction follow the pattern in (i) of [D-G-Z, Theorem II.6.7] and in (ii) of Lemma 2.3 above.

It is not difficult to establish the following characterisations for these stronger differentiability conditions.

THEOREM 2.7. *Consider a Banach space X*

(i) *X is VUG if and only if given $F \in S(X^{**})$ the mapping $x \mapsto F(f_x)$ for any selection $f_x \in \partial\|x\|$ is uniformly continuous on $S(X)$.*

(ii) *X is VEUG if and only if for all sequences $\{x_n\}$, $\{y_n\}$ in $S(X)$ such that $\|x_n - y_n\| \to 0$ as $n \to \infty$ there exists no $F \in S(X^{**})$ such that*

$$\limsup F(f_{y_n}) < 0 < \liminf F(f_{x_n})$$

where $f_{x_n} \in \partial\|x_n\|$ and $f_{y_n} \in \partial\|y_n\|$.

Hájek proved the significant result that a Banach space which is WUR is an Asplund space, [H, Theorem 1]. He showed that in a separable space X with non-separable dual X^* there exist sequences $\{x_n\}$ and $\{y_n\}$ in $S(X)$ where $\lim\limits_{n\to\infty} \|x_n + y_n\| = 2$ and a $\phi \in S(X^*)$ such that $\lim\limits_{n\to\infty} \phi(x_n) = 0$ and $\lim\limits_{n\to\infty} \phi(y_n) = \alpha > 0$. This result of Hájek's was extended to EWUR spaces in [G-V, Theorem 2] which we state as Theorem 2.8. A key step in proving this extension was modifying the Haar system (see [Di, p. 234ff]) used in Hájek's proof so that convex combinations of the elements would not converge weakly to 0, using the modified Haar system, one can show that the x_n's above do not converge weakly to 0. Then an application of Theorem 2.2 shows that a separable Banach space with nonseparable dual cannot be EWUR.

THEOREM 2.8. *A Banach space X which is EWUR is an Asplund space.*

A Banach space with separable dual can be equivalently renormed by be WUR, [D-G-Z, Corollary II.6.9(ii), p.66]. So despite Example 2.1, Theorem 2.8 has the following implication.

COROLLARY 2.9. *A separable Banach space X which is EWUR can be equivalently renormed to be WUR.*

We are not aware of a simple direct proof of the preceding result without assuming first that X is Asplund. The following questions arise naturally from the previous two results.

QUESTION 2.10. *(a) Is a Lipschitz separated Banach space an Asplund space?*
(b) Does an EWUR Banach space admit an equivalent WUR norm?

We should also mention that [F-H-Z] strengthens Hájek's result [H] by showing that if X admits a WUR norm, then X^* is a subspace of a weakly compactly generated space (such spaces have several nice renormings; see [D-G-Z]). It would also be interesting to know if this result remains valid for EWUR or even Lipschitz separated Banach spaces.

The fact that EWUR spaces are Asplund spaces highlights their importance. The nice duality properties of EWUR spaces also have implications for weaker variants of the Lipschitz separated property in the doulbe dual. Indeed, if X is EWUR, then X^{**} is EW*UR. Therefore, the proof of [B-G-V, Theorem 3.1], shows that given a proper weak*-closed linear subspace Y of X^{**}, every 1-Lipschitz function f on Y with $\sup\{|f(x)| : x \in Y\} \leq M$ has the property that, for each $x \in X^{**}\backslash Y$,

$$\sup \tilde{f}(x) > \inf \tilde{f}(x)$$

over all $\tilde{f}$ 1-Lipschitz extensions of f with $\sup\{|\tilde{f}(x)| : x \in X^{**}\} \leq M$. That says that X^{**} is in some sense weak*-Lipschitz separated, and according to [B-G-V, Theorem 2.1] any bounded 1-Lipschitz function on Y can be extended to a 1-Lipschitz function on X^{**} whose Clarke subdifferential is identically $B(X^{***})$. It is not clear to us whether this property remains valid if X is only assumed to be Lipschitz separated.

3. Characterisations of rotundity in the second dual

We now determine analytic characterisations of rotundity properties of the second dual, in terms of rotundity properties related to EWUR and the Lipschitz separated property.

A Banach space X is *weakly mid-point locally uniformly rotund* (WMLUR) if given $z \in S(X)$, for every pair of sequences $\{x_n\}$, $\{y_n\}$ in $S(X)$ where $\|x_n + y_n - 2z\| \to 0$ as $n \to \infty$ we have $\{x_n - y_n\}$ converges weakly to 0. It has been shown that a Banach space X is WMLUR if and only if every point of $S(\widehat{X})$ is an extreme point of $B(X^{**})$, [S2, p.663]. An adaptation of this proof shows that we have another characterisation of this rotundity property which is of an equatorial form of rotundity.

We say that a Banach space X is *equatorially weakly mid-point locally uniformly rotund* (EWMLUR) if given $z \in S(X)$, for every pair of sequences $\{x_n\}$, $\{y_n\}$ in $S(X)$ where $\|x_n + y_n - 2z\| \to 0$ as $n \to \infty$ there exists no $\phi \in S(X^*)$ such that

$$\limsup \phi(y_n) < 0 < \liminf \phi(x_n).$$

THEOREM 3.1. *A Banach space X is EWMLUR if and only if every point of $S(\widehat{X})$ is an extreme point of $B(X^{**})$.*

PROOF. Suppose there exist $F_1, F_2 \in S(X^{**}), F_1 \neq F_2$ such that $\hat{z} = \frac{1}{2}(F_1 + F_2) \in \widehat{X}$. Write $Z \equiv \operatorname{sp} \hat{z}$. Then there exist $\phi \in S(X^*)$ and $\delta > 0$ such that $\hat{\phi}(F_1) > \delta$ and $\hat{\phi}(F_2) < -\delta$ and $\phi(z) = 0$. Consider $W_1 \equiv \left\{F \in X^{**} : \hat{\phi}(F) > \delta\right\}$ and $W_2 \equiv \left\{F \in X^{**} : \hat{\phi}(F) < -\delta\right\}$ and the convex set $W \equiv \frac{1}{2}\left(W_1 \cap B(\widehat{X})\right) + \frac{1}{2}\left(W_2 \cap B(\widehat{X})\right)$. We show that $\hat{z} \in \overline{W}$. Consider the weak open convex neighbourhood $N \cap X$ of 0 in X where N is a weak$*$ open convex neighbourhood of 0 in X^{**}. By Goldstine's Theorem there exists $\hat{x} \in (F_1 + N) \cap W_1$ and $\hat{y} \in (F_2 + N) \cap W_2$ where $x, y \in B(X)$. Then $\frac{1}{2}(\hat{x} + \hat{y}) \in \frac{1}{2}(F_1 + F_2) + N$ so $\frac{1}{2}(\hat{x} + \hat{y}) \in \hat{z} + N \cap \widehat{X}$. So then $\hat{z} \in \overline{W}$ since W is convex. Therefore, given $n \in \mathbb{N}$ there exists $\hat{x}_n \in W_1 \cap B(\widehat{X})$ and $\hat{y}_n \in W_2 \cap B(\widehat{X})$ such that $\|x_n + y_n - 2z\| < \frac{1}{n}$. But $\phi(x_n) > \delta$ and $\phi(y_n) < -\delta$, so X is not EWMLUR.

Conversely, suppose there exists some $z \in S(X)$ and $\phi \in S(X^*)$ and $\delta > 0$ and sequences $\{x_n\}$, $\{y_n\}$ in $S(X)$ such that $\|x_n + y_n - 2z\| \to 0$ but $\phi(x_n) > \delta$ and $\phi(y_n) < -\delta$. Then $\phi(x_n - y_n) > 2\delta$ for all $n \in \mathbb{N}$ and X is not WMLUR. So then $\hat{z}$ is not an extreme point of $B(X^{**})$.

A Banach space X is *weakly locally uniformly rotund* (WLUR) if given $z \in S(X)$ and sequences $\{x_n\}$ in $S(X)$ where $\lim_{n\to\infty} \|x_n + z\| = 2$ we have that $\{x_n\}$ converges weakly to z. It has been shown [G-K-M-S, p.419] that a WLUR space X has the property that for $F \in S(X^{**})$ and $z \in S(X)$, $F \neq \hat{z}$ we have $\|F + \hat{z}\| < 2$.

We now characterise this type of rotundity of X^{**} in terms of an equatorial rotundity on X. We say that a Banach space X is *convex equatorially weakly locally uniformly rotund* (CEWLUR) if given $z \in S(X)$ and sequence $\{x_n\}$ in $S(X)$ where

$$d(z, \text{co}\,\{-x_k : k \geq n\}) \to 2, \text{ as } n \to \infty$$

there exists no $\phi \in S(X^*)$ such that

$$\limsup \phi(x_n) < 0 < \phi(z)$$

THEOREM 3.2. *A Banach space X is CEWLUR if and only if given $F \in S(X^{**})$ and $z \in S(X)$, $F \neq \hat{z}$ we have $\|F + \hat{z}\| < 2$.*

PROOF. Suppose X^{**} does not exhibit the rotundity property. Then there exist $F \in S(X^{**})$ and $z \in S(X)$, $F \neq \hat{z}$ such that $\|F + \hat{z}\| = 2$. Then there exists $\phi \in S(X^*)$ such that $\hat{\phi}(F + \hat{z}) = 0$ and for some $\delta > 0$ we have $\hat{\phi}(F) < -\delta$ and $\phi(z) > \delta$. For each $n \in \mathbb{N}$ choose $\phi_n \in S(X^*)$ such that $\hat{\phi}_n(F + \hat{z}) > 2 - \dfrac{1}{2^n}$. By Goldstine's Theorem, for each $n \in \mathbb{N}$ there exists $x_n \in S(X^*)$ such that

$$\phi_k(x_n) > \hat{\phi}_k(F) - \frac{1}{2^n} \text{ for all } 1 \leq k \leq n \text{ and } \phi(x_n) < -\delta.$$

Then

$$\phi_k(z) \geq 1 - \frac{1}{2^n} \text{ and } \phi_k(x_n) \geq 1 - \frac{2}{2^n} \text{ for } n \geq k,$$

and so

$$d(z, \text{co}\{-x_k : k \geq n\}) \geq 2 - \frac{3}{2^n}.$$

We conclude that X is not CEWLUR.

Conversely, suppose X is not CEWLUR. Then there exist $z \in S(X)$ and a sequence $\{x_n\}$ in $S(X)$ where for each $n \in \mathbb{N}$

$$d(z, \text{co}\,\{-x_k : k \geq n\}) > 2 - \frac{1}{n}$$

and $\phi \in S(X^*)$ and $\delta > 0$ such that

$$\phi(x_n) < -\delta < 0 < \delta < \phi(z).$$

For each $n \in \mathbb{N}$ we can separate z and co $\{-x_k : k \geq n\}$ by an $f_n \in S(X^*)$ such that

$$f_n(z - z_k) > 2 - \frac{1}{n} \text{ for all } z_k \in \text{co}\,\{-x_k : k \geq n\}.$$

Then $f_n(z + x_k) > 2 - \dfrac{1}{n}$ for all x_k where $k \geq n$. But $\{\hat{x}_n\}$ has a weak* cluster point $F_0 \in B(X^{**})$ and $\phi(F_0) \leq -\delta$. Then $\hat{f}_n(\hat{z} + F_0) \geq 2 - \dfrac{1}{n}$ for all $n \in \mathbb{N}$ which implies that $\|F_0 + \hat{z}\| = 2$ although $F_0 \neq \hat{z}$ and so X^{**} does not have the rotundity property.

It is well known that a non-reflexive Banach space X has X^{****} not R. Further, denoting by Q the natural embedding of X in X^{**} and Q_2 the natural embedding of X^{**} in X^{****}, for every $F \in X^{**} \backslash \widehat{X}$, $F \in S(X^{**})$ then $\|Q_2 F\| = \|Q^{**}F\| = 1$, $Q_2 F \neq Q^{**}F$ and $\|Q_2 F + Q^{**}F\| = 2$, [Da, p.70]. So we can make the following deduction from the proof of Theorem 3.2.

COROLLARY 3.3. *Given a non-reflexive Banach space X, for every element $F \in X^{**} \backslash \widehat{X}$, $F \in S(X^{**})$ there exists a sequence $\{G_n\}$ in $S(X^{**})$ such that*

$$d\,(F, co\{-G_k : k \geq n\}) \to 2 \text{ as } n \to \infty$$

and $\Phi \in S(X^{***})$ and $\delta > 0$ such that

$$\Phi(G_n) < -\delta < 0 < \delta < \Phi(F).$$

We are now in a position to give an analytic characterisation of rotundity in the second dual. We say that a Banach space X is *convex equatorially weakly uniformly rotund* (CEWUR) if for every pair of sequences $\{x_n\}, \{y_n\}$ in $S(X)$ where

$$d\,(\text{co}\{x_k : k \geq n\},\ \text{co}\ \{-y_k : k \geq n\}) \to 2 \text{ as } n \to \infty$$

there exists no $\phi \in S(X^*)$ such that

$$\limsup \phi(y_n) < 0 < \liminf \phi(x_n).$$

THEOREM 3.4. *A Banach space X is CEWUR if and only if X^{**} is R.*

PROOF. If X^{**} is not R there exist $F, G \in S(X^{**}), F \neq G$ such that $\|F + G\| = 2$. Then there exists $\phi \in S(X^*)$ such that $\hat{\phi}(F + G) = 0$ and for some $\delta > 0$ we have $\hat{\phi}(F) < -\delta$ and $\hat{\phi}(G) > \delta$. For each $n \in \mathbb{N}$ choose $\phi_n \in S(X^*)$ such that $\hat{\phi}_n(F + G) > 2 - \dfrac{1}{2^n}$. By Goldstine's Theorem, for each $n \in \mathbb{N}$ there exist $x_n, y_n \in S(X^*)$ such that for all $1 \leq k \leq n$,

$$\phi_k(x_n) > \hat{\phi}_k(F) - \frac{1}{2^n} \geq 1 - \frac{2}{2^n} \quad \text{and} \quad \phi_k(y_n) > \hat{\phi}_k(G) - \frac{1}{2^n} \geq 1 - \frac{2}{2^n},$$
$$\phi(y_n) < -\delta \quad \text{and} \quad \phi(x_n) > \delta,$$

Then $d\,(\text{co}\ \{x_k : k \geq n\},\ \text{co}\ \{-y_k : k \geq n\}) \geq 2 - \dfrac{4}{2^n}$ and we conclude that X is not CEWUR.

Conversely, suppose X is not CEWUR. Then there exist sequences $\{x_n\}$, $\{y_n\}$ in $S(X)$ where for each $n \in \mathbb{N}$

$$d\,(\text{co}\ \{x_k : k \geq n\},\ \text{co}\ \{-y_k : k \geq n\}) > 2 - \frac{1}{n}$$

and $\phi \in S(X^*)$ and $\delta > 0$ such that

$$\phi(y_n) < -\delta < 0 < \delta < \phi(x_n).$$

So for each $n \in \mathbb{N}$ we can separate co $\{x_k : k \geq n\}$ and co $\{-y_k : k \geq n\}$ by an $f_n \in S(X^*)$ such that $f_n(x_k + y_k) > 2 - \frac{1}{n}$ for all x_k and y_k where $k \geq n$. But $\{\hat{x}_n\}$ has a weak*-cluster point $F_0 \in B(X^{**})$ and $\hat{\phi}(F_0) \geq \delta$ and $\{\hat{y}_n\}$ has a weak*-cluster point $G_0 \in B(X^{**})$ and $\hat{\phi}(G_0) \leq -\delta$. Then $\hat{f}_n(F_0+G_0) > 2-\frac{1}{n}$ for all $n \in \mathbb{N}$ which implies that $\|F_0+G_0\| = 2$ although $F_0 \neq G_0$ and so X^{**} is not R.

For comparison with Theorem 3.3 it is evident that the rotundity characterisation of Lipschitz separation could be given as follows.

PROPOSITION 3.5. *A Banach space X is Lipschitz separated if and only if for every pair of sequences $\{x_n\}, \{y_n\}$ in $S(X)$ where*

$$d\left(\{x_k : k \geq n\}, \{-y_k : k \geq n\}\right) \to 2 \text{ as } n \to \infty$$

there exists no $\phi \in S(X^*)$ such that

$$\limsup \phi(y_n) < 0 < \liminf \phi(x_n).$$

Finally, the isometric differences between CEWUR norms, the Lipschitz separated property and EWUR norms have been illustrated in [B-G-V] and [G-V]. We also know that there are Banach spaces with equivalent CEWUR norms but no equivalent EWUR norm (for example the non-Asplund James-Tree space has an equivalent norm whose second dual is R, [H, Theorem 4]). Unfortunately, we do not know exactly where Lipschitz separated spaces fall between CEWUR spaces and EWUR spaces in the isomorphic sense. Along these lines, substantial progress has been made recently in the understanding of isomorphic implications of certain rotundity conditions. In particular, [M-O-T-V] proved that if a Banach space admits an equivalent WLUR norm, then it has an equivalent LUR norm, whereas [Ha] showed that Banach spaces with an equivalent MLUR norm need not have an equivalent LUR norm. We refer to [M-O-T-V] and its references for more information on this topic which may point to further questions of interest concerning isomorphic implications of the corresponding equatorial rotundity properties.

References

[B-G-V] Jon Borwein, John Giles, Jon Vanderwerff, Rotund norms, Clarke subdifferentials and extensions of Lipschitz functions. CECM Preprint 98:126.

[Da] M. M. Day, *Normed Linear Spaces* 3rd edn. Springer-Verlag, 1973.

[Di] J. Diestel, *Geometry of Banach spaces – Selected Topics*. Lecture Notes in Mathematics **485**, Springer-Verlag, 1975.

[D-G-Z] Robert Deville, Gilles Godefroy, Václav Zizler, *Smoothness and renormings in Banach spaces*. Pitman Monographs and Surveys in Pure and Applied Mathematics, **64**, 1993. Longman Scientific Technical.

[F-H-Z] M. Fabian, P. Hájek, V. Zizler, On uniform Eberlein compact and uniformly Gateaux smooth norms, *Serdica Math. J.* **23** (1997), 351–362.

[F-P] C. Franchetti, R. Payá, Banach spaces with strongly subdifferentiable norm, *Bollettino U.M.I.* **7-B** (1993), 45–70.

[G] J.R Giles, Uniformly weak differentiability of the norm and a condition of Vlasov, *J. Austral. Math. Soc.* **21** (Series A) (1976), 393–409.

[G-M-K-S] J.R. Giles, P.S. Kenderov, W.B. Moors, S.D. Sciffer, Generic differentiability of convex functions on the dual of a Banach space, *Pacific J. Math.*, **172** (1996), 413–431.

[G-V] J.R. Giles, J.D. Vanderwerff, Asplund spaces and a variant of weak uniform rotundity, *Bull. Austral. Math. Soc.*, to appear.

[H] Petr Hájek, Dual renormings of Banach spaces, *Comment. Math. U. Carolinae*, **37** (1996), 241–253.

[Ha] R. Haydon, Trees in renorming theory, *Proc. London Math. Soc.*, to appear.

[M-O-T-V] A. Moltó, J. Orihuela, S. Troyanski, M. Valdivia, On weakly locally uniformly rotund Banach spaces, *J. Funct. Anal.* **163** (1999), 252–271.

[S1] M. A. Smith, Banach spaces that are uniformly rotund in weakly compact sets of directions, *Canad. J. Math.*, **29** (1977), 963–970.

[S2] M. A. Smith, A curious generalisation of local uniform rotundity, *Comment. Math. U. Carolinae*, **25** (1984), 659–665.

[V] L.P. Vlasov, On Chebychev sets, *Soviet Math. Dokl.* **8** (1967), 401–404.

Department of Mathematics, The University of Newcastle, NSW 2308, Australia

E-mail address: jan@maths.newcastle.edu.au

Department of Mathematics, La Sierra University, Riverside, CA, 92515, USA

E-mail address: jvanderw@LaSierra.edu

Canadian Mathematical Society
Conference Proceedings
Volume **27**, 2000

Codirectional Compactness, Metric Regularity and Subdifferential Calculus

Alexander Ioffe

To Jonathan Borwein, on the occasion of his award of a Dhc

ABSTRACT. The paper is devoted to study of and interrelations between compactness properties of infinitesimal coobject of nonsmooth analysis (subdifferentials, normal cones, coderivatives), metric regularity properties of single-valued and set-valued mappings between Banach spaces and subdifferential calculus rules. The main results include (a) equivalence theorems showing that various known compactness properties for coderivatives of set-valued mappings form two groups of equivalence, and, for sets, properties of both groups reduce to the "compact epi-Lipschitz property" of Borwein-Strojwas; (b) more precise neighborhood and point sufficient criteria for metric regularity and necessary conditions for metric regularity for certain classes of mappings and (c) subdifferential calculus rules based on new "metric" qualification conditions.

1. Introduction

The paper is devoted to the theory of subdifferentials of lower semicontinuous (non-convex) functions with the main emphasis on a class of compactness properties which play a central part in the subdifferential calculus. So let us begin by briefly explaining the function and the meaning of the properties to be considered (for which we do not really need even the concept of subdifferential).

Suppose, for instance, that we have a mapping F from a Banach space into itself which is continuously Fréchet differentiable at a certain point $\bar{x}$ and the derivative $F'(\bar{x})$ is an invertible operator. The classical inverse function theorem says in this case that F has a Lipschitz continuous (even continuously differentiable) inverse defined in a neighborhood of $\bar{y} = F(\bar{x})$.

Assume now that F is Fréchet differentiable at every point of a neighborhood of $\bar{x}$ but its derivative is not continuous. Then we still can get the same conclusion (existence of a Lipschitz inverse) assuming in addition that, say $F'(x)$ is invertible

1991 *Mathematics Subject Classification.* Primary 49J52, Secondary 49J50, 58C20.

Key words and phrases. lower semicontinuous function, set-valued mapping, subdifferential, normal cone, coderivative.

This research was supported by CNRS during my stay in the Laboratory of Applied Mathematics URA 1204, Univ. of Pau in the second half of 1995, by the U.S. National Science Foundation under grant DMS-9404128 and by the US-Israel BSF fund under the grant 94-00237.

for every x near $\bar{x}$ and the norms of $F'(x)$ and $[F'(x)]^{-1}$ are uniformly bounded in a neighborhood of $\bar{x}$. The classical theorem is an easy corollary of this fact.

We can interpret the situation as follows: to be able to judge about the existence of a Lipschitz inverse by properties of the derivative at $\bar{x}$ alone (not at all points of a neighborhood) we need some extra assumption (say, continuity of the derivative).

Similar situations often appear in non-smooth analysis, with subdifferentials, coderivatives and/or normal cones rather than with derivatives (e.g. whenever we deal with inverse function theorems, metric regularity property or Lagrange multiplier rules): what we should require of these objects to be able to make conclusions based on the analysis of them at a given point rather than at every point of a neighborhood.

This question and the properties that make such "point-based" conclusion possible is the main object of our studies in this paper. In the context of non-smooth analysis the first such property was introduced by Ioffe in 1984 [19] under the name "finite codimension property" (motivation behind this name is explained in no. 3.2 – as well as more details concerning earlier developments). The first main result of the paper – Theorem 1 states that several recently introduced properties (e.g. the "partial normal compactness property" of [47,48]) are just reformulations of the finite codimension property.[1]

A similar theorem (Theorem 2) is then established for another group of properties in which the "partially compact epi-Lipschitz property" of Jourani and Thibault [**36**] was chronologically the first.

As we shall see, both groups can be characterized in very similar terms but with references to different topologies. We have chosen the words "codirectional compactness", as the common name for the properties to emphasize that they are actually compactness properties of certain sets of covectors: subdifferentials, normal cones and coderivatives (cf. the property of "directional compactness" studied in [**29**]). Perhaps "normal compactness" could be also a good name but it was already used in a different context by Mordukhovich and Shao [**48**], so we have prefered to introduce a new term to avoid confusion.

It turns out that applied to sets (interpreted as constant set-valued mappings) the two first theorems give (in Theorem 3) a surprising chain of equivalences for compactness properties of normal cones with the "compactly epi-Lipschitz property" of Borwein-Strojwas [**7**] as the most prominent representative. It turns out in particular that all implications established in Loewen [**42**] can be reversed and extended to general Banach space in both directions.

The three theorems are the main results of the paper. They are stated in § 3 and proved in § 4. Bur before that we offer in §2 a short survey of certain classes of subdifferentials that have been recognized as important for the theory and applications. This half introductory section also contains some new results which, we hope, may help to better understand interrelations between different classes of subdifferentials.

We then turn to study in § 5 the metric regularity property and prove two theorems containing a neighborhood (Theorem 5) and a point (Theorem 6) sufficient

[1] A condition of a different type, stated in terms of tangent cones rather than coderivatives was offered by Aubin and Frankowska in [3]. The connection between their condition called *uniform sleekness* and the finite codimension property is not completely clear. However, the finite codimension property is always satisfied if, say $\dim Y < \infty$, whereas the sleekness condition can fail to be valid even if both X and Y are finite dimensional.

criteria for the latter using the information acquired in the previous part of the paper. The theorems unify and generalize earlier available criteria (e.g. [**23, 36, 48**]). In particular, the point criterion of Theorem 6 can now be applied with any limiting subdiffferential in a "good" space and not just with the approximate subdifferential in any Banach space (as in [**23, 36**]) or the limiting Fréchet subdifferential in an Asplund space (as in [**48**]). We conclude the discussion on the regularity matters with theorems containing necessary and sufficient conditions for for a set-valued mapping with convex graph to be metrically regular (Theorem 7) and for a set-valued mapping into a finite dimensional space to be pseudo-Lipschitz (Theorem 9).

The last section is devoted to the "exact" calculus of approximate and limiting subdifferentials. It does not contain an exhaustive presentation of the subject but rather a somewhat new view of it demonstrated by two fundamental examples: normal cone to intersection of sets and subdifferential of a sum of l.s.c. functions. It is well known (e.g. [**25, 47, 38, 56**]) that every formula of the "exact"subdifferential calculus requires certain constraint qualification, given as a rule in terms of "singular" subdifferentials, and certain compactness conditions ensuring convergence of some specific nets of functionals. It turns out that all these results have a two – level structure. At the "lower level" there are calculus rules obtained under a very weak "linear-rate metric qualification condition" stated in terms of some distance estimates and without any additional compactness assumptions.It is interesting to note that the metric qualification condition is even less restrictive than the weakest qualification conditions used in convex analysis (e.g. [**1, 55**]). At the "upper level" it appears that these distant estimates result from metric regularity of some simple set-valued mappings. It turns out further that the sufficient conditions for metric regularity (provided by Theorem 6) applied to these mappings reduce precisely to the basic subdifferential constraint qualification (introduces in [**44, 57**] for in finite dimensions and then extended in [**25, 47**] to infinite dimensional cases) and compactness conditions needed for the existing calculus rules (in fact in a slightly more general form, as far as the latter are concerned).

Bibliographic comments and most of non-technical remarks are collected in "Discussion" subsections 3.2, 5.4 and 6.4.

Notation. Throughout the paper X, Y, Z are Banach spaces, B stands for the unit ball (it will always be clear in which space), X^* is the topological dual of X with $\langle x^*, x\rangle$ being the canonical pairing on $X \times X^*$. By f, g we denote functions which can be extended-real-valued in which case we set $\text{dom } f = \{x : |f(x)| < \infty\}$. We denote by $\text{epi } f = \{(x, \alpha) : \alpha \geq f(x)\}$ the epigraph of f. By capital letters F, $G, \ldots$ we denote mappings, single-valued or set-valued, say, from X into Y and write $\text{dom } F = \{x : F(x) \neq \emptyset\}$ for the domain and $\text{Graph}F = \{(x, y) : y \in F(x)\}$ for the graph of F.

Given a set $S \subset X$ we denote by $\rho(x, S)$ the distance from x to S and by

$$\chi_S(x) = \begin{cases} 0, & \text{if} \quad x \in S; \\ \infty, & \text{if} \quad x \notin S. \end{cases}$$

the indicator of S and by $T(S, x)$ the contingent cone to S at x. For a $Q \subset X^*$ we denote by $s_Q(x) = \sup\{\langle x^*, x\rangle : x^* \in Q\}$ the support function of Q.

*****

The paper was basically written 3 years ago, in 1996 and, in the present form, does not differ much from what existed at that time. In the subsequent developments there is one results which would affect the content of the paper had it been then known. This is the equivalence theorem for various types of fuzzy rules first proved by Zhu for viscosity β-subdifferentials [**60**] and then extended to arbitrary subdifferentials and bigger set of properties by Ioffe [**61**] and Lassonde [**62**]. It is also worth mentioning that the fact explained in the remark following Definition 5.1 was independently discovered by Henrion in his habilitation thesis [**63**].

2. Subdifferentials: an overview

2.1. An axiomatic approach. By a *subdifferential* we usually mean a set-valued mapping which associates with every function defined on a Banach space X and any $x \in X$ a set $\partial f(x) \subset X^*$ (possibly empty), called the *subdifferential of f at x* satisfying a certain set of properties which in various publications may vary depending on the specific purpose. The following is a sufficiently complete list of properties, valid for all important subdifferentials, which seems to cover the majority of "axiomatic" definitions:

(a) $\partial f(x) = \emptyset$ if $x \notin \mathrm{dom} f$;

(b) $\partial f(x) = \partial g(x)$ if f and g coincide in a neighborhood of x;

(c) if f is convex, then $\partial f(x)$ is the subdifferential of f in the sense of convex analysis;

(d) $0 \in \partial f(x)$ if f attains a local minimum at x;

(e) if f satisfies the Lipschitz condition near x with constant K, then $\|x^*\| \le K$ if $x^* \in \partial f(x)$;

(f) if $f(x,y) = f_1(x) + f_2(y)$, then $\partial f(x,y) \subset \partial f_1(x) \times \partial f_2(y)$[2];

(g) if $f(x) = \lambda g(Ax + v)$, where $A : X \mapsto Y$ is a bounded linear operator *onto* Y and $\lambda > 0$, then $\partial f(x) = \lambda A^* \partial g(Ax + v)$.

If $S \subset X$, then the *normal cone to S* at $x \in S$, associated with the given subdifferential is defined as the subdifferential of the indicator function of S at x:

$$N(S,x) = \partial \chi_S(x).$$

Finally, let F be a set-valued mapping from X into Y. Let $y \in F(x)$. The *coderivative* of F at (x,y) associated with ∂ is the set-valued mapping from Y^* into X^* defined as follows

$$D^*F(x,y)(y^*) = \{x^* : \ (x^*, -y^*) \in N(\mathrm{Graph} F, (x,y)\}.$$

Before passing to specific classes of subdifferentials, we mention a few more useful concepts of a general nature. We shall call a subdifferential ∂ *robust* on a class of functions if for any function f of the class the graph of the set-valued mapping $x \mapsto \partial f(x)$ is closed in $X \times X^*$, in the product of the norm and weak-star topologies. Given two subdifferentials ∂ and $\tilde\partial$, we shall say that ∂ is the *regularization* of $\tilde\partial$ on a class of functions if for any function f of the class the graph of ∂f is the closure of the graph of $\tilde\partial f$ in the above mentioned topology. Typically, we are interested in robustness and regularizations on the class of locally Lipschitz functions.

We can also consider sequential regularisation of a subdifferential. Namely, ∂ is the *sequential regularization* of $\tilde\partial$ on a class of functions if for any function f

[2]I am indebted to M. Lassonde for the observation that the equality that was written earlier in the this formula is not satisfied for generalized gradients of Clarke and should be replaced by inclusion

of the class the graph of ∂f coincides with the collection of all limits of sequences $\{(x_n, x_n^*)\}$, with norm converging first components and weak-star converging second components, such that $x_n^* \in \tilde{\partial} f(x_n)$. That kind of operation typically makes sense in spaces whose duals have weak-star sequentially compact unit balls.

2.2. Subdifferentials associated with bornologies. Recall that a *bornology* in X is a collection β of bounded subsets of X with the properties that the union of elements of β is the whole of X and for any two elements S_1, S_2 of β there is an $S \in \beta$ containing both S_1 and S_2. We shall always assume that elements of the bornology we consider are symmetric with respect to the origin and that the bornology is stable with respect to linear homeomorphisms of X, that is to say, $S \in \beta$ implies that $AS \in \beta$ for any linear homeomorphism $A: X \mapsto X$. The two most important bornologies are formed by all symmetric bounded sets (bounded, or Fréchet bornology) and symmetric compact sets (compact, or Hadamard bornology). Given a bornology β in X, there is a canonical way of introducing a bornology in $X \times \mathbb{R}^n$ consisting of all sets $S \times NB$ ($N > 0$ and B being the unit ball in $\mathbb{R}^n$)). We shall use the same symbol β to denote this bornology in $X \times \mathbb{R}^n$.

Every bornology generates a certain concept of derivative (e.g. [**4**]). There are also several types of subdifferentials that can be associated with every bornology β. We shall call them *β-associated.*

The two basic β-associated subdifferentials are defined as follows. We say that x^* belongs to the *canonical β-subdifferential* of f at x if for any $S \in \beta$

$$\liminf_{t \to +0} t^{-1} \inf_{h \in S} (f(x+th) - f(x) - t\langle x^*, h\rangle) \geq 0.$$

For the canonical Dini-Hadamard subdifferential there is a representation through a directional derivative, similar to what we have in convex analysis. Namely, let

$$d^- f(x;h) = \liminf_{(t,u) \to (+0,h)} t^{-1}(f(x+tu) - f(x))$$

be the *Hadamard or contingent directional derivative* of f at x. Then the canonical Dini-Hadamard subdifferential is defined as the collection of all x^* such that $\langle x^*, h\rangle \leq d^- f(x;h)$ for all h. If $f(x) = \chi_S(x)$ is the indicator function of a set S, then $d^- f(x;h)$ is the indicator function of a cone $T(S,x)$ known as the *contingent cone* to S at x (see [**3**]).

Following [**58**], we say that x^* belongs to the *variational β-subdifferential* of f at x if there is a function $\varphi(\cdot)$, β- differentiable at x with $x^* = \varphi'(x)$, satisfying the Lipschitz condition near x and such $f - \varphi$ attains a local minimum at x.

It is an easy matter to verify that all the axioms (a) – (g) are satisfied for the two subdifferentials introduced.

REMARK. There is a difference between the variational β-subdifferentials appeare and the "viscosity subdifferentials" considered in [**11**] (without using this named) and [**9**]: we require that φ be differentiable only at x, not in a neighborhood of x.[3]

PROPOSITION 2.1. *Consider the following statements:*

(a) x^* belongs to the variational β-subdifferential of f at x;

(b) there is a $\lambda \geq 1$ such that $(x^*, -1)$ belongs to the canonical β-subdifferential of $\lambda\rho(\cdot, \mathrm{epi} f)$ at $(x, f(x))$;

(c) x^* belongs to the canonical β-subdifferential of f at x;

[3]I am thankful to the referee to drawing my attention to this difference.

(d) $(x^*, -1)$ belongs to the canonical β-subdifferential of $R(u,\alpha) = (f(u) - \alpha)^+$ at $(x, f(x))$.

Then the implications (a) $\Rightarrow$ (b) $\Rightarrow$ (c) $\Rightarrow$ (d) hold for any bornology β and any function f.

If β is the Fréchet bornology, then (c) $\Rightarrow$ (a), so that the properties (a), (b) and (c) are equivalent.

If f satisfies the Lipschitz condition near x, then (d) $\Rightarrow$ (b), so that the properties (b), (c) and (d) are equivalent.

PROOF. (a) $\Rightarrow$ (b). By definition $x^* = \varphi'(x)$ for a function φ satisfying the Lipschitz condition near x and such that $f - \varphi$ attains at x a local minimum. We may assume that $f(x) = \varphi(x)$ in which case for any $S \in \beta$ and any $N > 0$ the inequality

$$\rho((x+th, f(x)+t\alpha), \mathrm{epi} f) \geq \rho((x+th, \varphi(x)+t\alpha), \mathrm{epi}\varphi)$$

holds for all $(h,\alpha) \in S \times [-N, N]$ if t is sufficiently small. Therefore it is sufficient to show that there is a $\lambda \geq 1$ such that

$$\liminf_{t\to+0} \inf_{h\in S, |\alpha|\leq N} t^{-1}(\lambda\rho((x+th, \varphi(x)+t\alpha), \mathrm{epi}\varphi) - t\langle x^*, h\rangle + t\alpha) \geq 0,$$

To this end we shall make use of the following fact (see [**14, 25**]): *let K be the Lipschitz constant of φ. Consider the norm $\|(x,\alpha)\|_K = K\|x\| + |\alpha|$ in $X \times I\!R$. Then*

$$\rho_K((u,\alpha), \mathrm{epi}\varphi) = (\varphi(u) - \alpha)^+$$

for all (u,α) sufficiently close to $(x, \varphi(x))$.

It follows that there is a $\lambda \geq 1$ such that

$$\lambda\rho((x+th, \varphi(x)+t\alpha), \mathrm{epi}\varphi) \geq (\varphi(x+th) - \varphi(x) - t\alpha)^+.$$

As x^* is the β-derivative of φ at x, we have (given an $S \in \beta$)

$$\varphi(x+th) - \varphi(x) = t(\langle x^*, h\rangle + r(t))$$

for all $h \in S$, where $r(t) \to 0$ as $t \to 0$.

It follows that for any $h \in S$

$$\lambda\rho((x+th, \varphi(x)+t\alpha), \mathrm{epi}\varphi) \geq t(\langle x^*, h\rangle - \alpha + r(t))^+ \geq t(\langle x^*, h\rangle - \alpha + r(t))$$

and therefore

$$\lim_{t\to+0} \inf_{h\in S, |\alpha|\leq N} t^{-1}(\lambda\rho((x+th, \varphi(x)+t\alpha), \mathrm{epi}\varphi) - t\langle x^*, h\rangle + t\alpha) \geq \lim_{t\to+0}(-r(t)) = 0.$$

(b) $\Rightarrow$ (c). If (b) holds, then for any $S \in \beta$, $N > 0$

$$\inf_{h\in S, |\alpha|\leq N} t^{-1}(\lambda\rho((x+th, f(x)+t\alpha), \mathrm{epi} f) - t\langle x^*, h\rangle + t\alpha) \geq r(t) \to 0. \tag{2.1}$$

Let K be the upper bound of the norms of $h \in S$. We shall first show that (2.1) implies that

$$\liminf_{t\to+0} \inf_{h\in S} t^{-1}(f(x+th) - f(x)) = K_1 \geq -(K\|x^*\| + 1) > -\infty.$$

Indeed, assuming the contrary, we shall find, for any sufficiently small t, an $h = h(t) \in S$ such that $f(x+th) - f(x) < -t(K\|x^*\|+1)$. Then, taking $N = K\|x^*\|+1$, we conclude that $f(x) + t\alpha > f(x+th)$ for $\alpha = -N$, so that

$$\lambda\rho((x+th, f(x) - tN), \mathrm{epi} f) - t\langle x^*, h\rangle - tN = -t(N + \langle x^*, h\rangle) \leq -t$$

which implies that the liminf in (2.1) as $t \to 0$ is not greater than -1 in contradiction with the assumption.

Take $N > |K_1| + K\|x^*\| + 1$, and set $\alpha(t,h) = \min\{t^{-1}(f(x+th) - f(x), N\}$. Then $|\alpha(t,h)| \le N$. If now $t^{-1}(f(x+th) - f(x)) \le N$, then by (2.1)

$$\begin{aligned} &-\langle x^*, h\rangle + t^{-1}(f(x+th) - f(x)) \\ &\qquad = t^{-1}(\lambda\rho((x+th, \varphi(x) + t\alpha(t,h)), \mathrm{epi}\varphi) - t\langle x^*, h\rangle + t\alpha(t,h)) \ge r(t). \end{aligned}$$

Otherwise

$$t^{-1}(f(x+th) - f(x)) \ge N \ge \langle x^*, h\rangle.$$

Thus, for all $h \in S$ we have

$$t^{-1}((f(x+th) - f(x)) - t\langle x^*, h\rangle) \ge \min\{r(t), 0\} \to 0.$$

(c) $\Rightarrow$ (d). This is trivial.

A proof of the implication (c)$\Rightarrow$(a) in case when β is the Fréchet bornology can be found in [**11**], and the implication (d)$\Rightarrow$(b) in case when f satisfies the Lipschitz condition with constant K is immediate from the mentioned fact that $(f(u) - \alpha)^+$ is the distance to epi$f$ corresponding to the K-norm in $X \times I\!\!R$, □

The proposition suggests one more possibility to introduce a subdifferential associated with the given bornology. Namely, we shall say that x^* belongs to the *metric β-subdifferential* of f at x if there is a $\lambda \ge 1$ such that $(x^*, -1)$ belongs to the canonical β-subdifferential of $\lambda\rho(\cdot, \mathrm{epi}f)$ at $(x, f(x))$.

It follows from Proposition 2.1 that the metric β-subdifferential coincides with the canonical β-subdifferential if either the function satisfies the Lipschitz condition near the point or if β is the Fréchet bornology. It follows in particular that the canonical Fréchet normal cone to S at x is generated by the Fréchet subdifferential of the distance function $\rho(\cdot, S)$ at x (cf [**26**]).

We shall call the three subdifferentials *elementary.* The following proposition describes some fundamental relationship between the elementary subdifferentials and the corresponding normal cones.

PROPOSITION 2.2. *For any lower semicontinuous function f the equality*

$$\partial f(x) = \{x^* : \ (x^*, -1) \in N((x, f(x)), \mathrm{epi}f)\} \tag{2.2}$$

is valid whenever ∂ stands for one of the three elementary β-subdifferentials and N is the corresponding normal cone.

If, furthermore, ∂ is the metric β-subdifferential, then, in addition,

$$N(S, x) = \bigcup_{\lambda > 0} \lambda\partial\rho(x, S) \tag{2.3}$$

for any closed set S.

The proof easily follows from definitions and properties (a) – (e) of subdifferentials.

The proposition shows a certain advantage of the metric β-subdifferential over the two other elementary subdifferentials, namely, the possibility to define the normal cone by means of the distance function. (We observe that both equalities are valid for convex functions and convex subdifferentials.) Otherwise, the properties of the three elementary subdifferentials seem to be very similar and most of statements can be made in terms of "an elementary" subdifferential for they equally apply to any of them.

Finally for any elementary β-subdifferential ∂ we can consider its sequential regularization ∂ on the class of lower semicontinuous functions which is called the *limiting ∂-subdifferential.*

For further information about elementary subdifferentials see e.g. [**6, 9, 11, 18, 21, 42**]. The limiting Fréchet subdifferential in spaces with Fréchet differentiable norms and, more generally, in Asplund spaces has been thoroughly studied in [**40, 47, 48, 49**].

2.3. Trustworthiness and fuzzy calculus. To be able to apply a subdifferential to analysis on a given space or a class of spaces, we have to be sure that $\partial f(x)$ really carries useful information about functions on the space. This would at least require that $\partial f(x) \neq \emptyset$ on a sufficiently rich set of points of the domain of f. The concept of trustworthiness introduced in [**18**] serves this purpose by describing a class of spaces suitable for application of a given subdifferential. The concept was first introduced for (canonical) Dini-Hadamard and Fréchet subdifferentials (see also [**9**]) where it was adjusted for other elementary β-subdifferentials). Here we slightly modify the concept and extend it to arbitrary subdifferentials.

Let $f_1, \dots, f_k$ be l.s.c. functions on X. It is said that f_i is *uniformly lower semicontinuous* at x if they are finite at x and there is a $c > 0$ such that for any k sequences $\{x_{in}\}$, $i = 1, \dots, k$ such that $f_i(x_{in}) < \infty$, $\|x_{in} - x\| \le c$, $\|x_{in} - x_{jn}\| \to 0$ as $n \to \infty$, there is a sequence of x_n with $\|x_n - x\| \le c$ such that $\|x_n - x_{in}\| \to 0$ and

$$\liminf_{n\to\infty} \sum (f_i(x_{in}) - f_i(x_n)) \ge 0. \tag{2.4}$$

DEFINITION 2.1. We shall say that X is a *∂-trustworthy* space (or trustworthy with respect to ∂), provided the following *basic fuzzy principle* holds in X: if the functions $f_1, \dots, f_k$ are lower semicontinuous and uniformly lower semicontinuous at a certain x at which the sum of the functions $f_1 + \dots + f_k$ has a local minimum then for any $\varepsilon > 0$ there are x_i, x_i^* with

$$\|x_i - x\| \le \varepsilon, \quad |f_i(x) - f_i(x_i)| \le \varepsilon, \quad x_i^* \in \partial f_i(x_i)$$

and such that

$$\|\sum x_i^*\| \le \varepsilon.$$

Finally, given a bornology β, we say that X is *β-trustworthy* if it is trustworthy with respect to the variational β-subdifferential (and hence to any β-associated subdifferential, as the variational β-subdifferential is the smallest among β-associated subdifferentials).

It was shown in [**6**] that any space with an equivalent β-differentiable norm is β-trustworthy. In fact, *any space on which there is a β-differentiable Lipschitz bump function is β-trustworthy.* A proof of this statement can be easily obtained from the proof of Lemma 2 of [**18**] if we use in it the variational principle of Deville-Godefroy-Zissler [**11**] (not yet known when [**18**] appeared) instead of the variational principle of Ekeland.

Furthermore, it is known that *a space is Fréchet trustworthy if and only if it is Asplund.* The "only if" part of this result follows from [**18**] (actually it was shown that a weaker property is sufficient for the space to be Asplund) and the crucial "if" part was established by Fabian [**13**]. (It has to be remarked that in [**13**] the basic fuzzy principle in Asplund spaces was proved under a seemingly additional assumption that all functions f_i but possibly one of them are Lipschitz near x (in which case the uniform lower semicontinuity condition is trivially satisfied). However, the general form of the basic fuzzy principle, as stated above, for the Fréchet subdifferential in Asplund spaces follows from this result.) The following proposition gives a new metric characterization of uniform lower semicontinuity.

PROPOSITION 2.3. *The functions $f_1, \dots, f_k$ are uniformly lower semicontinuous at x if and only if there is a nondecreasing function $\omega(t)$ on $I\!R_+$, continuous and equal to zero at zero such that*

$$\rho((u,\alpha), \mathrm{epi}(\sum f_i)) \le \omega(\sum \rho((u,\alpha_i), \mathrm{epi} f_i) \tag{2.5}$$

for all u of a neighborhood of x and all α, α_i with $\sum \alpha_i = \alpha$.

PROOF. Suppose there is a $c > 0$ such that $f_i(u) > \infty$ and (2.5) holds whenever $\|u - x\| \le c$ and $\sum \alpha_i = \alpha$. Let x_{in}, $i = 1, ..., k$, $n = 1, 2, ...$, be such that $\|x_{in} - x\| < c$, $\|x_{in} - x_{jn}\| \to 0$ as $n \to \infty$ and $f_i(x_{in}) < \infty$. Set $\alpha_{in} = f_i(x_{in})$ and $u_n = x_{1n}$. Then for all $i = 1, ..., k$ we have $\rho((u_n, \alpha_{in}), \mathrm{epi} f_i) \le \|x_{in} - x_{1n}\| \to 0$. By (2.5), $\rho((u_n, \alpha_n), \mathrm{epi}(\sum f_i)) \to 0$ (where $\alpha_n = \sum \alpha_{in}$). This means that there are $(x_n, \beta_n) \in \mathrm{epi}(\sum f_i)$ such that $\|u_n - x_n\| \to 0$, $\alpha_n - \beta_n \to 0$. The first of these two relations shows that $\|x_n - x_{in}\| \to 0$ for every i and the second that

$$0 = \lim_{n\to\infty} (\sum_i f_i(x_{in}) - \beta_n) \le \liminf_{n\to\infty} (\sum_i f_i(x_{in}) - \sum_i f_i(x_n)).$$

Thus (2.5) implies uniform lower semicontinuity.

To prove the converse, we set $f = \sum f_i$ and

$$\omega(t) = \sup\{\rho((u,\alpha), \mathrm{epi} f) : \sum \rho((u,\alpha_i), \mathrm{epi} f_i) \le t,\ \|u - x\| \le c/2,\ \sum \alpha_i = \alpha\}.$$

This function is obviously nonnegative and equal to zero at zero. Therefore we have to check that it is continuous at zero.

So let $t_n \to 0$. Let (u_n, α_{in}) be such that $\sum \rho((u_n, \alpha_{in}), \mathrm{epi} f_i) \le t_n$ and u_n are all within $c/2$ of x. Then there are $(x_{in}, \beta_{in}) \in \mathrm{epi} f_i$ such that $\|x_{in} - x\| \le c$, $|\beta_{in} - \alpha_{in}| \le 2t_n$ and $\|x_{in} - u_n\| \le 2t_n$. Then $\|x_{in} - x_{jn}\| \to 0$ and by uniform lower semicontinuity we can find x_n with $\|x_n - x_{in}\| \to 0$ such that (2.4) is satisfied. Set $\alpha_n = \sum \alpha_{in}$, $\beta_n = \sum \beta_{in}$. Then by (2.4)

$$\liminf(\beta_n - \sum f_i(x_n)) \ge \liminf(\sum (f_i(x_{in}) - f_i(x_n)) \ge 0$$

which means (as $\|(x_n, \beta_n) - (u_n, \alpha_n)\| \to 0$) that

$$0 = \lim \rho((x_n, \beta_n), \mathrm{epi}(\sum f_i)) = \lim \rho((u_n, \alpha_n), \mathrm{epi}(\sum f_i)).$$

□

Proposition 2.3 justifies the following

DEFINITION 2.2. We say that the functions $f_1, ..., f_k$ satisfy the *general metric qualification condition* at x if there is a nonnegative continuous function $\omega(t)$ on $I\!R_+$ such that (2.5) holds for all u of a neighborhood of x and all α, α_i with $\sum \alpha_i = \alpha$.

In other words: the functions are uniformly lower semicontinuous at x if and only they satisfy the general metric qualification condition.

The proof of the first part of the proposition below can be obtained exactly as it was done in [**18**] for Dini-Hadamard subdifferential (with the additional help of the renorming lemma in [**25**]); for Fréchet subdifferentials the result was considered in [**13, 25**], and the proof of the second part of the proposition is very similar; its precise wording can be obtained from the proof of Proposition 4 in [**6**] applied to the Fréchet subdifferential.

PROPOSITION 2.4. *Suppose that ∂ is an elementary subdifferential and X is a ∂-trustworthy space. Then the following weak fuzzy calculus rule is valid for any l.s.c. functions $f_1, ..., f_k$ which are finite at x:*

if $x^* \in \partial(f_1 + ... + f_k)(x)$, then for any $\varepsilon > 0$ and any weak-star neighborhood V of the origin in X^*, there are x_i, x_i^* such that

$$\|x_i - x\| \le \varepsilon, \quad x_i^* \in \partial f_i(x_i), \quad \sum x_i^* \in x^* + V.$$

In an Asplund space the following strong fuzzy calculus rule (with Fréchet subdifferential) is valid for any l.s.c. functions $f_1, ..., f_k$, provided they are finite at x and satisfy the general metric qualification condition:

if $x^* \in \partial(f_1 +, ..., +f_k)(x)$, then for any $\varepsilon > 0$ there are x_i, x_i^* such that

$$\|x_i - x\| \le \varepsilon, \quad x_i^* \in \partial f_i(x_i), \quad \|\sum x_i^* - x^*\| \le \varepsilon.$$

For more information on the subject see [**9, 13, 18, 24**].

2.4. Approximate subdifferentials. Unlike elementary subdifferentials and limiting elementary subdifferentials, approximate subdifferentials make sense and have good analytic properties in any Banach space. It can be shown that topological (non-sequential) limits of all elementary (hence limiting elementary) subdifferentials in trustworthy spaces coincide with the approximate subdifferential for Lipschitz functions and are not smaller than the approximate subdifferential for arbitrary l.s.c. functions. Moreover, in certain most important cases (as, say, the cases of a separable or a reflexive or, more generally, weakly compactly generated X) limiting subdifferentials themselves are topologically closed and therefore coincide with the approximate subdifferential if the function is Lipschitz.

This determines the special role of the approximate subdifferential in nonsmooth analysis and allows to view various limiting subdifferentials as "almost realizations" of the approximate subdifferential in certain specific classes of spaces. In certain cases a limiting subdifferential can be smaller than the approximate subdifferential. A crucial question is whether these differences can be used in analysis to get better information about a function or a mapping. So far, to our knowledge, no confirmation to such a possibility has been found.

The structure of the definition of the approximate subdifferential is more complicated (and resembles the structure of the original definition of the generalized gradient given by Clarke (see [**10**]).

Let L be a subspace of X. Set

$$\partial_L^- f(x) = \{x^* \in X^* : \ d^- f(x;h) \ge \langle x^*, h\rangle, \ \forall h \in L\}.$$

If $L = X$, this is of course the Dini-Hadamard subdifferential. If now f satisfies the Lipschitz condition near x, we define the approximate subdifferential of f at x as follows:

$$\partial f(x) = \bigcap_{L \in \mathcal{F}} \limsup_{u \to x} \partial_L^- f(u),$$

where the limsup, as usual, is taken in the norm–to–weak-star topology and $\mathcal{F}$ is the collection of all finite dimensional subspaces of X. This is equivalent to the formula below which is slightly longer but substantially more convenient in applications

$$\partial f(x) = \bigcap_L \limsup_{u \to x} [\partial_L^- f(u) \bigcap kB], \tag{2.6}$$

where k is a Lipschitz constant of f in a neighbourhood of x.

For an l.s.c. function which is not necessarily Lipschitz continuous, we define the approximate subdifferential through the approximate subdifferential of the distance to the epigraph (which is already a Lipschitz function with Lipschitz constant one) using the relations (2.3) and (2.2) as *the definition* of the approximate subdifferential and the approximate normal cone. This definition is correct in the sense that for a Lipschitz function f we obtain the same result as from (2.6), on the one hand (which is an easy consequence of Proposition 2.1), and that the result does not depend on the specific choice of an equivalent norm in X (see [**25**]).

REMARK. What we have just defined as the approximate subdifferential coincides with what was called in [**25**] the *nucleus of the G-subdifferential.* We shall, however, as was initiated in [**6**], use for this object the name *approximate subdifferential* and shall not work with its topological closure as in [**25**]. The reasons are the following. As was shown in [**25**], in most cases the object just defined displays better analytic properties than its weak-star closure. Furthermore, a very convenient representations in terms of β-derivatives in spaces with β-differentiable renorms was found in [**6**]. The two definitions are equivalent at least if either the function is Lipschitz continuous or the space is finite dimensional (and the function lower semicontinuous).

There is an alternative way to define the approximate subdifferential for Lipschitz functions by means of so called *separable reduction.* Let L be a closed subspace of X, and let f_L be the restriction of f to L, considered as a function on L. Likewise, for any $x^* \in X^*$ we denote by x^*_L the restriction of x^* to L considered as an element of L^*. (We emphasize the difference between the notations $\partial^-_L f(x)$ and $\partial^- f_L(x)$. In particular, the first is a subset of X^* whereas the second lies in L^*.)

PROPOSITION 2.5. (Separable reduction theorem [**28**]). *Assume that X is a Banach space and f a function on X satisfying the Lipschitz condition in a neighborhood of a given point $x \in X$. Then the following three conditions are equivalent:*

(a) $x^* \in \partial f(x)$;

(b) for any weak-star neighborhood $U \subset X^*$ of the origin there is a separable subspace $\tilde{L}$ (depending on x and x^*) such that for any closed separable subspace $L \subset X$ containing x and $\tilde{L}$ we have $x^*_L \in \partial f_L(x) + U_L$;

(c) for any weak-star neighborhood $U \subset X^*$ of the origin and any separable subspace $\tilde{L}$ there is a closed separable subspace $L \subset X$ containing x and $\tilde{L}$ and such that $x^*_L \in \partial f_L(x) + U_L$.

It should be mentioned in connection with Proposition 2.5 that if X itself is a separable space (or, more generally, a space with a Gâteaux differentiable renorm) then, as was shown in [**6**], for the approximate subdifferential of a Lipschitz (near x) function f we have the following simple representation

$$\partial f(x) = \limsup_{u \to x} \partial^- f(u). \tag{2.7}$$

It can also be shown that in an Asplund weakly compactly generated space the approximate subdifferential coincides with the limiting Fréchet subdifferential for any lower semicontinuous function.

The following proposition allows to apply Separable reduction theorem to functions which are just lower semicontinuous, not necessarily satisfying the Lipschitz condition.

PROPOSITION 2.6. *Let $\{S_i\}$ be a countable collection of subsets of X. Then for any separable subspace $M \subset X$ there is a bigger closed separable subspace L such that $\rho(x, S_i) = \rho(x, S_i \bigcap L)$ for any $x \in L$.*

PROOF. Set $M_1 = M$ and for any $k = 1, 2, ...$ define M_{k+1} as the closed subspace spanned by M_k and the union of all $\{x_{imn}\}$, where $x_{imn} \in S_i$ and $\rho(x_n, S_i) = \lim_{m\to\infty} \|x_n - x_{imn}\|$ for any x_n of a dense countable subset of M_k. Then for any i and any $x \in M_k$ we have $\rho(x, S_i) = \rho(x, S_i \bigcap M_{k+1})$. Define L as the closure of $\bigcup M_k$. □

The proof of Proposition 2.5 is based on the fundamental minimality property of approximate subdifferentials. Before stating it, we notice that, as follows from (2.6) and the weak fuzzy calculus rule of Proposition 2.4, the inclusion

$$\partial(f_1 + f_2)(x) \subset \partial f_1(x) + \partial f_2(x) \tag{2.8}$$

is always valid for approximate subdifferentials if both functions (or at least one of them) are Lipschitz continuous. To emphasize the difference between (2.8) and the fuzzy calculus rules discussed above, it is usually said that the approximate subdifferential has the *exact calculus* on the class of Lipschitz continuous functions (actually on a much broader - see Section 6). The minimality property of the approximate subdifferential is now stated as follows.

PROPOSITION 2.7 ([**21, 25**]). *Suppose that ∂ stands for the approximate subdifferential and $\tilde{\partial}$ is any robust subdifferential with exact calculus on the class of Lipschitz continuous functions. Then for any such function*

$$\partial f(x) \subset \tilde{\partial} f(x), \quad \forall\, x.$$

We conclude the section with a proposition which emphasizes universality of the approximate subdifferential.

PROPOSITION 2.8. *Every Banach space is approximate trustworthy.*

PROOF. The proof basically follows the lines of standard proofs for the basic fuzzy principle [**6, 9, 21, 22, 30**]. Let $f_1, ..., f_k$ be l.s.c.functions on a Banach space X which are finite and satisfy the general metric qualification condition at a certain $x \in X$ at which the sum $f_1 + ... + f_k$ attains a local minimum. Fix a corresponding $c > 0$ (from the qualification condition) and consider for any $r > 0$ the function

$$p_r(x_0, ..., x_k) = \sum_{i=0}^{k} (f_i(x_i) + r\|x_i - x_0\|^2) + \|x_0 - x\|^2,$$

where $f_0(\cdot) \equiv 0$. Set

$$\alpha_r = \inf\{p_r(x_0, ..., x_k) : \ \|x_i - x\| \le c, \ i = 1, ..., k\}$$

and choose u_{ir} such that $p_r(u_{1r}, ..., u_{kr}) \le \alpha_r + 1/r^2$. Arguing as in [**18, 21, 30**], we conclude that $\|u_{ir} - u_{jr}\| \to 0$ as $r \to \infty$ and then, using the general metric qualification condition, that all of them actually converge to x.

Applying Ekeland's variational principle, we find x_{ir} with $\|u_{ir} - x_{ir}\| \le 1/r$ such that the function

$$q_r(x_0, ...x_k) = p_r(x_0, ..., x_k) + (1/r)\sum_{i=1}^{k} \|x_i - x_{ir}\|$$

attains its minimum at $(x_{0r}, ...x_{kr})$ on the set $\{(x_0, ...x_k) : \ \|x_i - x\| \le c\}$. As all x_{ir} converge to x, the point is interior to the set for large r, so we have $(0, ..., 0) \in \partial q_r(x_{0r}, ...x_{kr})$.

We can rewrite q_r as follows:

$$q_r(x_0,...x_k) = f(x_0,...,x_k) + r\sum_{i=1}^{k}\|x_i - x_0\|^2 + \|x_0 - x\|^2 + (1/r)\sum_{i=1}^{k}\|x_i - x_{ir}\|,$$

where $f(x_0,...,x_k) = \sum f_i(x_i)$. The first term in this sum is a lower semicontinuous function, the rest are convex continuous, hence locally Lipschitz. It follows that we can apply (2.8) and conclude that there are $(x_0^*,...,x_k^*) \in \partial f(x_0,...,x_k)$, $u_{ir}^* \in X^*$ in the subdifferential of $r\|\cdot\|^2$ at $x_{ir} - x_{0r}$, $i = 1,...,k$, and $v_r^* \in X^*$ in the subdifferential of $\|\cdot\|$ at $x_{0r} - x$ such that

$$\|(x_0^*,...,x_k^*) - (-\sum_{i=1}^{k} u_{ir}^* + v_r^*, u_{1r}^*,...,u_{kr}^*)\| \le 1/r.$$

But as f is a separable function, by the property (f) of subdifferentials, the relation $(x_0^*,...,x_k^*) \in \partial f(x_0,...,x_k)$ is equivalent to $x_i^* \in \partial f_i(x_i)$, $i = 0,...,k$. It follows in particular that $x_{0r}^* = 0$.

Thus, we have: $\|\sum u_{ir}^*\| \le \|v_r^*\| + 1/r \le 2\|x_{0r} - x\| + 1/r = \delta_r \to 0$ and $\|x_{ir}^* - u_{ir}^*\| \le 1/r \to 0$, so that $\|\sum x_{ir}^*\| \le k(\delta_r + 1/r) \to 0$.

This completes the proof of the proposition. □

3. Codirectional compactness

3.1. Definitions and statements of the main theorems. Everywhere in this section and in what follows F is a set-valued mapping from X into Y with closed graph (unless otherwise is specified). We also fix throughout a point $(\bar{x},\bar{y}) \in \text{Graph}F$, that is to say, with $\bar{y} \in F(\bar{x})$ ($\bar{y} = F(\bar{x})$ if F is single-valued). Let ∂ be a certain subdifferential – we shall specify its nature whenever necessary. Let finally $\mathcal{N}$ be a class of bounded nets in X^*, e.g. the class of norm converging nets or the class of bounded weak-star converging nets etc.

DEFINITION 3.1. We say that F is ∂-codirectionally compact at $(\bar{x},\bar{y})$ on the class $\mathcal{N}$ (of nets in X^*) if for any net $\{(x_\alpha, y_\alpha, x_\alpha^*, y_\alpha^*)\}$ such that

(a) $y_\alpha \in F(x_\alpha)$, $(x_\alpha, y_\alpha) \to (\bar{x},\bar{y})$;
(b) $x_\alpha^* \in D^*F(x_\alpha, y_\alpha)(y_\alpha^*)$, $\{x_\alpha^*\} \in \mathcal{N}$;
(c) y_α^* are uniformly (norm) bounded,

weak-star convergence of y_α^* to zero implies norm convergence:

$$y_\alpha^* \to 0 \text{ (weak}^*) \Rightarrow \|y_\alpha^*\| \to 0.$$

It is clear that the richer is the class $\mathcal{N}$, the stronger is the corresponding codirectional compactness property. In particular, codirectional compactness on the class of nets weak-star converging to zero implies codirectional compactness on the class of nets norm converging to zero.

The first two theorems stated in this section contain characterizations of codirectional compactness properties for the two mentioned types of convergence. Together the theorems imply a fairly complete characterization for codirectional compactness of sets (defined below as codirectional compactness of the constant mapping whose unique value is the given set) in Theorem 3.

THEOREM 1. *The following properties are equivalent for any subdifferential* ∂*:*

(CP1) there are a finite dimensional subspace $L \subset Y$, and positive numbers $\varepsilon > 0$, $\gamma > 0$ such that the inequality

$$\|x^*\| \ge \gamma\|y^*\| \tag{3.1}$$

holds whenever

$$\|(x,y)-(\bar{x},\bar{y})\| \le \varepsilon,\ y \in F(x),\ x^* \in D^*F(x,y)(y^*); \tag{3.2}$$

and

$$\rho(y^*, L^\perp) \le \varepsilon\|y^*\| \tag{3.3}$$

(CP2) there are a compact set $Q \subset Y$ with $Q = -Q$, $\varepsilon > 0$, $\gamma > 0$ such that (3.1) holds whenever (3.2) is satisfied together with

$$s_Q(y^*) \le \varepsilon\|y^*\|; \tag{3.4}$$

(CP3) there are a compact set $Q' \subset Y$, $\varepsilon > 0$, $\gamma > 0$ such that the inequality

$$\|x^*\| + s_{Q'}(y^*) \ge \gamma\|y^*\| \tag{3.5}$$

holds whenever (3.2) is satisfied;
(CP4) there are two compact sets Q, $Q' \subset Y$ with $Q = -Q$, $\varepsilon > 0$, $\gamma > 0$ such that (3.5) holds whenever (3.2) is satisfied together with (3.4);
(CP5) F is ∂-codirectionally compact at $(\bar{x},\bar{y})$ on the class of bounded norm converging (to zero) nets in X^*.

THEOREM 2. *The implications (CP6) ⇒(CP7)⇒(CP8) are always valid for the properties (CP6) – (CP8) below with ∂ being the approximate subdifferential in (CP7) and (CP8). If in addition F is pseudo-Lipschitz at $(\bar{x},\bar{y})$ then (CP8) implies (CP6), so that the three properties are actually equivalent:*

(CP6) there are compact sets $P \subset X$, $Q \subset Y$, $\varepsilon > 0$, $\gamma > 0$ such that

$$F(x)\bigcap(\bar{y}+\varepsilon B) + t\gamma B \subset F(x+tP) - tQ$$

for any $t \in [0,\gamma]$ and $x \in \bar{x} + \varepsilon B$;
(CP7) there are compact sets $P \subset X$, $Q \subset Y$, $\varepsilon > 0$, $\gamma > 0$ such that

$$s_P(-x^*) + s_Q(y^*) \ge \gamma\|y^*\|,$$

provided (3.2) is satisfied;
(CP8) F is ∂-codirectionally compact at $(\bar{x},\bar{y})$ on the class of (norm) bounded nets in X^* weak-star converging to zero.

We have to add a few short comments to the statements. First of all, we notice that the conditions (CP3) and (CP7) are very similar: the second is obtained from the first if we replace $\|x^*\|$ by $s_P(-x^*)$. The same replacement in (CP2) also gives a condition equivalent to (CP7) - this can be easily verified as in the proof of Theorem 1 in the next section. But with (CP1) this trick does not work: we get a property which does not even follow from (CP6) as is seen from the following example.

EXAMPLE 1. Let $X = Y = H$ be a Hilbert space and $A : H \to H$ a closed self-adjoint unbounded operator with Im $A = H$ and a compact inverse A^{-1}. (For instance, fix an orthonormal basis $\{e_1, e_2, \dots\}$ in H and define A by setting for $x = \sum x^i e_i$, $Ax = (x^1, 2x^2, 3x^3, \dots)$.) Then A satisfies (CP6) (e.g. with P equal to $A^{-1}B = \{x :\ \sum(nx^n)^2 \le 1\}$) and $Q = \{0\}$ but for any two finite dimensional subspaces M and L we can always find a $u \in H$ such that $u \in L^\perp$ and $Au \in M^\perp$. (Indeed, this reduces to finding a solution of a finite system of homogeneous linear equations with infinitely many unknowns.) This means that the property which we get from (CP1) if we replace $\|x^*\|$ by $\rho(x^*, M^\perp)$ with a finite dimensional M is not satisfied for A.

We next show that (CP8) is not, in general, equivalent to (CP7).

EXAMPLE 2. Again, let $X = Y = H$, and let the mapping $F : H \mapsto H$ be defined by

$$[F(x)]^n = \sqrt{|x^n|} \cdot \text{sign} x^n.$$

This is a mapping with closed graph and for $u \in D^*F(x)(v)$ we have

$$u^n = \frac{v^n}{\sqrt{|x^n|}}, \text{ if } x^n \neq 0, \text{ or } v^n = 0, \ u^n \text{ arbitrary if } x^n = 0. \tag{3.6}$$

It can be easily seen that F satisfies (CP8) at zero: if v_α are norm bounded, $u_\alpha \in D^*F(x_\alpha)(v_\alpha)$ for some $x_\alpha \to 0$ (in the norm topology) and u_α are norm bounded, then $\|v_\alpha\| \to 0$ by (3.6). On the other hand, F does not satisfy (CP7). To see this, fix $\varepsilon > 0$, $\gamma > 0$ and compact sets $P, Q \subset H$.(We may assume without loss of generality that both contain zero.) We then can find a number n^0 such that $s_{P+Q}(x) < \gamma/2$ if $x^k = \delta_{nk}$ for some $n \geq n^0$, where δ_{nk} is the symbol of Kroneker; (Obviously, such an n^0 exists as $P + Q$ is a compact set.) Now take an $n > n^0$ and an x with $x^k = \varepsilon^2\delta_{nk}$.Then $\|x\| = \varepsilon^2 < \varepsilon = \|F(x)\|$ and by (3.6) the vector $u : u^k = \delta_{nk}$ belongs to $D^*F(x)(v)$ where $v^k = \varepsilon\delta_{nk}$. But for these u and v we have

$$s_P(u) + s_Q(v) < \gamma.$$

In other words, for any $\varepsilon > 0, \gamma > 0$ and any pair of compact sets in H, there is an x such that (CP7) is not satisfied.

These examples emphasize that the conditions of Theorem 2 are substantially stronger than the conditions of Theorem 1 and more sensitive to variations. It seems appropriate to mention in this connection that the identity mapping satisfies the equivalent conditions of Theorem 1 but does not satisfy any of conditions of Theorem 2.

Applying together the two theorems to the simplest albeit important case when $F(x) \equiv S$ is a constant mapping we arrive at the following (rather surprising) list of equivalent compactness properties for a set.

THEOREM 3. *Let $S \subset X$ be a closed set containing $\bar{x}$, and let ∂ stand for the approximate subdifferential. Then the following properties are equivalent:*

(S1) there are a finite dimensional subspace $L \subset X$ and positive numbers $\varepsilon > 0$, $\gamma > 0$ such that

$$\rho(x^*, L^\perp) \geq \gamma\|x^*\|$$

whenever

$$x \in S, \ \|x - \bar{x}\| \leq \varepsilon, \ x^* \in N(S, x); \tag{3.7}$$

(S2) there are a compact set $P \subset X$, $\varepsilon > 0$, $\gamma > 0$ such that

$$s_P(x^*) \geq \gamma\|x^*\|$$

whenever x, x^* satisfy (3.7);

(S3) there are a weak-star locally compact cone $K \subset X^*$ and $\varepsilon > 0$ such that $N(S, x) \subset K$ whenever $x \in S$, $\|x - \bar{x}\| \leq \varepsilon$;

(S4) any bounded net $\{x^*_\alpha\}$ such that $x^*_\alpha \in N(S, x_\alpha)$ for some $x_\alpha \in S$, $x_\alpha \to \bar{x}$ norm converges to zero, provided it converges to zero in the weak-star topology;

(S5) there are a compact set $P \subset X$, $\varepsilon > 0$, $\gamma > 0$ such that

$$S \bigcap (\bar{x} + \varepsilon B) + tB \subset S + tP, \quad \forall \ t \in [0, \gamma].$$

Moreover, if X is a β-trustworthy space, then the theorem remains valid if ∂ is any β-associated subdifferential.

Observe that, unlike the other four properties in Theorem 3, (S5) is intrinsic - it does not involve any subdifferential. This property is known as the *compactly epi-Lipschitz* property of Borwein and Strojwas [**7**]. It seems however reasonable for the sake of uniformity of terminology to call a set satisfying the five equivalent properties of Theorem 3 near $\bar{x}$ *codirectionally compact at* $\bar{x}$.

We further note an obvious fact that in case when Y is finite dimensional, the five properties of Theorem 1, (CP7) and (CP8) are automatically satisfied as well as the five properties of Theorem 3 for sets in finite dimensional spaces. More interesting are consequences of Theorems 1 and 2 for set-valued mappings from a finite dimensional space.

THEOREM 4. *Assume that dim $X < \infty$. Then properties (CP1) – (CP8) are equivalent and, moreover, all of them are equivalent to*

(CP9) the graph of F is codirectionally compact (= compactly epi-Lipschitz).

3.2. Discussion. Condition (CP1) is a version of the finite codimension property first introduced in [**19**] for generalized gradients of Clarke and later modified and extended to approximate subdifferentials in [**14, 23**]. There is no difficulty in extending the definition of finite codimension property to any subdifferential ∂. Namely, we say following [**14**] that a (single-valued) mapping Φ from X into Y has the *finite codimension property with respect to* a certain set S at $\bar{x} \in S$ if there are a finite dimensional subspace $L \subset Y$, $\varepsilon > 0$, $\gamma > 0$ such that $\|x^* + u^*\| \geq \gamma$ whenever $x^* \in D^*F(x)(y^*)$, $u^* \in N(S,x)$ for some $x \in S$ with $\|x - \bar{x}\| \leq \varepsilon$ and some y^* of norm one satisfying $\rho(y^*, L^\perp) \leq \varepsilon$.

(To briefly trace back the source of this definition, we consider a bounded linear operator $A: X \mapsto Y$ whose image V is a closed subspace of Y of finite codimension. Let W be a finite dimensional complement of W in Y. Then Y^* is the direct sum of $V^\perp$ and $W^\perp$ and the latter is isomorphic to V^*. As A is onto V, we have (by the Banach open mapping theorem) that there is a $\gamma > 0$ such that $\|A^*w^*\| \geq \gamma\|w^*\|$ for any $w^* \in W^*$. We note further that for $y^* = v^* + w^*$, we have $\rho(y^*, W^\perp) \leq \|v^*\|$. So choosing a small $\varepsilon > 0$ we have for $y^* = v^* + w^*$ with $\|y^*\| = 1$, $\|v^*\| \leq \varepsilon$:

$$\rho(y^*, W^\perp) \leq \varepsilon, \quad \|A^*y^*\| = \|A^*w^*\| \geq \gamma\|w^*\| \geq \gamma(1-\varepsilon)$$

and it remains to remember that A^* is precisely the coderivative of A.)

The definition of finite codimension property as stated above applies to restrictions of single-valued mappings to closed sets. But it was observed in [**23**] that properties of set-valued mappings relating to openness can be equivalently expressed in terms of the restriction of the projection onto the image space to the graph of the set-valued mapping. The finite codimension property is no exception as the following proposition shows.

PROPOSITION 3.1. *The (set-valued) mapping F satisfies (CP1) at $(\bar{x}, \bar{y})$ with a certain subdifferential ∂ if and only if the projection* $\mathrm{Pr}_Y : X \times Y \mapsto Y$ *has the finite codimension property at $(\bar{x}, \bar{y})$ with respect to* Graph$F$ *(with the same subdifferential).*

PROOF. We have ${\mathrm{Pr}_Y}^*(y^*) = (0, y^*)$, so the finite codimension property with respect to Graph$F$ means that

$$\|(0, y^*) + (u^*, -v^*)\| \geq \gamma\|y^*\|,$$

provided $(u^*, -v^*) \in N(\mathrm{Graph}F, (x,y))$ for $(x,y) \in \mathrm{Graph}F$ sufficiently close to $(\bar{x}, \bar{y})$ and y^* satisfying $\rho(y^*, L^\perp) \leq \varepsilon\|y^*\|$. Suppose $\rho(v^*, L^\perp) \leq \varepsilon\|v^*\|$. Then taking $y^* = v^*$, we get $\|u^*\| \geq \gamma\|v^*\|$, that is to say,

$$x^* \in D^*F(x,y)(y^*),\ \rho(y^*, L^\perp) \leq \varepsilon\|y^*\| \Rightarrow \|x^*\| \geq \gamma\|y^*\|$$

which is exactly (CP1). The converse is trivial for if (CP1) holds for F, then, taking the sum norm in $X^* \times Y^*$, we see that

$$\|(u^*, y^* - v^*)\| \geq \|u^*\| \geq \gamma\|v^*\|$$

if $u^* \in D^*F(x,y)(v^*)$.

□

Condition (CP4) reduces to the *partial normal compactness* condition of Mordukhovich and Shao [**48**] if we specify Q as $L \bigcap B$, where L is a finite dimensional subspace of Y (in which case (3.4) reduces to (3.3)), but stated in the general situation, not only for limiting Fréchet subdifferentials in Asplund spaces as in [**47**]. As a matter of fact, one of the starting points for this study was that in [**47**] (CP4) was not recognized as a mere paraphrase of the finite codimensional property.

A stronger form of condition (CP5) which is the codirectional compactness property with respect to norm converging nets was first mentioned in [**38, 52**][4] where it was applied to all nets, not necessarily satisfying the boundedness condition (c) of Definition 3.1. We shall see in the proof that the boundedness condition is essential to the equivalence (and later – that it is sufficient for applications). Note that in [**52**] the condition ((CP5) minus boundedness) is studied in a more general context of homogeneous mappings between dual spaces, not necessarily coderivatives. To my understanding, in both cases the condition was distilled from proofs of "point" openness (or regularity) criteria. We shall see later, however, that for that purpose a weaker sequential form of the condition (see Definition 5.4) is sufficient.

Condition (CP6) is a reformulation of the "partly compactly epi-Lipschitz" property introduced by Jourani and Thibault [**36**]; the implications (CP6) ⇒ (CP7) ⇒ (CP5) for approximate subdifferentials were also proved in this paper. This circle of results was largely motivated by the study of Loewen [**42**] where in particular the implications (S5) ⇒ (S2) ⇒ (S3) were established for Fréchet and limiting Fréchet subdifferentials in reflexive spaces, although the proofs easily carry over to the case of arbitrary spaces with β-differentiable renorms. In turn, Loewen's study was partly influenced by the ealier work by Borwein and Strojwas [**7**] where the compactly epi-Lipschitz property (S5) was first introduced.

Conditions (CP3) and (CP8) seem to have not been mentioned earlier.

I am indebted to L. Thibault for the observation that the equivalence between (CP6) and (CP9) under the assumptions of Theorem 4 follows from Theorem 4.7 of the already mentioned concurrent paper [**38**].

It should be further noted that the conditions (S1) – (S4) in Theorem 3 are equivalent for any subdifferential and only the equivalence of them with the last condition (S5) (under the pseudo-Lipschitz assumption) is specific for approximate and limiting β-associated subdifferentials. The condition (S5) is not equivalent to the other four conditions, say, for the generalised gradient of Clarke (see [**5**] for an

[4]There was some priority dispute concerning (CP5) (see [**49, 52**]). It seems, however, that credit should be evenly shared between the parties. Indeed, in the published version of [**49**] a weaker sequential version of the property appears under the name *sequential partial normal compactness*, similar to sequential coderivative compactness of Definition 5.4 in this paper. But preprint versions of [**38, 52**] appeared before the published version of [**49**] and in all earlier versions of [**49**] the latter property appeared in a stronger (non-infinitesimal) form and as a corollary of partial normal compactness rather than in a special definition.

example of a compactly epi-Lipschitz set with the Clarke normal cone not being locally compact).

Finally, we note that Theorem 4 means in particular that for mappings from a finite dimensional space all "partial" compactness properties are equivalent and can be expressed in terms of a "compact epi-Lipschitz" property of the graph. So replacing the latter by the first does not make a result stronger.

4. Proofs of the theorems

4.1. Proof of Theorem 1. Verification of equivalence of the first four properties is not difficult. To prove that (CP1) $\Rightarrow$ (CP2) we only have to note that $\rho(y^*, L^\perp) = s_Q(y^*)$ with $Q = L \bigcap B$. If (CP2) holds then the obvious inequality (valid as (3.1) holds if $s_Q(y^*) \leq \varepsilon \|y^*\|$)

$$\|x^*\| + s_Q(y^*) \geq \max\{\|x^*\|, s_Q(y^*)\} \geq \min\{\gamma, \varepsilon\}\|y^*\|$$

proves (CP3). The implication (CP3) $\Rightarrow$ (CP4) is trivial. Let finally (CP4) hold. Denote by m the upper bound of the norms of elements of $Q \bigcup Q'$. Choosing a $\delta < \min(\varepsilon, \gamma/2)$ and taking a finite $\delta/2m$-net $\{y_1, ..., y_k\}$ in $Q \bigcup Q'$, we have (for $L = \text{span}\ \{y_1, ..., y_k\}$)

$$\rho(y^*, L^\perp) \geq (1/m)\max\{s_Q(y^*), s_{Q'}(y_*)\} - (\delta/2m)\|y^*\|,$$

so that if $\rho(y^*, L^\perp) \leq (\delta/2m)\|y^*\|$, then $s_Q(y^*) \leq \delta\|y^*\| \leq \varepsilon\|y^*\|$ and, consequently, by (3.5)

$$\begin{aligned}\|x^*\| &\geq \gamma\|y^*\| - s_{Q'}(y_*)\\ &\geq \gamma\|y^*\| - m\rho(y^*, L^\perp) - (\delta/2)\|y^*\| \geq \gamma\|y^*\| - \delta\|y^*\| \geq (\gamma/2)\|y^*\|\end{aligned}$$

which gives (CP1).

The implication (CP3) $\Rightarrow$ (CP5) is again easy to verify: if we have a net $\{x_\alpha, y_\alpha, x^*_\alpha, y^*_\alpha\}$ satisfying the properties (a) - (c) of Definition 3.1 and such that $\|x^*_\alpha\| \to 0$, then (as $\|y^*_\alpha\|$ are uniformly norm bounded and $s_{Q'}$ is weak-star continuous on bounded subsets of Y^*) we have $s_{Q'}(y^*_\alpha) \to 0$ if y^*_α go to zero in the weak-star topology. Therefore by (3.5) the norms of y^*_α go to zero.

It remains to prove that (CP5) implies, say (CP1). Assuming that the implication does not hold, we, for every finite dimensional $L \subset Y$, every $\varepsilon > 0$ and every $\gamma > 0$, can find a pair $(x, y) \in \text{Graph}F$ with $x^* \in D^*F(x, y)(y^*)$ such that $\|(x, y) - (\bar{x}, \bar{y})\| < \varepsilon$, $\rho(y^*, L^\perp) \leq \varepsilon\|y^*\|$ and $\gamma\|y^*\| > \|x^*\|$. The latter means in particular that $y^* \neq 0$, so we may assume that $\|y^*\| = 1$ (as all relations above are homogeneous with respect to (x^*, y^*)).

Consider the collection of all pairs $\alpha = (L, \varepsilon)$, where L is a finite dimensional subspace of Y and $\varepsilon > 0$ with the natural order: $\alpha = (L, \varepsilon) \succ (L', \varepsilon') = \alpha'$ if and only if $L' \subset L$ and $\varepsilon' \geq \varepsilon$. With any such pair we shall associate a set $U_\alpha = L^\perp + \varepsilon B$. Clearly, these sets form a basis for the weak-star topology in Y^*. As follows from the discussion in the preceding paragraph, there is a net $\{x_\alpha, y_\alpha, x^*_\alpha, y^*_\alpha\}$ such that $y_\alpha \in F(x_\alpha)$, $\|(x_\alpha, y_\alpha) - (\bar{x}, \bar{y})\| \leq \varepsilon_\alpha$, $\|{y_\alpha}^*\| = 1$, $x^*_\alpha \in D^*F(x_\alpha, y_\alpha)(y^*_\alpha)$, $\|x^*_\alpha\| \leq \varepsilon_\alpha$ and $y^*_\alpha \in U_\alpha$ (Here ε_α stands for the ε-component of α.) It follows that $\|x^*_\alpha\| \to 0$ and $y^*_\alpha \to 0$ in the weak-star topology. By (CP5) this means that the norms of y^*_α must also go to zero. The contradiction completes the proof.

4.2. Proof of Theorem 2. (CP6) $\Rightarrow$ (CP7). Let $L \subset X$ and $M \subset Y$ be separable subspaces containing $\bar{x}$, P and $\bar{y}$, Q respectively. Take $u \in L$ and $v \in M$ within $\varepsilon/3$ of $(\bar{x}, \bar{y})$ and let

$$(w^*, z^*) \in \lambda\partial^- \rho_{L\times M}((u, v), \text{Graph}F) \tag{4.1}$$

for some $\lambda > 0$, where $\rho_{L\times M}$ is the restriction of $\rho(\cdot, \mathrm{Graph}F)$ to $L \times M$ (not the distance in $L \times M$!). Let $(u_t, v_t) \in \mathrm{Graph}F$ be such that

$$\|(u,v) - (u_t, v_t)\| = \rho((u,v), \mathrm{Graph}F) + o(t).$$

Then $\|(u_t, v_t) - (\bar{x}, \bar{y})\| \leq \varepsilon$ for sufficiently small t and, given a $b \in (\gamma B) \bigcap M$, we can find $p_t \in P$, $q_t \in Q$ such that

$$v_t + t(b + q_t) \in F(u_t + tp_t).$$

It follows that

$$\begin{aligned}
&\rho((u + tp_t, v + t(b + q_t)), \mathrm{Graph}F) \\
&\qquad \leq \|(u,v) - (u_t, v_t)\| + \rho((u_t + tp_t, v_t + t(b + q_t)), \mathrm{Graph}F) \\
&\qquad = \rho((u,v), \mathrm{Graph}F) + o(t).
\end{aligned}$$

Let now $p \in P$, $q \in Q$ be certain limit points of p_t and q_t respectively as $t \to +0$. As $u + tp_t \in L$, $v + t(b + q_t) \in M$, the inequality above implies by (4.1) that

$$\begin{aligned}
&\langle w^*, p\rangle + \langle z^*, b + q\rangle \\
&\quad \leq \lambda d^- \rho_{L\times M}(\cdot, \mathrm{Graph}F)((u,v); (p, b+q)) \\
&\quad \leq \lambda \limsup_{t\to+0} t^{-1}(\rho((u + tp_t, v + t(b + q_t)), \mathrm{Graph}F) - \rho((u,v), \mathrm{Graph}F)) \leq 0.
\end{aligned}$$

It follows that

$$\langle z^*, b\rangle \leq s_P(-w^*) + s_Q(-z^*)$$

whenever $(u, v, w^*, z^*) \in L \times M \times L^* \times M^*$ satisfies (4.1), $b \in (\gamma B) \bigcap M$ and (u, v) is within $\varepsilon/3$ of $(\bar{x}, \bar{y})$. Therefore for such (u, v, w^*, z^*)

$$\gamma\|z^*\| \leq s_P(-w^*) + s_Q(-z^*). \tag{4.2}$$

As the functions on the right are weak-star continuous on bounded sets and (2.7) is valid for separable spaces, (4.2) holds for all (u, v, w^*, z^*) satisfying the above conditions with (4.1) replaced by

$$(w^*, z^*) \in \lambda\partial\rho_{L\times M}((u,v), \mathrm{Graph}F). \tag{4.3}$$

Let now $(x, y) \in \mathrm{Graph}F$, $\|(x,y) - (\bar{x}, \bar{y})\| \leq \varepsilon/3$ and $x^* \in D^*F(x,y)(y^*)$. The latter means that

$$(x^*, -y^*) \in \lambda\partial\rho((x,y), \mathrm{Graph}F)$$

for some $\lambda > 0$. By the separable reduction theorem (Proposition 2.5), given a weak-star neighborhood $W \subset X^* \times Y^*$, there is a separable subspace $\tilde{N} \subset X \times Y$ containing (x, y) such that for any bigger subspace N

$$(x^*, -y^*)_N \in \lambda\rho_N((x,y), \mathrm{Graph}F) + W_N.$$

We can take $L \subset X$, $M \subset Y$ big enough to make sure that $\tilde{N} \subset L \times M$, $P \subset L$ and $Q \subset M$. Then the last inclusion means that there is $(w^*, z^*) \in L^* \times M^*$ satisfying $(x_L^*, -y_M^*) - (w^*, z^*) \in W_{L\times M}$ along with (4.3) (with (u, v) replaced by (x, y)) and (4.2). We can moreover assume that $(L \times M)^\perp$ is contained in W in which case choosing $u^* = u^*(W) \in X^*$ and $v^* = v^*(W) \in Y^*$ such that $u_L^* = w^*$, $v_M^* = z^*$ with $\|u^*\| = \|w^*\|$, $\|v^*\| = \|z^*\|$ we shall have $(x^* - u^*, y^* - v^*) \in W$. Furthermore, $s_P(-u^*) = s_P(-w^*)$ and $s_Q(-v^*) = s_Q(-z^*)$ as $P \subset L$ and $Q \subset M$. Therefore, when W ranges through a base of weak-star neighborhoods in $X^* \times Y^*$, we shall have the nets $\{u^*(W)\}$ and $\{v^*(W)\}$ weak-star converging to x^* and y^* respectively, and, as these nets are norm bounded (by (4.3)), it follows from (4.2) that

$$\begin{aligned}
\gamma\|x^*\| &\leq \liminf\nolimits_W \|u^*(W)\| \\
&\leq \liminf\nolimits_W (s_P(-u^*(W)) + s_Q(-v^*(W))) = s_P(-x^*) + s_Q(-y^*).
\end{aligned}$$

(CP7) $\Rightarrow$ (CP8). This is obvious.

So it remains to prove that (CP8) implies (CP6) if F is pseudo-Lipschitz near $(\bar{x}, \bar{y})$. The latter means that there are $K > 0,\ \delta > 0$ such that

$$F(x) \bigcap (\bar{y} + \delta B) \subset F(x') + K\|x - x'\|B$$

if $x,\ x' \in \bar{x} + \delta B$. It follows (e.g. [**14**], Proposition 1) that, taking the "K-norm" $\|(x, y)\|_K = K\|x\| + \|y\|$ in $X \times Y$, we have

$$\rho^K((x, y), \text{Graph}F) = \rho(y, F(x)). \tag{4.4}$$

for (x, y) close to $(\bar{x}, \bar{y})$. (Here ρ^K is the distance in the K-norm!)

In the remaining part of the proof we use exclusively this K-norm in $X \times Y$. (Recall [**25**] that the approximate normal cone and the approximate coderivative do not depend on specific choice of the norm.)

Consider the collection of all triples $\alpha = (L, M, \varepsilon)$ with L and M being finite dimensional subspaces of X and Y respectively and $\varepsilon \in (0, 1)$ and endow this collection with the natural order $\alpha = (L, M, \varepsilon) \succ (L', M', \varepsilon') = \alpha'$ if and only if $L' \subset L$, $M' \subset M$ and $\varepsilon' \geq \varepsilon$ (as it was done in the proof of Theorem 1). Set

$$P_\alpha = L_\alpha \bigcap B, \qquad Q_\alpha = M_\alpha \bigcap B,$$

where L_α, M_α and ε_α are the components of the triple α.

If (CP6) does not hold, then for any α there are $(x_\alpha, y_\alpha) \in \text{Graph}F$ with $\|(x_\alpha, y_\alpha) - (\bar{x}, \bar{y})\| \leq \varepsilon_\alpha$, $b_\alpha \in \varepsilon_\alpha B$ and $t_\alpha \in (0, \varepsilon_\alpha)$ such that

$$(y_\alpha + t_\alpha(b_\alpha + Q_\alpha)) \bigcap F(x_\alpha + t_\alpha P_\alpha) = \emptyset. \tag{4.5}$$

Set $\tau_\alpha = \sqrt{\varepsilon_\alpha} t_\alpha$ and consider the function

$$\varphi_\alpha(p, q) = \rho((y_\alpha + t_\alpha(b_\alpha + q)), F(x_\alpha + t_\alpha p)) + \tau_\alpha(\|p\| + \|q\|).$$

In what follows, we fix an α and suppress it from the notation – up to a special notice. By (4.5), $\varphi(p, q) > 0$ for any $p \in P,\ q \in Q$. On the other hand,

$$\varphi(0, 0) = \rho(y + tb, F(x)) \leq t\|b\| \leq t\varepsilon,$$

hence

$$0 < \mu = \min\{\varphi(p, q) :\ p \in P,\ q \in Q\} \leq t\varepsilon.$$

The minimum is attained (as φ is l.s.c. and P and Q are compact sets) at certain $p' \in P$, $q' \in Q$ with $\|p'\| + \|q'\| \leq t\varepsilon/\tau = \sqrt{\varepsilon}$.

Set

$$\sigma = \rho(y + t(b + q'), F(x + tp')) > 0$$

and choose a $\lambda > 0$ to satisfy

$$\lambda < \min\{\tau,\ \sigma/(1 + t),\ 1 - \sqrt{\varepsilon}\}. \tag{4.6}$$

By (4.4) μ is the minimal value in the problem of minimizing

$$\psi(u, v, p, q) = K\|u - (x + tp)\| + \|v - (y + t(b + tq)\| + \tau(\|p\| + \|q\|)$$

over $(u, v) \in \text{Graph}F$, $p \in P$, $q \in Q$. Take a $v' \in F(x + tp')$ with

$$\|v' - (y + t(b + tq'))\| \leq \rho(y + t(b + q'), F(x + tp')) + \lambda^2.$$

Then the vector $(x + tp', v', p', q')$ is admisssible for the problem and

$$\psi(x + tp', v', p', q') \leq \mu + \lambda^2.$$

By the variational principle of Ekeland there are $(\tilde{u},\tilde{v}) \in \mathrm{Graph}F$, $\tilde{p} \in P$, $\tilde{q} \in Q$ such that

$$\|\tilde{u} - (x + tp')\| \le \lambda,\ \|\tilde{v} - v'\| \le \lambda,\ \|\tilde{p} - p'\| \le \lambda,\ \|\tilde{q} - q'\| \le \lambda \tag{4.7}$$

and the function

$$\psi(u,v,p,q) + \lambda(\|u - \tilde{u}\| + \|v - \tilde{v}\| + \|p - \tilde{p}\| + \|q - \tilde{q}\|)$$

attains at $(\tilde{u},\tilde{v},\tilde{p},\tilde{q})$ its minimum on $\mathrm{Graph}F \times P \times Q$. The function satisfies the Lipshitz condition, so taking a sufficiently large C (say $C = K + 1 + \lambda + \tau$) we can be sure that the function

$$\begin{aligned} g(u,v,p,q) \quad = \quad & K\|u - (x + tp)\| + \|v - (y + t(b + q))\| + \tau(\|p\| + \|q\|) \\ & +\lambda(\|u - \tilde{u}\| + \|v - \tilde{v}\| + \|p - \tilde{p}\| + \|q - \tilde{q}\|) \\ & +C(\rho((u,v), \mathrm{Graph}F) + \rho(p,P) + \rho(q,Q)) \end{aligned}$$

also attains an unconditional minimum at $(\tilde{u},\tilde{v},\tilde{p},\tilde{q})$.

This means that zero belongs to $\partial g(\tilde{u},\tilde{v},\tilde{p},\tilde{q})$ and applying the standard calculus rules, we find

$$\begin{aligned} & u^* \in K\partial\|\tilde{u} - (x + t\tilde{p})\|,\ v^* \in \partial\|\tilde{v} - (y + t(b + \tilde{q}))\|, \\ & (x^*,y^*) \in C\partial\rho((\tilde{u},\tilde{v}), \mathrm{Graph}F), \\ & p^* \in C\partial\rho(\tilde{p},P), \quad q^* \in C\partial\rho(\tilde{q},Q) \end{aligned}$$

such that

$$\left.\begin{aligned} & \|x^* + u^*\| \le \lambda, \quad \|y^* + v^*\| \le \lambda \\ & \|p^* - tu^*\| \le \lambda + \tau \le 2\tau, \quad \|q^* - tv^*\| \le \lambda + \tau \le 2\tau \end{aligned}\right\} \tag{4.8}$$

We note further that by (4.6), (4.7)

$$\|\tilde{v} - (y + t(b + \tilde{q}))\| \ge \|v' - (y + t(b + q))\| - \|\tilde{v} - v'\| - t\|\tilde{q} - q'\| \ge \sigma - (1 + t)\lambda > 0.$$

This implies that $\|v^*\| = 1$. We further see that

$$\|\tilde{p}\| \le \lambda + \sqrt{\varepsilon} < 1,\ \|\tilde{q}\| \le \lambda + \sqrt{\varepsilon} < 1$$

which means by (4.6) that $\tilde{p}$ and $\tilde{q}$ belong to relative interiors of P and Q respectively and, consequently, that $p^* \in L^{\perp}$, $q^* \in M^{\perp}$. This allows us to deduce from (4.8) that

$$\begin{aligned} & \|x^*\| \le K + \lambda, \quad 1 - \lambda \le \|y^*\| \le 1 + \lambda, \\ & x^* \in L^{\perp} + (3\tau/t)B \subset L^{\perp} + 3\sqrt{\varepsilon}B = U, \\ & y^* \in M^{\perp} + (3\tau/t)B \subset M^{\perp} + 3\sqrt{\varepsilon}B = V. \end{aligned}$$

Returning to the original notation, we conclude that for any α we can find $(x_\alpha, y_\alpha) \in \mathrm{Graph}F$ (e.g. $x_\alpha = \tilde{u}$, etc.) with (x_α, y_α) converging to $(\bar{x},\bar{y})$ and $(x^*_\alpha, y^*_\alpha) \in C\partial\rho((x_\alpha,y_\alpha), \mathrm{Graph}F)$ (note that C does not depend on α !) with

$$\|x^*_\alpha\| \le K + \varepsilon_\alpha, \qquad 1 - \varepsilon_\alpha \le \|y^*_\alpha\| \le 1 + \varepsilon_\alpha \tag{4.9}$$

and $x^*_\alpha \in U_\alpha$ $(= L^{\perp}_\alpha + 3\sqrt{\varepsilon_\alpha}B)$, $y^*_\alpha \in V_\alpha$.

As in the proof of Theorem 1, we see that the sets U_α and V_α form bases for weak-star topologies in the corresponding spaces. Therefore both x^*_α and y^*_α weak-star converge to zero which by (CP8) means that the norms of y^*_α must also go to zero in contradiction with (4.9). This completes the proof of Theorem 2 for approximate subdifferentials.

Very little change is needed to extend the proof to β-associated subdifferentials if the spaces are β-trustworthy. The proof of (CP6) $\Rightarrow$ (CP7) is even simpler in this case as no restrictions to separable subspaces is any more needed. The proof that (CP8) implies (CP6) under the assumption that F is pseudo-Lipschitz near $(\bar{x},\bar{y})$ also follows the same lines and actually remains the same till the end if ∂ is a limiting subdifferential. If ∂ is not a limiting subdifferential but one of elementary subdifferentials, then we have to apply the basic fuzzy calculus rule to the function $g(u, v, p, q)$ to obtain an analogue of (4.8) and not the exact calculus as in the proof.

4.3. Proof of Theorem 3. As we mentioned, Theorem 3 is a result of joint application of Theorems 1 and 2 to the constant mapping $F(x) \equiv S$, say from X into X. Then for the approximate (and any β-associated if available) coderivative we have the expression

$$D^*F(x,y)(y^*) = \begin{cases} \{0\}, \text{ if } y^* \in N(S,y) \\ \emptyset, \text{ if } \ y^* \notin N(S,y), \end{cases}$$

where $N(S, y)$ is the normal cone to S at y. It follows that the codirectional compactness properties (CP5) and (CP8) of such a mapping on the classes of norm and weak-star converging nets coincide and reduce to (S4). The property (CP1) trivially reduces to (S1) while (CP2), (CP3) and (CP7) reduce to (S2) and (CP6) to (S5). We further note that a constant mapping is pseudo-Lipshitz at every point of its graph. Therefore Theorems 1 and 2 give equivalence of (S1), (S2), (S4) and (S5). The implication (S2) $\Rightarrow$ (S3) was proved by Loewen in [**42**]. So we only need to show that (S3) implies (S2).

As K is weak-star locally compact, there is a finite collection $\{x_1, ..., x_k\}$ of points of X such that for $U = \{x^* \in X^* : |\langle x^*, x_i\rangle| \le 1, \ i = 1, ..., k\}$, the set $U \bigcap K$ is norm bounded, say $\|x\| \le r$ if $x^* \in U \bigcap K$. Now set $P = \{\pm x_i, \ i = 1, ..., k\}$. Let $x^* \in K$ and $\lambda = s_P(x^*)$. If $\lambda = 0$, then $tx^* \in U \bigcap K$ for any $t > 0$ which may be only if $x^* = 0$. If $\lambda > 0$, then $x^*/\lambda \in U \bigcap K$, that is $\|x^*/\lambda\| \le r$ which means that $\|x^*\| \le r s_P(x^*)$ and the proof is completed.

4.4. Proof of Theorem 4. As X is finite dimensional, the ∂-codirectional compactness properties on the classes of norm and weak-star converging nets coincide. This gives equivalence of (CP5) and (CP8). Likewise, (CP7) is equivalent to (CP3). To see it, it sufficed to take $P = B$ in (3.7).

Furthermore, (CP8) is equivalent to (S4) for Graph$F$. Indeed , the implication (S4) $\Rightarrow$ (CP8) is valid for S being the graph of F even if X is not finite dimensional. On the other hand, if $(x^*_\alpha, y^*_\alpha) \in N((x_\alpha, y_\alpha), \text{Graph}F)$ weak-star converge to zero and $\|y^*_\alpha\|$ go to zero along a subsequence of indices, the same is true for $\|(x^*_\alpha, y^*_\alpha)\|$ as X is finite dimensional.

Thus, (CP8) is equivalent for (S5) for the graph of F. And the latter easily implies (CP6) (which is a known fact - see e.g. [**36**]).

5. Metric regularity and related properties

5.1. A general sufficient condition for metric regularity. We start with the well-known definitions of metric regularity for single-valued and set-valued mappings.

DEFINITION 5.1. A single-valued mapping F is said to be (metrically) regular near $\bar{x}$ with respect to a set S containing $\bar{x}$ at rate $K > 0$ if for all $x \in S$ of a neighborhood of $\bar{x}$ and all y of a neighborhood of $\bar{y} = F(\bar{x})$

$$\rho(x, F^{-1}(y) \bigcap S) \le K\|y - F(x)\|.$$

If F is a set-valued mapping then we call it (metrically) regular near $(\bar{x}, \bar{y})$ at rate K if for all (x, y) sufficiently close to $(\bar{x}, \bar{y})$

$$\rho(x, F^{-1}(y)) \le K\rho(y, F(x)) \tag{5.1}$$

(where we set $\rho(y,\emptyset)=\infty$). The lower bound of all such K is called the rate of regularity of F at $\bar{x}$ with respect to S in the first case or at $(\bar{x},\bar{y})$ in the second. We say that F is regular at a linear rate if it is regular at a certain rate K.

REMARK. In definitions of metric regularity many authors (e.g. [**9, 36, 48, 58**]) require that (5.1) be valid only for (x,y) sufficiently close to $(\bar{x},\bar{y})$, and such that the distance from $\bar{y}$ to $F(x)$ is sufficiently small. It is an easy matter to verify that the two definitions are equivalent.

Indeed, let (5.1) holds if $\|x-\bar{x}\|<\varepsilon$, $\|y-\bar{y}\|<\varepsilon$, $\rho(\bar{y},F(x))<\varepsilon$. Set $U=\{(x,y):\ \|\bar{y}-y\|+K^{-1}\|\bar{x}-x\|<\varepsilon/2\}$. If $\rho(y,F(x))<\varepsilon/2$, then $\rho(\bar{y},F(x))<\varepsilon$ and, therefore (5.1) is satisfied. If $\rho(y,F(x))\geq\varepsilon/2$, then (as $\rho(\bar{y},F(\bar{x}))=0$)

$$\begin{aligned}\rho(x,F^{-1}(y)) &\leq \rho(\bar{x},F^{-1}(y))+\|x-\bar{x}\|\\ &\leq K\rho(y,F(\bar{x}))+\|x-\bar{x}\|\\ &\leq K\|y-\bar{y}\|+\|x-\bar{x}\|\leq K\varepsilon/2\leq K\rho(y,F(x)).\end{aligned}$$

The definition of metric regularity implies solvability on S of the equation $F(x)=y$, $x\in S$ or the inclusion $y\in F(x)$ for all y sufficiently close to $\bar{y}$. In fact, it turns out that the regularity property is actually equivalent to a stronger opennes property of F (on S in the single-valued case). The following is the precise statement of this remarkable fact.

PROPOSITION 5.1. *A set-valued mapping F is metrically regular near $(\bar{x},\bar{y})$ at rate $K>0$ if and only if there is an $\varepsilon>0$ such that for any $(x,y)\in \mathrm{Graph}F$ with $\|x-\bar{x}\|<\varepsilon$, $\|y-\bar{y}\|<\varepsilon$ and any $t\in(0,\varepsilon)$ we have $y+(t/K)B\subset F(x+tB)$ or, equivalently,*

$$F(x)\bigcap(\bar{y}+\varepsilon B)+(t/K)\overset{\circ}{B}\subset F(x+tB).$$

If the latter property holds, we say that F is *open near* $((\bar{x},\bar{y}))$ *at the rate* $c=K^{-1}$. The upper bound of all such c is called the *rate of openness of* F *at* $((\bar{x},\bar{y})$. Here $\overset{\circ}{B}$ is the interior of B.

PROOF. If F is metrically regular near $(\bar{x},\bar{y})$, then there is a $\delta>0$ such that (5.1) holds for $\|x-\bar{x}\|<\delta$, $\|y-\bar{y}\|<\delta$, $\rho(\bar{y},F(x))<\delta$. Set $\varepsilon=\delta/2$ and let $\|x-\bar{x}\|<\varepsilon$, $\|y-\bar{y}\|<\varepsilon$, $y\in F(x)$. If now $\|v-y\|<t/K$, $t\leq\varepsilon$, then by (5.1),

$$\rho(x,F^{-1}(v))\leq K\rho(v,F(x))\leq K\|v-y\|<t$$

which means that there is a u with $\|u-x\|<t$ such that $v\in F(u)$. In other words, $v\in F(x+tB)$. This proves the "only if" part of the proposition.

The proof of the "if" part is equally simple. Take (x,y) such that $\|x-\bar{x}\|<\varepsilon/2$, $\|y-\bar{y}\|<\varepsilon/2K$, $\rho(\bar{y},F(x))<\varepsilon/2K$. Then $\rho(y,F(x))<\varepsilon/K$ that is to say, $\rho(y,F(x))=\tau/K$ for some $\tau<\varepsilon$. Take a $t\in(\tau,\varepsilon)$. Then there is a $v\in F(x)$ with $\|y-v\|<t/K$ so that $y\in V+(t/K)\overset{\circ}{B}\subset F(x+t\overset{\circ}{B})$. This means that for a certain u with $\|u-x\|<t$ we have $y\in F(u)$. Therefore $\rho(x,F^{-1})\leq\|x-u\|<t$. As this is true for any $t>\tau$, we conclude that $\rho(x,F^{-1})\leq\tau=K\rho(y,F(x))$ an the proof is completed. □

We have had already a chance to refer to a remark in [**23**] that, as far as the openness (or surjection) property is concerned, a set-valued mapping can always be equivalently replaced by the restriction of the projection into Y to the graph of the mapping. The same must be true for the regularity property as well (due to equivalence). The following proposition gives an explicit quantitative representation of this fact.

PROPOSITION 5.2. *Let F be a set-valued mapping with closed graph. It is regular near $(\bar{x}, \bar{y}) \in \text{Graph}F$ at rate K if and only if the projection $Pr_Y : X \times Y \mapsto Y$ is regular near $(\bar{x}, \bar{y})$ with respect to* Graph$F$ *at rate $K + 1$, provided $X \times Y$ is endowed with the sum norm $\|(x, y)\| = \|x\| + \|y\|$.*

PROOF. If $X \times Y$ is endowed with the sum norm, then

$$\begin{aligned} &\rho((x, v), Pr_Y^{-1}(y) \bigcap \text{Graph}F) \\ &\qquad = \inf\{\|x - u\| + \|y - v\| : (u, y) \in \text{Graph}F\} \\ &\qquad = \rho(x, F^{-1}(y)) + \|y - v\|. \end{aligned} \tag{5.2}$$

Thus, if F is regular at rate K and $(x, v) \in \text{Graph}F$, then (by definition and (5.2))
(5.3)

$$\rho((x, v), Pr_Y^{-1}(y) \bigcap \text{Graph}F) \leq K\rho(y, F(x)) + \|y - Pr_Y(x, v)\| \leq (K + 1)\|y - v\|$$

which means that the projection is regular with respect to Graph$F$ at rate $K + 1$.

Conversely, assume that there is an $\varepsilon > 0$ such that (5.3) holds for all (x, y, v) such that $\|x - \bar{x}\| < \varepsilon$, $\|y - \bar{y}\| < \varepsilon$, $\|v - \bar{y}\| < \varepsilon$, $v \in F(x)$. Then by (5.2), (5.3) for such (x, y, v)

$$\rho(x, F^{-1}(y)) \leq K\|v - y\| \tag{5.4}$$

This will obviously imply (5.1) if x is within ε of $\bar{x}$, y is within ε of $\bar{y}$ and the distance from $\bar{y}$ to $F(x)$ is less than $\varepsilon/2$.

□

The theorem below contains the main subdifferential criterion for metric regularity. To state the theorem we need some additional notation.

DEFINITION 5.2. [17]. Consider a set-valued mapping $\mathcal{D}$ from Y^* into X^*. The *Banach constant* of the mapping is

$$C^*(\mathcal{D}) = \inf\{\|x^*\| : x^* \in \mathcal{D}(y^*), \ \|y^*\| = 1\}$$

(setting $C^*(\mathcal{D}) = \infty$ if the intersection of the domain of $\mathcal{D}$ with the unit sphere is empty).

In what follows we consider only the sum norm in $X \times Y$. (This is a matter of convenience only).

THEOREM 5. *Assume that $X \times Y$ is a ∂-trustworthy space, F is a set-valued mapping from X to Y with closed graph and*

$$C^*(D^*F(x, y)) \geq m \quad \text{if } \|(x, y) - (\bar{x}, \bar{y})\| < \varepsilon,$$

Then F is regular on the open ε-ball around $(\bar{x}, \bar{y})$ with a rate not greater than m^{-1}.

PROOF. We shall proof a slightly stronger result. Define for $r < \infty$ the "r-coderivative" of F at (x, y) (associated with ∂) by

$$D_r^*F(x, y)(y^*) = \{x^* : (x^*, -y^*) \in \lambda\partial\rho((x, y), \text{Graph}F) \text{ for some } \lambda \in [0, r]\}$$

and set $D_\infty^*F(x, y)(y^*) = D_\infty^*F(x, y)(y^*)$. What we shall prove is that F is regular near $(\bar{x}, \bar{y})$, provided there are $m > 0$, $r \in (2(1 + m), \infty]$ and $\varepsilon > 0$ such that

$$C^*(D_r^*F(x, y)) \geq m \quad \text{if } \|(x, y) - (\bar{x}, \bar{y})\| < \varepsilon, \tag{5.5}$$

We shall give the proof only for the case $r < \infty$. The argument for $r = \infty$ is precisely the same with the only difference that instead of $\rho((x, y), \text{Graph}F)$, we have to conside the indicator function of Graph$F$.

Suppose F is not regular near $(\bar{x},\bar{y})$ at rate K. We shall show that $K < m^{-1}$ in which case application of this to every point of the ε-ball around $(\bar{x},\bar{y})$ (instead of $(\bar{x},\bar{y})$ itself) concludes the proof.

So assume the contrary: $K \geq m^{-1} = a$. By Proposition 4.3, projection Pr_Y is not regular near $(\bar{x},\bar{y})$ at rate $K+1$ with respect to Graph$F$. This means that there are sequences $\{(u_n,v_n)\} \to (\bar{x},\bar{y})$, $(u_n,v_n) \in \text{Graph}F$ and $\{y_n\} \to \bar{y}$ such that

(5.6) $$\gamma_n = \rho((u_n,v_n), Pr_Y^{-1}(y_n) \bigcap \text{Graph}F) > (K+1)\|y_n - v_n\| \geq (a+1)\|y_n - v_n\|.$$

Set $\varepsilon_n = \|y_n - v_n\|$. As $\gamma_n > 0$ by (5.6), and $(u_n,v_n) \in \text{Graph}F$, we have $(u_n,v_n) \notin Pr_Y^{-1}(y_n)$ which means that $\varepsilon_n = \|y_n - v_n\| > 0$. Choose positive λ_n and μ_n to make sure that for any n

$$\gamma_n > \lambda_n > (a+1+\mu_n)\varepsilon_n,$$

and consider the function

$$f_n(x,y) = \|y - y_n\|.$$

By (5.6), $f(u_n,v_n) = \varepsilon_n \leq \inf f_n + \varepsilon_n$. Applying the variational principle of Ekeland, we shall find $(w_n,z_n) \in \text{Graph}F$ such that

(5.7) $$\|w_n - u_n\| + \|z_n - v_n\| \leq \lambda_n < \gamma_n$$

and the function

$$\|y - y_n\| + (\varepsilon_n/\lambda_n)(\|x - w_n\| + \|y - z_n\|)$$

attains minimum on Graph$F$ at (w_n,z_n). As $\varepsilon_n \leq \lambda_n$, this function satisfies the Lipschitz condition with constant 2, so the function

$$g_n(x,y) = \|y - y_n\| + 2\rho((x,y), \text{Graph}F) + (\varepsilon_n/\lambda_n)(\|x - w_n\| + \|y - z_n\|)$$

attains at (w_n,z_n) an unconditional minimum.

Observe that $y_n \neq z_n$. Indeed, if $y_n \in F(w)$, then $\|w - u_n\| + \|y_n - v_n\| \geq \gamma_n$ by (5.6), so (5.7) shows that $y_n \notin F(w_n)$.

As ∂ satisfies the basic fuzzy principle on $X \times Y$, we can choose (w'_n, z'_n) and (w''_n, z''_n) sufficiently close to (w_n,z_n) to make sure that both pairs go to $(\bar{x},\bar{y})$ as $n \to \infty$, $z'_n \neq y_n$ and

$$0 \in \{0\} \times \partial\|z'_n - y_n\| + 2\partial\rho((w''_n, z''_n), \text{Graph}F) + (a+1+\mu_n)^{-1}B.$$

This means that there are $v^*_n \in \partial\|z'_n - y_n\|$ and $(w^*_n, -z^*_n) \in 2\partial\rho((w''_n,z''_n), \text{Graph}F)$ such that

(5.8) $$\|w^*_n\| \leq (a+1+\mu_n)^{-1}, \quad \|z^*_n - v^*_n\| \leq (a+1+\mu_n)^{-1}.$$

As $z'_n \neq y_n$, we have $\|v^*_n\| = 1$, so that $1+(a+1+\mu_n)^{-1} \geq \|z^*_n\| \geq 1-(a+1+\mu_n)^{-1}$.

Set $x^*_n = w^*_n/\|z^*_n\|$, $y^*_n = z^*_n/\|z^*_n\|$. Then $\|y^*_n\| = 1$ and $(x^*_n, -y^*_n) \in r_n\partial\rho((w''_n,z''_n), \text{Graph}F)$, where

$$r_n = 2/\|z^*_n\| \leq 2(1-(a+1+\mu_n)^{-1})^{-1} = 2\frac{a+1+\mu_n}{a+\mu_n} = 2\frac{1+m+m\mu_n}{1+m\mu_n}.$$

As $\mu_n \to 0$, we have $r > r_n$ for sufficiently large n which means that $x^*_n \in D^*_r F(w''_n,z''_n)(y^*_n)$. By the assumptions, $\|x^*_n\| \geq m$ as $\|y^*_n\| = 1$. So we get from (5.8)

$$\begin{aligned} m \leq \|x^*_n\| &\leq (a+1+\mu_n)^{-1}/\|z^*_n\| \\ &\leq (a+1+\mu_n)^{-1}(1-(a+1+\mu_n)^{-1})^{-1} = \tfrac{m}{1+m\mu_n} < m. \end{aligned}$$

The contradiction completes the proof. □

5.2. Point criterion for metric regularity. Theorem 5 gives a fairly precise sufficient criterion for regularity. But in certain important situations we need a criterion entirely in terms of infinitesimals relating to a given specified point, not to all points in a neighborhood. First order necessary conditions in optimization is the most graphic demonstration of such a situation.

In other words, we need a "point" criterion for regularity. As we mentioned in Introduction, it was for the sake of such criteria that many of the compactness conditions discussed above were introduced and studied . Taking some earlier results into account (e.g. [**23, 36, 48**]), we may conclude that assuming in addition that F is ∂-codirectionally compact on the class of norm converging nets, we shall have a point criterion instead of (5.5). In fact a somewhat different and possibly weaker assumption is sufficient.

DEFINITION 5.3. We shall say that the sequential ∂-codirectional compactness property holds for F at $(\bar{x}, \bar{y})$ if whenever the sequence of $\{(x_n, y_n, x_n^*, y_n^*)\}$ is such that $y_n \in F(x_n)$, $(x_n, y_n) \to (\bar{x}, \bar{y})$, $x_n^* \in D^*(x_n, y_n)(y_n^*)$ for some r, the same for all n, $\|x_n^*\| \to 0$ and $y_n^* \to 0$ in the weak-star topology, we actually have $\|y_n^*\| \to 0$.

Sequential ∂-codirectional compactness is not stronger than ∂-codirectional compactness on the class of norm converging nets, it actually looks as being strictly weaker but it may well be of no difference in typical situations. On the other hand, verification of this property may be simpler and therefore it would be reasonable to look for other equivalent characterizations for it.

We can now state and prove the theorem. Set

$$\mathrm{Ker} D^* F(x,y) = \{y^* : \ 0 \in D^* F(x,y)(y^*)\}.$$

THEOREM 6. *Let ∂ be the regularization (or sequential regularization) of a subdifferential $\tilde{\partial}$, and let X be a $\tilde{\partial}$-trustworthy space. Suppose that F has the sequential $\tilde{\partial}$-codirectional compactness property at $(\bar{x}, \bar{y})$. Then the condition*

$$\mathrm{Ker} D^* F(\bar{x}, \bar{y}) = \{0\} \tag{5.9}$$

is sufficient for F to be metrically regular near $(\bar{x}, \bar{y})$ at a linear rate.

(Here, of course, D^* is the coderivative associated with ∂).

PROOF. All we need is to verify that under the assumptions of the theorem the conditions of Theorem 5 are satisfied for F and $\tilde{\partial}$. Assuming the contrary, we have to conclude that for any $m > 0$ and any $\varepsilon > 0$ there is a $(x,y) \in \mathrm{Graph} F$ within ε of $(\bar{x}, \bar{y})$ such that $C^*(\tilde{D}^* F(x,y)) < m$.

If so, then there is a sequence of $(x_n, y_n) \in \mathrm{Graph} F$ converging to $(\bar{x}, \bar{y})$ and such that for some (x_n^*, y_n^*) with $x_n^* \in \tilde{D}^* F(x_n, y_n)(y_n^*)$ we have $\|y_n^*\| = 1$, $\|x_n^*\| \to 0$. Any weak-star limit point of $\{(x_n^*, y_n^*)\}$ has the form $(0, y^*)$, so by (5.9) $y^* = 0$ which means that zero is the only weak-star limit point of the sequence $\{y_n^*\}$ and, consequently, the sequence weak-star converges to zero. As F has the sequential $\tilde{\partial}$-codirectional compactness property at $(\bar{x}, \bar{y})$, it follows that $\|y_n^*\| \to 0$ and we arrive at a contradiction. □

The last theorem, unlike Theorem 5, does not offer an estimate for the rate of regularity. Again , we need to assume that dim $X < \infty$ to be able to add the statement that the rate of regularity at $(\bar{x}, \bar{y})$ is not greater than $(C^*(D^* F(\bar{x}, \bar{y})))^{-1}$.

The following obvious corollary lists some specific situations in which the theorem applies.

COROLLARY 7. *The condition of the theorem is satisfied if either*

(a) X and Y are arbitrary Banach spaces and ∂ is the approximate subdifferential; or

(b) X and Y are β-trustworthy spaces and ∂ is any of limiting β-associated subdifferentials.

A particular realization of (b) is the case when in both X and Y there exist β-differentiable Lipschitz bump functions or when X and Y are both Asplund and ∂ is the limiting Fréchet subdifferential. So the theorem contains the Asplund space results of [**48**].

Thus, we have two types of conditions which imply metric regularity: (5.5) and two conditions of Theorem 5, namely, (5.9) and sequential codirectional compactness. The two latter conditions imply (5.5) as we have seen. On the other hand, we observe that the condition $C^*(D^*F(x,y)) \geq m > 0$ for all $(x,y) \in \text{Graph}F$ of a neighborhood of $(\bar{x},\bar{y})$ implies "full" codirectional compactness on the class of norm converging nets.

5.3. Metric regularity and properness. Characterizations of metric regularity for convex multifuctions and of the pseudo-Lipschitz property of set-valued mappings with finite dimensional ranges. We shall next discuss the question of how necessary the conditions of Theorems 5 and 6 are for metric regularity. In two simple cases the conditions do appear to be necessary: if $F = A$ is a linear bounded operator (in which case, by the Banach open mapping theorem, (5.5) holds if A is metrically regular) and if X and Y are both finite dimensional (in which case F is automatically codirectionally compact and, as was shown in [**46**], (5.9) holds if F is regular at $(\bar{x},\bar{y})$). So the question is what kind of weaker assumptions on F may make the conditions of Theorems 4 and/or 5 also necessary for metric regularity.

We start the discussion on this subject with the following definition. Set $F_\varepsilon(x) = F(x + \varepsilon B)$.

DEFINITION 5.4. We say that F is directionally proper near $(\bar{x},\bar{y})$ if there are $K > 0$ and $\alpha \in [0,1)$ with the following property:

(DP1) for any $(x,y) \in \text{Graph}F$ of a neighborhood of $(\bar{x},\bar{y})$ there is an $\varepsilon > 0$ such that for any $v \in T(F_\varepsilon(x),y)$ there are $h \in X, w \in Y$ satisfying the following conditions:

$$\|h\| \leq K\|v\|,\ \|v - w\| \leq \alpha\|v\|,\ (h,w) \in T(\text{Graph}F,(x,y)). \tag{5.10}$$

(Recall that $T(S,x)$ is the contingent cone to S at $x \in S$.)

PROPOSITION 5.3. *A bounded linear operator $A: X \mapsto Y$ is directionally proper near zero (hence everywhere) if and only if $\text{Im}A$ is a closed subspace of Y.*

PROOF. The "if" part of the proposition follows from the statement of the Banach open mapping theorem, and the proof of the "only if" part is more or less a part of its proof theorem as we shall see. So let A be directionally proper. We first show that for any $y \in \text{Im}\ A$ there is an $x \in X$ such that $\|x\| \leq (K/(1-\alpha))\|y\|,\ A(x) = y$.

Indeed, set $y_1 = y$, and for every $k = 1,2,\ldots$ find an h_k such that $\|h_k\| \leq K\|y_k\|,\ \|Ah_k - y_k\| \leq \alpha\|y_k\|$ and set $y_{k+1} = y_k - Ah_k,\ x_k = \sum_{i=1}^k h_i$. Then

$$\|y_{k+1}\| = \|Ah_k - y_k\| \leq \alpha\|y_k\| \leq \ldots \leq \alpha^k\|y_1\| \to 0$$

as $k \to \infty$. It follows that x_k form a Cauchy sequence converging to a certain x with

$$\|x\| \le \sum \|h_k\| \le \sum K\|y_k\| \le (K/(1-\alpha))\|y\|$$

and

$$\begin{aligned} Ax_k - y &= \sum_{i=1}^{k} Ah_i - y_1 = \sum_{i=2}^{k} Ah_i + Ah_1 - y_1 \\ &= \sum_{i=2}^{k} Ah_i - y_2 = \ldots = Ah_k - y_k = -y_{k+1} \to 0. \end{aligned}$$

Set $K_1 = K/(1-\alpha)$. Let a sequence $\{y_k\} \subset \mathrm{Im}A$ converges to a certain y. Taking if necessary a subsequence we can guarantee that the series $\sum \|y_{k+1} - y_k\|$ also converges. Take an x_1 such that $Ax_1 = y_1$, $\|x_1\| \le K_1\|y_1\|$ and, for any k, a u_k such that $Au_k = y_{k+1} - y_k$ with $\|u_k\| \le K_1\|y_{k+1} - y_k\|$. Then $x_k = x_1 + \sum h_i$ converge to a certain x such that $Ax = y$. □

It immediately follows from the proposition that a bounded metrically regular linear operator is directionally proper.

PROPOSITION 5.4. *Assume as usual that F is a set-valued mapping from X into Y. If dim $X < \infty$ and F is metrically regular (at a linear rate) near $(\bar{x}, \bar{y})$, then F is directionally proper near $(\bar{x}, \bar{y})$ with K arbitrarily close to the constant of openness of F at $(\bar{x}, \bar{y})$.*

PROOF. As F is regular at a linear rate near $(\bar{x}, \bar{y})$, it is open at a linear rate near the point. This means that there is an $m > 0$ such that for every $(x, y) \in \mathrm{Graph}F$ sufficiently close to $(\bar{x}, \bar{y})$ there is an $\varepsilon > 0$ such that for all $\delta \in (0, \varepsilon)$ the ball $y + \delta B$ belongs to $F_{m\delta}(x)$. In other words, for any nonzero $v \in Y$ and any positive $t \le \delta/\|v\|$ there is an $h_t \in X$ with $\|h_t\| \le m\|v\|$ such that $y + tv \in F(x + th_t)$. As X is finite dimensional, we may assume that h_t converge to some h for which we have by definition $(h, v) \in T(\mathrm{Graph}F, (x, y))$, $\|h\| \le m\|v\|$ which means that F is directionally proper with $\alpha = 0$, hence concluding the proof. □

If $(x^*, -y^*) \in \partial^- \rho(\mathrm{Graph}F, (x, y))$, then, of course

$$\langle x^*, h \rangle - \langle y^*, v \rangle \le 0, \quad \forall\, (h, v) \in T(\mathrm{Graph}F, (x, y)). \tag{5.11}$$

Therefore if F is directionally proper near $(\bar{x}, \bar{y})$, then there is an $\varepsilon > 0$ such that for all $(x, y) \in \mathrm{Graph}F$ sufficiently close to $(\bar{x}, \bar{y})$ and all $x^* \in D^*_- F(x, y)(y^*)$, we have

$$\sup\{-\langle y^*, v \rangle : \ v \in T(F_\varepsilon(x), y),\ \|v\| \le 1\} \le \alpha\|y^*\| + K\|x^*\|.$$

In particular, if F is regular (hence open) near $(\bar{x}, \bar{y})$, then $T(F_\varepsilon(x), y) = Y$ and we get the inequality

$$(1 - \alpha)\|y^*\| \le K\|x^*\|.$$

Therefore we can state the following

PROPOSITION 5.5. *Assume that F is directionally proper near $(\bar{x}, \bar{y})$ with constants K and α. Then the condition*

$$C^*(D^*_- F(x, y)) \ge m = (1 - \alpha)/K$$

for all $(x, y) \in \mathrm{Graph}F$ sufficiently close to $(\bar{x}, \bar{y})$ is necessary for metric regularity of F near $(\bar{x}, \bar{y})$.

The proposition, in turn, leads to the following characterization of metric regularity for convex set-valued mapping (i.e. set-valued mappings with convex graphs).

THEOREM 8. *Let F be a convex set-valued mapping from X into Y with closed graph. Let $F(X)$ be the image of X under F. Let finally, $\bar{y} \in F(\bar{x})$. Then the following properties are equivalent:*

(a) $\bar{y} \in \text{int } F(X)$;
(b) *F is metrically regular at $(\bar{x}, \bar{y})$ at a linear rate;*
(c) *F is open at $(\bar{x}, \bar{y})$ at a linear rate;*
(d) *F^{-1} is pseudo Lipschitz at $(\bar{y}, \bar{x})$;*
(e) *there is an $m > 0$ such that $C^*(D^*F(x, y)) \geq m$*

for all $(x, y) \in \text{Graph}F$ of a neighborhood of $(\bar{x}, \bar{y})$.

Moreover, if the equivalent conditions (a) – (e) are satisfied then the rate of openness of F at $((\bar{x}, \bar{y}))$ is equal to $C^*(D^*F(x, y))$.

Here D^* refers to the subdifferential in the sense of convex analysis with which all other subdifferentials (of a convex function) coincide.

PROOF. Equivalence of (b), (c) and (d) is a general fact already mentioned. Implication (a) $\Rightarrow$ (c) is the famous Robinson-Ursescu extension of the Banach open mapping theorem [**53, 59**] and the inverse implication is, of course, a trivial fact. Implication (e) $\Rightarrow$ (b) follows from Theorem 5. So we only need to verify that, say, (c) $\Rightarrow$ (e) and the Banach constant of $D^*F(\bar{x}, \bar{y})$ coincides with the constant of openness of F.

So we assume that (c) holds. Let $c > 0$ be the rate of openness of F at $(\bar{x}, \bar{y})$. This means that for any $\delta \in (0, c)$ there is an $\varepsilon = \varepsilon(\delta)$ such that

$$y + t(c - \delta)\overset{\circ}{B} \subset F(x + tB) \tag{5.12}$$

when $(x, y) \in \text{Graph}F$, $\|x - \bar{x}\| + \|y - \bar{y}\| < \varepsilon$, $0 < t < \varepsilon$. It follows that $T(F_\varepsilon(x), y) = Y$ for any such (x, y). Take a $v \in Y$ with $\|v\| = 1$ and a $t \in (0, \varepsilon)$ and set $v_1 = (1 + \delta)^{-1}v$. Then $v_1 \in \overset{\circ}{B}$ and by (5.12) there is an h with $\|h_1\| \leq 1$ such that $y + t(c - \delta)v_1 \in F(x + th_1)$. As $\text{Graph}F$ is a convex set, we get from here that $(x + \lambda th_1, y + \lambda(c - \delta)tv_1) \in \text{Graph}F$ for all $\lambda \in [0, 1]$ which means that $(h_1, (c - \delta)v_1) \in T(\text{Graph}F, (x, y))$.

Set $w = v_1$, $h = h_1/(c - \delta)$. Then $(h, w) \in T(\text{Graph}F, (x, y))$ and

$$\|h\| \leq (c - \delta)^{-1} = (c - \delta)^{-1}\|v\|, \ \|w - v\| \leq \delta(1 + \delta)^{-1} = \delta(1 + \delta)^{-1}\|v\|.$$

These relations are homogeneous with respect to (h, v, w), so they mean that F is directionally proper near $((\bar{x}, \bar{y}))$ with $K = (c - \delta)^{-1}$ and $\alpha = \delta(1 + \delta)^{-1}$.

By Proposition 5.9 (as D^*F coincides with the Dini-Hadamard coderivative due to convexity), $C^*(D^*F(x, y)) \geq (c - \delta)(1 + \delta)^{-1}$ if $(x, y) \in \text{Graph}F$, and $\|x - \bar{x}\| + \|y - \bar{y}\| < \varepsilon(\delta)$. Therefore, for a sufficiently large r we have (as $C^*(D_r^*F) \geq C^*(D^*F)$)

$$c' = \lim_{\varepsilon \to 0} \inf\{C^*(D_r^*F(x, y)) : (x, y) \in \text{Graph}F, \ \|x - \bar{x}\| + \|y - \bar{y}\| < \varepsilon\} \geq c. \tag{5.13}$$

As the graph of F is convex, D^*F at any point coincides with the approximate coderivative of F at the same point. Applying Theorem 5 along with Proposition 5.2, we conclude that the constant of openness of F at $(\bar{x}, \bar{y})$ is not greater than c'. Together with (5.13) this gives $c' = c$ and it remains to note that (as the convex subdifferential of a continuous convex functions is an u.s.c. set-valued mapping from the norm to the weak-star topology)

$$\liminf C^*(D_r^*F(x_n, y_n)) \geq C^*(D_r^*F(\bar{x}, \bar{y})),$$

whenever $(x_n, y_n) \to (\bar{x}, \bar{y})$, $y_n \in F(x_n)$ which implies that $C^*(D_r^*F(\bar{x},\bar{y})) = c'$ for any sufficiently large r and therefore $C^*(D^*F(\bar{x},\bar{y})) = c' = c$. This completes the proof.

□

Proposition 5.9 gives a necessary condition for metric regularity in terms of the Dini-Hadamard subdifferential. We shall obtain a necessary condition for metric regularity in terms of approximate subdifferential under a stronger version of (DP1), namely

(DP2) there are a compact set $P \subset X$ and an $\alpha \in [0,1)$ such that for any $(x,y) \in \text{Graph}F$ of a neighborhood of $(\bar{x},\bar{y})$ there is an $\varepsilon > 0$ such that for any $v \in T(F_\varepsilon(x), y)$ with $\|v\| = 1$ there is a pair $(h,u) \in T(\text{Graph}F, (x,y))$ satisfying $\|u - v\| \le \alpha\|v\|$, $h \in P$.

PROPOSITION 5.6. *Suppose that both X and Y are Hadamard trustworthy spaces and F satisfies (DP2) near $(\bar{x},\bar{y})$. Suppose further that F is regular at a linear rate near $(\bar{x},\bar{y})$. Then there are a compact set $P \in X$ and a $\lambda > 0$ such that the inequality*

$$\|y^*\| \le s_P(x^*)$$

holds whenever $(x,y) \in \text{Graph}F$, $\|(x,y) - (\bar{x},\bar{y})\| < \lambda$ and $x^* \in D^*F(x,y)(y^*)$, D^*F being the approximate coderivative of F at (x,y).

In particular, F is approximate codirectionally compact on the class of weak-star converging nets and (5.9) is satisfied.

PROOF. As F is regular near $(\bar{x},\bar{y})$ at a linear rate, $T(F_\varepsilon, y) = Y$ for all $(x,y) \in \text{Graph}F$ sufficiently close to $(\bar{x},\bar{y})$ and all positive ε. By (DP2) this means that there are a compact set $P' \subset X$ and an $\alpha \in [0,1)$ such that for any $v \in Y$, $\|v\| = 1$ we can find a pair $(h,u) \in T(\text{Graph}F, (x,y))$ with $h \in P'$, $\|u - v\| \le \alpha$.

Let $(x^*, -y^*) \in r\partial^-\rho((x,y), \text{Graph}F)$. Then as follows from (5.11)

$$\begin{aligned} -\langle y^*, v\rangle &\le -\langle x^*, h\rangle + \langle y^*, u - v\rangle \\ &\le s_{P'}(-x^*) + \alpha\|y^*\|. \end{aligned}$$

As this is true for any v of the unit sphere, we have

$$(1-\alpha)\|y^*\| \le s_{P'}(-x^*)$$

or, setting $P = -(1-\alpha)^{-1}P'$,

$$\|y^*\| \le s_P(x^*). \tag{5.14}$$

The desired result follows now exactly in the same way as in the proof of Theorem 2 (the separable case) with the account of the facts that the approximate subdifferential of a Lipschitz function (the distance function in our case) is the upper limit of Dini-Hadamard subdifferentials if the space is Hadamard trustworthy and s_P is weak-star continuous on bounded subsets of X^*. □

Again, it is not clear whether the result can be extended to arbitrary Banach spaces for (DP2) can hardly be inherited by separable subspaces. So we need to further modify the condition to make it work in general situations:

(DP3) there exist a compact set $P' \subset X$, $\varepsilon > 0$, $\delta > 0$ and $\alpha \in [0,1)$ such that the inclusion

$$(y + tb) \bigcap (\bar{y} + \varepsilon B) \bigcap F_\delta(x) \subset F(x + tP') + t\alpha B$$

holds whenever $(x,y) \in \text{Graph}F$, $\|(x,y) - (\bar{x},\bar{y})\| < \varepsilon$ and $0 \le t < \varepsilon$.

THEOREM 9. *Let X and Y be arbitrary Banach spaces, and let F satisfy (DP3). If F is regular at a linear rate near $(\bar{x}, \bar{y})$, then the conclusion of Proposition 5.10 is valid, namely there are a compact set $P \in X$ and a $\lambda > 0$ such that (5.13) holds whenever $(x, y) \in \text{Graph}F$, $\|(x, y) - (\bar{x}, \bar{y})\| < \lambda$ and $x^* \in D^*F(x, y)(y^*)$, D^*F being the approximate coderivative of F at (x, y).*

In particular, F is approximate codirectionally compact on the class of weak-star converging nets and (5.9) is satisfied.

PROOF. We can rewrite (DP3) in the following way: if $(x, y) \in \text{Graph}F$ is within ε of $(\bar{x}, \bar{y})$, then for any v with $\|v\| < \varepsilon$ such that $y + v \in F_\delta(x)$, there are $h \in X$, $w \in Y$ satisfying the following relations:

$$h \in \|v\|P', \ \|w\| \leq \alpha\|v\|, \ y + v \in F(x + h) + w. \tag{5.15}$$

On the other hand, as F is regular at a linear rate, there is an $m > 0$ such that (taking a smaller ε if necessary) for $\delta = m\varepsilon$ the set $F_\delta(x)$ contains the ε-ball around y.

Set $M = \max\{\|u\| : u \in P'\}$ and take $(x, y) \in \text{Graph}F$ and v satisfying

$$\left.\begin{array}{l} \|x - \bar{x}\| + (1 - \alpha)^{-1}M\|v\| < \varepsilon, \\ \|y - \bar{y}\| + (1 - \alpha)^{-1}(1 + \alpha)M\|v\| < \varepsilon, \\ \|v\| < \varepsilon. \end{array}\right\} \tag{5.16}$$

Choose (h, w) according to (5.15) and consider the sequence $\{(x_k, y_k, v_k, h_k, w_k)\}$ defined recoursively as follows:

$$x_1 = x, \ y_1 = y, \ v_1 = v, \ h_1 = h, \ w_1 = w$$

....

$$x_{k+1} = x_k + h_k, \ y_{k+1} = y_k + v_k - w_k, \ v_{k+1} = w_k,$$
$$h_{k+1} \in \|v_{k+1}\|P', \ \|w_{k+1}\| \leq \alpha\|v_{k+1}\|, \ y_{k+1} + v_{k+1} \in F(x_{k+1} + h_{k+1}) + w_{k+1}$$

....

Such a sequence can be constructed, provided at each step we have (x_k, y_k) within ε of $(\bar{x}, \bar{y})$ and $\|v_k\| \leq \varepsilon$. We shall see that these conditions are satisfied. Indeed, they are definitely satisfied for $k = 1$. Suppose they are satisfied for first k elements of the sequence ($k \geq 1$). We have

$$\left.\begin{array}{l} \|v_{k+1}\| = \|w_k\| \leq \alpha\|v_k\| \leq \ldots \leq \alpha^k\|v\|, \\ \|h_{k+1}\| \leq M\|v_{k+1}\| \leq M\alpha^k\|v\|, \\ \|x_{k+1} - x_k\| = \|h_k\| \leq M\alpha^{k-1}\|v\|, \\ \|y_{k+1} - y_k\| = \|v_k - w_k\| \leq \|v_k\| + \|v_{k+1}\| \leq (1 + \alpha)\alpha^k\|v\|, \end{array}\right\} \tag{5.17}$$

so that (taking (5.16) into account)

$$\|v_{k+1}\| < \varepsilon,$$
$$\|x_{k+1} - \bar{x}\| \leq \|x_{k+1} - x\| + \|x - \bar{x}\| \leq \|x - \bar{x}\| + M(1 - \alpha)^{-1}\|v\| < \varepsilon,$$
$$\|y_{k+1} - \bar{y}\| \leq \|y_{k+1} - y\| + \|y - \bar{y}\| \leq \|y - \bar{y}\| + (1 + \alpha)(1 - \alpha)^{-1}\|v\| < \varepsilon,$$

and we conclude that the sequence is indeed well defined.

We see further from (5.17) that $v_k, w_k \to 0$, $h_k \to 0$ and $\{x_k\}$ and $\{y_k\}$ are Cauchy sequences. Let $\tilde{x}$ and $\tilde{y}$ be their limits. Then $(\tilde{x}, \tilde{y})$ are within ε of $(\bar{x}, \bar{y})$. Furthermore, as the graph of F is closed, we conclude that $\tilde{y} \in F(\tilde{x})$. From the definition of y_k we see that at each step $y_{k+1} + v_{k+1} = y_k + v_k$, so that $\tilde{y} = y + v$. Finally, for $\tilde{x}$ we have setting $P = (1 - \alpha)^{-1}\text{cl conv}(P \bigcup\{0\})$ (clearly this is a compact set):

$$\tilde{x} = x + \sum h_k \in x + (1 - \alpha)\sum \|v_k\|P \subset x + (1 - \alpha)\|v\|\sum \alpha^k P = x + \|v\|P.$$

We have therefore proved the following: under the assumptions there is a compact set $P \subset Y$ such that for any $(x,y) \in \text{Graph}F$ within ε of $(\bar{x},\bar{y})$ and any $v \in Y$, $\|v\| \leq \varepsilon$ we have $y+v \in F(x+\|v\|P)$. In other words, we see that F satisfies (CP6) with $Q = \{0\}$, so by Theorem 2,

$$\|y^*\| \leq s_P(-x^*)$$

if $x^* \in D^*F(x,y)(y^*)$ which implies that F is both ∂-codirectionally compact and satisfies Ker $D^*F(x,) = \{0\}$. This completes the proof of the theorem. □

We observe further that condition (DP3) is automatically satisfied if X is finite dimensional and F is regular, as immediately follows from Proposition 5.2. Taking Theorem 4 and 6 into account we arrive to the following characterization of metric regularity for mappings from a finite dimensional space.

PROPOSITION 5.7. *Suppose that* dim $X < \infty$. *Then F is metrically regular at a linear rate near $(\bar{x},\bar{y})$ if and only if the following two conditions hold:*

(a) the graph of F is codirectionally compact at $(\bar{x},\bar{y})$;

(b) Ker $D^*F(\bar{x},\bar{y}) = \{0\}$,

∂ being the approximate subdifferential and D^* the approximate coderivative.

More important however seems to be a reformulation of this proposition for the pseudo-Lipschitz property of set-valued mappings into finite dimesional spaces. Recall [**8, 50**] that a set valued mapping is metrically regular near a certain point if and only if the inverse mapping is pseudo-Lipschitz near the point. We therefore have the following characterization for the pseudo-Lipschitz property of mappings into $I\!R^n$.

THEOREM 10. *A set-valued mapping F with closed graph into a finite dimensional space is pseudo-Lipschitz near $(\bar{x},\bar{y})$ if and only if*

(a) the graph of F is codirectionally compact (= compactly epi-Lipschitz) at $(\bar{x},\bar{y})$ and

(b) $D^*F(\bar{x},\bar{y})(0) = \{0\}$,

∂ being the approximate subdifferential and D^* the corresponding coderivative

If, moreover, X is a β-trustworthy space, then the statement is true for ∂ being any of the limiting β-subdifferentials.

5.4. Discussion. For the genesis of the definitions of metric regularity see e.g. [**8, 12, 15, 17, 31, 50, 53, 54**]. The role of A.A. Miljutin must be especially emphasized. Although his first publication on the subject appeared only in 1980 [**12**], he was most probably the first to understand the importance of the property, for the theory of necessary optimality conditions in particular, as early as in mid-sixties. It is also interesting to observe that for the first time when the term "regularity" appeared, probably in [**31, 53, 54**], it was applied to the property which is now characterized as "openness" of local approximations.

A weaker property in which, say, (5.1) is required for x in a neighborhood of $\bar{x}$ only with $y = \bar{y}$ has also been considered. This property can be characterized as "regularity *at* a point" (vs. metric regularity as regularity *near* a point). This is actually a property playing a fundamental role in the theory of necessary optimality conditions. When applied to constraint mappings in optimization problems, it allows to lift them into the cost function (through so called "exact penalties", always non-differentiable in this case, by the way) and by that to reduce a constraint problem to a problem without constraints (see e.g. [**14, 19**])

Proofs of Proposition 5.2 first appeared in [**8, 50**] but the fact itself was known and mentioned much earlier [**12, 16**] (including the relationship between the rates). Moreover, for set-valued mapping with convex graph and set-valued mappings of the form

$$F(x) = \begin{cases} G(x) + K, & \text{if} \quad x \in C \\ \emptyset, & \text{if} \quad x \notin C, \end{cases}$$

where $C \subset X$ is a closed convex set, $K \subset Y$ a closed convex cone and G is a C^1-mapping (single-valued) proofs of the "if" part of the proposition was given even earlier in [**53, 54**] and, in the second case, the "only if" part was also demonstrated under a certain additional assumption. Furthermore, it was shown in [**8, 50**] (and mentioned earlier in [**23**]) that both properties are equivalent to Aubin's pseudo-Lipschitz property of the inverse mapping.

Theorem 5 is a slightly modified (and adjusted for metric regularity rather than openness) version of the "dual surjection criterion" of [**23**]. It is also very close to recent results by Jourani and Thibault [**38**]. The main difference is that in [**23**] it was assumed that ∂ has a version of the strong fuzzy calculus and a "full" coderivative was used in (5.5). Neither affects the scheme of the proof very much but the stronger fuzzy calculus assumption on ∂ sorts out all non-limiting β-associated subdifferentials, except for the Fréchet subdifferential, whereas Theorem 4 can be applied with all practically important classes of subdifferentials.

The dual Banach constant was introduced by the author in [**17**] for homogeneous mappings between arbitrary Banach spaces (not necessarily dual). A similar characteristic was introduced also in [**12**] for coderivatives associated with generalized gradients of Clarke. It was also in [**17**] and then in [**23**] that the Banach constant was first used to get estimates for rates of regularity and openness. Variants and particular cases of these estimates can be found in a number of subsequent publications [**46, 48, 38**].

Theorem 6 is a modified version of the point criterion established in [**23**] where a (generally stronger and equivalent if $\dim Y < \infty$) condition

$$C^*(D^*F(\bar{x}, \bar{y})) > 0$$

was used instead of (5.9). The latter was first introduced in [**46**] in the finite dimensional context in which case it is also necessary for metric regularity [**41**], and in [**48**] in the context of the limiting Fréchet subdifferential in Asplund spaces.

It can be shown that the requirement in the definition of sequential codirectional compactness is equivalent to that of the existence of a subsequence of $\{y_n^*\}$ norm converging to zero. In this latter form the property was actually used in [**48**] (in the context of limiting Fréchet subdifferentials in Asplund spaces) although the explicit definition given in the original version of [**48**], dealt with a bigger collection of sequences $\{(x_n, y_n)\}$ which were only required to be sufficiently close to $(\bar{x}, \bar{y})$ and not necessarily converge to it. However, whereas sequential weak-star convergence is a natural assumption in spaces with sequentially weak-star compact dual balls, in particular in Asplund spaces, the fact that only sequential weak-star convergence need to be considered also in the general situation of Theorem 6 seems to be rather surprising.

There is a definite similarity between the directional properness property (DP1) and the one used by Aubin [**2**] partly for the same purpose of proving an open mapping theorem for set-valued mappings. Namely, we shall get Aubin's condition if we assume that the property holds for all $v \in Y$, not just for $v \in T(F_\varepsilon(x), y)$. Further properties (DP2) and (DP3) take their roots in the classical definition of a proper mapping (i.e. such that the preimage of any compact set is compact). All three seem to appear here for the first time as well as most of the results of the preceding subsection. All necessary references concerning Theorem 7 can be found in the proof: the implication (e) $\Rightarrow$ (c) and the concluding statement are new, the rest is well known.

It was observed by Thibault that Proposition 5.11 can also be deduced from the results of the concurrent paper [**38**]. Necessity of the condition (b) of Theorem 9 for metric regularity of mappings from Asplund spaces (which are Fréchet trustworthy) and ∂ being the limiting Fréchet subdifferential was established in [**48**].

6. Application to subdifferential calculus

The purpose of this section is to demonstrate the connection between metric regularity and some basic exact calculus rules. We shall consider here two types of qualification conditions. Conditions of the first type are stated in terms of certain distance estimates which follow from metric regularity of certain set-valued mappings. Conditions of the second type which are the standard subdifferential qualification conditions turn out to be precisely the realization of the sufficient conditions for metric regularity given by Theorem 6 for the above mentioned set-valued mappings.

We shall mainly speak about the approximate subdifferential but all arguments in 6.2 and 6.3 without any change apply to limiting Fréchet subdifferentials in Asplund spaces (even when the latter does not coincide with the approximate subdifferential), whereas Proposition 6.1 for that case requires a much simpler proof.

6.1. Calculus rules under the linear-rate qualification condition. We shall say that the the functions $f_1, ..., f_k$ satisfy the *linear-rate qualification condition* at x if they are finite at x and there are $K > 0$ and $\varepsilon > 0$ such that

$$\rho((u,\alpha), \text{epi}(f_1 + ... + f_k)) \le K \sum_{i=1}^{n} \rho((u,\alpha_i), \text{epi} f_i) \tag{6.1}$$

whenever $\|u - x\| \le \varepsilon$, $|\alpha - f(x)| \le \varepsilon$, $\sum \alpha_i = \alpha$ and α_i are sufficiently close to $f_i(x)$. (The latter requirement can be dropped without changing the property.)

PROPOSITION 6.1. *Let $f_1, ..., f_k$ be lower semicontinuous functions satisfying the linear-rate qualification condition at x. Then*

$$\partial(f_1 + ... + f_k)(x) \subset \partial f_1 + ... + \partial f_k(x). \tag{6.2}$$

PROOF. Set $f = \sum f_i$. Let $x^* \in \partial f(x)$. Then there is a $\lambda > 0$ such that $(x^*, -1) \in \lambda\partial\rho((x, f(x)), \text{epi} f)$.

Assume first that X is a separable space. Then, as follows from [**6**], Proposition 2, there is a sequence $\{(x_n, \alpha_n)\} \subset \text{epi} f$ converging to $(x, f(x))$ as $n \to \infty$ such that for certain $(x_n^*, -\gamma_n) \in \lambda\partial^-\rho((x_n, \alpha_n), \text{epi} f)$ we have $\gamma_n \to 1$ and $x_n^* \to x^*$ in the weak-star topology. The norms of all x_n^* are uniformly bounded and $\alpha_n = f(x_n)$ for all sufficiently large n (for the second component of any elementary subdifferential to the distance function to the epigraph at a point (u, α) with $\alpha > f(u)$ must be zero). We therefore have

$$\langle x_n^*, h\rangle - \gamma_n\alpha \le \lambda d^-\rho(\cdot, \text{epi} f)((x_n, f(x_n)); (h, \alpha)), \quad \forall h,\ \alpha$$

and, by the linear-rate qualification condition: for any $(h, \alpha_1, ..., \alpha_n)$

$$\langle x_n^*, h\rangle - \gamma_n \sum \alpha_i \le \lambda K d^-(\sum \rho(\cdot, \text{epi} f_i))((x_n, f_1(x_n), ..., f_k(x_n)); (h, \alpha_1, ..., \alpha_k)). \tag{6.3}$$

Consider the following functions on $X \times \mathbb{R}^k$:

$$\varphi_i(u, \alpha_1, ..., \alpha_k) = \lambda K \rho((u, \alpha_i), \text{epi} f_i).$$

Then (6.3) means that

$$(x_n^*, -\gamma_n, ..., -\gamma_n) \in \partial^-(\sum \varphi_i)(x_n, f_1(x_n), ..., f_k(x_n)). \tag{6.4}$$

Observe that, as every φ_i does not depend on α_j with $j \neq i$, the corresponding components of elements of its subdifferential are zeros.

Let $\{U_n\}$ be a base of neighborhood of the bounded weak-star topology of X^*. Using the weak fuzzy calculus rule of Proposition 2.4, we find for any n some x_{in}, x_{in}^*, α_{in}, γ_{in} such that

$$\left.\begin{array}{ll} \|x_{in} - x\| \leq 1/n, & |\alpha_{in} - f(x_n)| \leq 1/n; \\ \sum x_{in}^* \in x_n^* + U_n, & |\gamma_{in} - \gamma_n| \leq 1/n; \\ (x_{in}^*, -\gamma_{in}) \in K\lambda\partial^- \rho((x_{in}, \alpha_{in}), \text{epi} f_i). & \end{array}\right\} \tag{6.5}$$

It remains to observe again that (as follows from the last relation in (6.5)) all x_{in}^* are uniformly norm bounded and to pass to the limit as $n \to \infty$.

Thus, the proposition holds for functions on separable spaces. Let now X be an arbitrary Banach space. To complete the proof it is sufficient to show that for any weak-star neighborhood U of zero in X^* there are $x_i^* \in X^*$ such that

$$(x_i^*, -1) \in \lambda \partial \rho((x, f_i(x)), \text{epi} f_i), \qquad x^* - \sum x_i^* \in U.$$

So fix a U. By Proposition 2.5 there is a separable subspace $\tilde{L} \subset X$ such that

$$(x^*, -1)_{L \times \mathbb{R}} \in \lambda \partial \rho_{L \times \mathbb{R}}((x, f(x), \text{epi} f) + U_L$$

for any separable L containing $\tilde{L}$. (Here ρ_L means the restriction of the distance in $X \times \mathbb{R}$ to $L \times \mathbb{R}$, not the distance in the subspace.) By Proposition 2.6 we can choose L to make sure that

$$\begin{array}{l} \rho((u, \alpha), \text{epi} f) = \rho((u, \alpha), (\text{epi} f \bigcap (L \times \mathbb{R})), \\ \rho((u, \alpha), \text{epi} f_i) = \rho((u, \alpha), (\text{epi} f_i \bigcap (L \times \mathbb{R})) \end{array}$$

for all $x \in L$, all α and all $i = 1, ..., k$. Then the restrictions of f_i to L satisfy the linear-rate qualification condition (with the same K) and ρ_L coincides with the distance (to the corresponding epigraph) in $L \times \mathbb{R}$. So by what has been just proved, there are vectors $w_i^* \in L^*$ such that $(w_i^*, -1) \in K\lambda\partial\rho_L((x, f(x)), \text{epi} f_i)$ and $x_L^* - \sum w_i^* \in U_L$. But then there are $x_i^* \in X^*$ such that $(x_i^*, -1) \in K\lambda\partial\rho((x, f(x), \text{epi} f_i)$ and $(x_i^*)_L = w_i^*$. As we can always take L big enough to make sure that $L^\perp \subset U$, the inclusion $x_L^* - \sum (x_i^*)_L \in U_L$ implies that $x^* - \sum x_i^* \in U$ and the proof is completed. □

Applied to indicator functions of sets the proposition gives the following result.

PROPOSITION 6.2. *Let $S_1, ..., S_k$ be closed subsets of X, and let $\bar{x} \in \bigcap S_i$. Assume that the following qualification condition holds: for all x of a neighborhood of $\bar{x}$*

$$\rho(x, \bigcap S_i) \leq K(\rho(x, S_1) + ... + \rho(x, S_k)).$$

Then

$$N(\bigcap S_i, \bar{x}) \subset N(S_1, \bar{x}) + ... + N(S_k, \bar{x}). \tag{6.6}$$

6.2. Normal cone to intersection – standard qualification condition. Let again $S_1, ..., S_k$ be closed sets in X, and let $\bar{x} \in \bigcap S_i$. Consider the set-valued mapping

$$F(x) = (S_1 - x) \times ... \times (S_k - x) = S_1 \times ... \times S_k - (x, ..., x) \tag{6.7}$$

from X into X^k. Suppose this mapping is metrically regular at $(\bar{x}, 0, ..., 0)$. What does this say about the sets S_i themselves?

Taking the sum norm in X^k, we have for any $x \in X$ and any $(x_1, ..., x_k) \in X^k$:

$$\rho((x_1, ..., x_k), F(x)) = \rho(x + x_1, S_1) + ... + \rho(x + x_k, S_k). \tag{6.8}$$

On the other hand, $F^{-1}(x_1, ..., x_k) = \{x : x + x_i \in S_i, \ i = 1, ..., k\}$, so that

$$\rho(x, F^{-1}(x_1, ..., x_k)) = \rho(x, \bigcap(S_i - x_i)).$$

If we now take $x_i = 0$, then the metric regularity condition for F at $(\bar{x}, 0, ...,)$ implies the inequality

$$\rho(x, \bigcap S_i) \leq K \sum \rho(x, S_i)$$

for all x sufficiently close to $\bar{x}$ which is precisely the linear-rate qualification condition of Proposition 6.2.

This means that any sufficient condition for F to be metrically regular at $(\bar{x}, 0, ...0)$ autimatically becomes a qualification condition for the calculus rule for normal cones stated in Proposition 6.2.

Theorems 5 and 6 offer two such sufficient conditions. The condition of Theorem 5, though weaker than that of Theorem 6, can hardly be of a particular interest as long as we are talking about exact calculus rules like (6.2) or (6.6). On the other hand, fuzzy calculus rules need much weaker (than the linear-rate) qualification conditions as we have seen in Section 2.

So let us examine what kind of conditions can be offered by Theorem 6. There are two of them in the statement of the theorem. One says that $\mathrm{Ker} D^*F(x, y) = \{0\}$ and the other that F satisfies the sequential codirectional compactness property.

The coderivative of the mapping F defined by (6.1) is easy to calculate. We already mentioned that for a pseudo-Lipschitz mapping F the equality (4.4) holds if we choose a suitable norm in $X \times Y$. Our mapping F is even Lipschitz continuous, so by (6.2)

$$\rho((x, x_1, ...x_k), \mathrm{Graph} F) = \rho((x_1, ...x_k), F(x)) = \sum \rho(x + x_i, S_i)$$

if an appropriate norm has been chosen in $X \times X^k$.

Now, as coderivatives do not depend on the specific choice of a norm and as the mapping $(x, x_1, ...x_k) \mapsto (x + x_1, ..., x + x_i)$ is obviously onto, property (g) of subdifferential (see 2.1) leads to the conclusion that for $y^* = (x_1^*, ..., x_k^*)$,

$$D^*F(\bar{x}, 0)(-y^*) = \begin{cases} x_1^* + ... + x_k^* & , \text{ if } x_i^* \in N(S_i, x_i); \\ \emptyset & , \text{ otherwise.} \end{cases}$$

So the condition $\mathrm{Ker}\ D^*F(\bar{x}, 0) = \{0\}$ reduces to

$$x_1^* + ... + x_k^* = 0, \quad x_i^* \in N(S_i, x_i), \ i = 1, ..., k \ \Rightarrow \ x_1^* = ... = x_k^* = 0. \tag{6.9}$$

This is a well known qualification condition traditionally used in theorems about normal cones to intesections (e.g. [**25, 44, 57**]).

The sequential codirectional compactness property in our case reads (if we write x_i instead of $x + x_i$):

if for every $i = 1, ..., k$, we have $x_{in} \in S_i$, $x_{in} \to \bar{x}$, $x_{in}^* \in r\partial\rho(x_{in}, S_i)$ for some $r \geq 1$,

$\|x^*_{1n}+...+x^*_{kn}\| \to 0$ and $(0,...,0)$ is a weak-star limit point of $\{(x^*_{1n},...,x^*_{kn})\}$, then $\sum \|x_{in_s}\| \to 0$ along a subsequence $\{n_s\}$ of indices.

This property is actually what is needed in proofs of calculus rules. Similar but somewhat stronger properties were introduced in [**32**] (strong independence) and [**51**] (alliedness).

Following the latter, we shall say that the sets $S_1,...,S_k$ are *generically allied* at $\bar{x}$ if the above formulated property holds. We observe that in this case the qualification condition (6.9) is automatically satisfied.

Summarising, we conclude that Theorem 6 and Proposition 6.2 imply together the following result.

PROPOSITION 6.3. *Let $S_1,...,S_k$ be closed subsets of X which are generically allied at $\bar{x} \in \bigcap S_i$. Then*

$$N(\bigcap S_i, \bar{x}) \subset N(S_1,\bar{x}) + ...N(S_k,\bar{x}).$$

This is in particular true if all sets S_i but at most one of them are codirectionally compact (compactly epi-Lipschitz) at $\bar{x}$.

In slightly different terms the last statement was proved in [**39**] (see also [**32**]); for an earlier result under a stronger assumptions that all sets but one are epi-Lipschitz at $\bar{x}$ see [**25**].

6.3. Subdifferential of a sum – standard qualification condition. Here we shall slightly modify the previous example to obtain the fundamental subdifferential inclusion for sums of functions. It has to be observed that this inclusion can be also obtained as a Corollary of Proposition 6.1 in the same way as it was done in [**25**]. But we prefer to give a direct proof here to demonstrate the connection between metric regularity and subdifferential calculus.

To begin with, we recall the concept of *singular subdifferential* first introduced for Clarke's generalized gradients in [**56**]:

$$\partial^{\infty} f(x) = \{x^* : \ (x^*,0) \in N(\mathrm{epi} f, (x, f(x))).$$

Let $f_1,...,f_k$ be lower semicontinuous functions on X which are finite at $\bar{x}$. Our first purpose will be to construct a set-valued mapping F such that metric regularity of this mapping reduces to the linear-rate qualification condition (6.1).

The construction is almost obvious in view of the previous section. Namely, let F be the set-valued mapping from $W = X \times I\!R^k$ into $Z = (X \times I\!R)^k$ defined as follows:

$$F(x,\alpha_1,...,\alpha_k) = (\mathrm{epi} f_1 - (x,\alpha_1)) \times ... \times (\mathrm{epi} f_k - (x,\alpha_k)).$$

Writing w for $(x,\alpha_1,...,\alpha_k) \in W$ and z for $((x_1,\alpha_1),...,(x_k,\alpha_k)) \in Z$ (and taking again the sum norms in the product spaces), we get

$$\rho(z,F(w)) = \sum \rho((x+x_i,\alpha_i+\beta_i), \mathrm{epi} f_i) \tag{6.10}$$

and

$$\rho(w,F^{-1}(z)) = \inf\{\|x-u\| + \sum |\alpha_i - \gamma_i| : \gamma_i + \beta_i \geq f_i(x+x_i)\}.$$

In particular for $z=0$ (i.e. $x_i = 0, \ \beta_i = 0$)

$$\begin{aligned} \rho(w,F^{-1}(0)) &= \inf\{\|x-u\| + \textstyle\sum |\alpha_i - \gamma_i| : \ \gamma_i \geq f_i(u)\} \\ &\geq \inf\{\|x-u\| + |\textstyle\sum \alpha_i - \sum \gamma_i| : \ \sum \gamma_i \geq \sum f_i(u)\} \\ &= \rho((x,\textstyle\sum \alpha_i), \mathrm{epi} \sum f_i), \end{aligned}$$

so that if F is regular at $(\bar{w}, \bar{z})$ $(\bar{w} = (\bar{x}, f_1(\bar{x}), ..., f_k(\bar{x})),\ \bar{z} = 0)$, then

$$\rho((x,\alpha), \text{epi} \sum f_i) \leq K \sum \rho((x, \alpha_i), \text{epi} f_i)$$

for $\sum \alpha_i = \alpha$ and x close to $\bar{x}$. This is precisely the linear-rate qualification condition (6.1).

We have furthermore $F(w) = D - Aw$, where $D = \text{epi} f_1 \times ... \times \text{epi} f_k$ and the operator $A:\ W \mapsto Z$ is defined by $A(x, \alpha_1, ..., \alpha_k) = ((x, \alpha_1), ..., (x, \alpha_k))$. Then A^* sends every $((x_1^*, \gamma_1), ..., (x_k^*, \gamma_k)) \in Z^*$ into $(\sum x_i^*, \gamma_1, ..., \gamma_k) \in W^*$.

Applying property (g) of subdifferential, we find that $w^* = (x^*, \delta_1, ..., \delta_k) \in D^*F(w,z)(z^*)$ for some $z^* = ((x_1^*, \gamma_1), ..., (x_k^*, \gamma_k)$ only if $(x_i^*, \gamma_i) \in N(\text{epi} f_i, (x_i, \alpha_i)$, $x^* = \sum x_i^*$ and $\delta_i = \gamma_i$ for all i.

Therefore the condition $\text{Ker} D^*F(\bar{x}, 0) = \{0\}$ reduces to the following statement: if $(x_i^*, 0) \in N(\text{epi} f_i, (\bar{x}, f_i(\bar{x}))$ for all $i = 1, ..., k$ and $\sum x_i^* = 0$, then all x_i^* must be equal to zero, or in other words,

$$(6.11) \qquad x_i^* \in \partial^\infty f_i(\bar{x}), \quad \sum x_i^* = 0 \quad \Rightarrow \quad x_1^* = ... = x_i^* = 0.$$

which is exactly the standard qualification condition.

Finally taking into account that $\gamma \leq 0$ whenever $(x^*, \gamma) \in N(\text{epi} f(x, f(x))$, we conclude as in the previous case that sequential codirectional compactness of F at $(\bar{w}, \bar{z})$ (where $\bar{w} = (\bar{x}, f_1(\bar{x}), ..., f_k(\bar{x}))$ and $\bar{z} = ((\bar{x}, f_1(\bar{x})), ..., (\bar{x}, f_k(\bar{x}))))$ is equivalent to the following *sequential alliedness* condition for functions f_i:

> if $(x_{in}^*, \gamma_{in}) \in r\partial\rho(\text{epi} f_i, (x_{in}, f_i(x_{in}))$ for some $r \geq 1$ and x_{in} converging to $\bar{x}$ with $f_i(x_{in})$ converging to $f_i(\bar{x})$, $\| \sum_i (x_{in}^* \| \to 0$, $\sum_i \gamma_{in} \to 0$ as $n \to \infty$ and zero is a weak-star cluster point of $\{(x_{1n}^*, ..., x_{kn}^*)\}$, then actually the norms of all (x_{in}^*) go to zero along a certain subsequence n_s of indices.

Thus, the following theorem is true.

THEOREM 11. *Let $f_1, ...f_k$ be lower semicontinuous functions on X which are finite and sequentially allied at $\bar{x}$. Suppose the qualification condition (6.11) holds. Then*

$$\partial(f_1 + ... + f_k)(\bar{x}) \subset \partial f_1(\bar{x}) + ... + \partial f_k(\bar{x}).$$

This is in particular true if the epigraphs of all f_i with a possible exception of one of them are codirectionally compact (compactly epi-Lipschitz) at $(\bar{x}, f_i(\bar{x}))$.

The last statement for approximate subdifferential was obtained in [**32**] and for the limiting Fréchet subdifferential in Asplund spaces in [**47**]. For earlier versions of the theorem see [**22, 24, 25, 40, 44, 45**].

6.4. Discussion. The implication of Proposition 6.2 was first established in the proof of Theorem 5.4 in [**25**] under the additional assumption that all sets save at most one are epi-Lipschitz. A corresponding implication with a weaker topological assumption was recently established in [**37, 39**] for coderivatives of compositions. The novelty of Propositions 6.1 and 6.2 therefore lies in the understanding that all these implications (equivalent to each other as they are) do not require any compactness assumptions in addition to the linear-rate metric qualification condition.

The first parts of no. 6.2 and 6.3 contain the essence of our approach to subdifferential calculus. It seems to be important to note that the linear-rate metric qualification conditions are not equivalents of metric regularity of the corresponding set-valued mappings but only follow from them. In fact the conditions are equivalent to regularity of the mappings *at* the corresponding points (see (5.4)).

Calculus rules for various subdifferentials were in focus of attention in nonsmooth analysis, perhaps since the beginning. Finite dimensional versions of Proposition 6.3 and Theorem 10 were first announced in [**44**]. Very close results for limiting Fréchet subdifferentials in Asplund spaces was proved in [**47**] and the second parts of the results, as stated, were obtained in [**39**]. For versions of the results obtained under somewhat stronger assumptions see [**22**] (finite dimensional case), [**40**] (limiting Fréchet subdifferential, spaces with Fréchet differentiable renorms) and [**24, 25**] (approximate subdifferentials in Banach and locally convex spaces).

References

[1] H. Attouch and H. Brezis, Duality for the sum of convex functions in general Banach spaces, in *Aspects of Mathematics and its Applications*, J. Barroso, editor, North Holland Elesevier 1986, pp. 125-133.

[2] J.-P. Aubin, Lipschitz behavior of solutions to convex minimization problems, *Math. Oper. Res.* **9** (1984), 87-111.

[3] J.-P. Aubin and H. Frankowska, *Set-Valued Analysis*, Burkhaüser 1990.

[4] V.I. Averbuch and O.G. Smolyanov, The theory of differentiation in linear topological spaces, *Russian Math. Surveys* **22** (1967), 201–258.

[5] J. Borwein, Epi-Lipschitz like sets in Banach spaces: theorems and examples, *Nonlinear Analysis. Theory, Methods, Appl.* **11** (1987), 1207–1217.

[6] J. Borwein and A. Ioffe, Proximal analysis in smooth spaces, *Set-Valued Analysis*, **4** (1996), 1-24.

[7] J. Borwein and H.Strojwas, Tangential approximations, *Nonlinear Analysis TMA* **9** (1985), 1347-1366.

[8] J. Borwein and D.M. Zhuang, Verifiable necessary and sufficient conditions for regularity of set-valued and single-valued maps, *J. Math. Anal. Appl.* **134** (1988), 441-459.

[9] J. Borwein and Q.J. Zhu, Viscosity solutions and viscosity subderivatives in smooth Banach spaces with applications to metric regularity, *SIAM J. Control Optimization*, **134** (1996), 1576-1991.

[10] F.H. Clarke, *Optimization and Nonsmooth Analysis*, Wiley 1983.

[11] R. Deville, G. Godefroy and V. Zissler, *Smoothness and Renormings in Banach Spaces*, Longman 1993.

[12] A. Dmitruk, A.A. Miljutin and N. Osmolovskii, Ljusternik's theorem and the theory of extrema, *Russian Math. Surveys* **35:6** (1980), 11-51.

[13] M. Fabian, Subdifferentiability and trustworthiness in the light of the new variational principle of Borwein and Preiss, *Acta Univ. Carolinae* **30** (1989), 51-56

[14] B. Ginsburg and A. Ioffe, The maximum principle in optimal control of systems governed by semilinear equations, Proceedings of the IMA Workshop on Nonsmooth Analysis and Geometric Methods in Deterministic Optimal Control, Minneapolis 1993, B. Mordukhovich and H. Sussman eds, Springer 1995, to appear.

[15] A. Ioffe, Regular points of Lipschitz functions, *Trans. Amer. Math. Soc.* **251** (1979), 61-69

[16] A. Ioffe, Sous-differentielles approchées de fonctions numériques, *C.R. Acad. Sci. Paris* **292** (1981), 675-678

[17] A. Ioffe, Nonsmooth analysis: differential calculus of non-differentiable mappings, *Trans. Amer. Math. Soc.* **266** (1981), 1-56.

[18] A. Ioffe, On subdifferentiability spaces, *Ann. N.Y. Acad. Sci* **410** (1983), 107-119.

[19] A. Ioffe, Necessary conditions in nonsmooth optimization, *Math. Oper. Res.* **9** (1984), 159-189.

[20] A. Ioffe, Calculus of Dini subdifferentials of functions and contingent derivatives of set-valued maps, *Nonlinear Analysis TMA* **8** (1984), 517–539.

[21] A. Ioffe Subdifferentiability spaces and nonsmooth analysis, *Bull Amer. Math Soc.* **10** (1984), 87-89.

[22] A. Ioffe, Approximate subdifferentials and applications 1. The finite dimensional theory. *Trans. Amer. Math Soc.* **281** (1984), 289–316.

[23] A. Ioffe, On the local surjection property, *Nonlinear Analysis TMA* **11** (1987), 565-592.

[24] A. Ioffe, Approximate subdifferentials and applications 2. *Mathematika* **33** (1986), 111-128.

[25] A. Ioffe, Approximate subdifferentials and applications 3. The metric theory, *Mathematika* **36** (1989), 1–38.

[26] A. Ioffe, Proximal analysis and approximate subdifferentials, *J. London Math. Soc* **41** (1990), 175–192.

[27] A. Ioffe, Nonsmooth subdifferentials: their calculus and applications, in *World Congress of Nonlinear analists '92*, Proceedings of the 1st Congress of Nonlinear Analists, Tampa 1991, Walter de Gruyter 1996, pp. 2299-2310.

[28] A. Ioffe, Separable reduction theorem for approximate subdifferentials, C.R. Acad. Sci. Paris, **323** (1996), 107-112.

[29] A. Ioffe, Directional compactness, scalarization and semi-Fredholm mappings, *Nonlinear Analysis. Theory, Methods, Appl.*, **29** (1997), 201-219.

[30] A. Ioffe and R.T. Rockafellar, The Euler and the Weierstrass condition for nonsmooth variational problems, *Calculus of Variationa and Partial Diff. Equations*, **4** (1996), 60-87.

[31] A. Ioffe and V.M. Tikhomirov, Theory of Extremal Problems, Nauka, Moscow 1974, English translation: North Holland 1979.

[32] A.Jourani, Intersection formulae and marginal functions in Banach spaces *J. Math. Anal. Appl.* **192** (1995), 867-891.

[33] A. Jourani and L. Thibault, The approximate subdifferential of a composite function, *Bull. Aust. Math. Soc.* **47** (1993), 443-455.

[34] A. Jourani and L. Thibault, Metric regularity and strongly compactly Lipschitzean mappings, *Nonlinear Analysis TMA* **24** (1995), 229-240.

[35] A. Jourani and L. Thibault, A note on Fréchet and approximate subdifferentials of composite functions, *Bull. Austr. Math. Soc.* **49**, 111-116.

[36] A. Jourani and L. Thibault, Verifiable conditions for openness and regularity of multivalued mappings, *Trans. Amer. Math. Soc.* **347** (1995), 1255-1268.

[37] A. Jourani and L. Thibault, Metric regularity and subdifferential calculus in Banach spaces, *Set-Valued Analysis* **3**(1995), 87-100.

[38] A. Jourani and L. Thibault, Coderivatives of multivalued mappings, locally compact cones and metric regularity, preprint 1995.

[39] A. Jourani and L. Thibault, Extensions of subdifferential calculus rules in Banach spaces and applications, *Canadian J. Math.*, to appear.

[40] A. Kruger, Properties of generalized differentials, Siberian J. Math. **26** (1985), 822–832.

[41] A covering theorem for set-valued mappings, *Optimization* **19** (1988), 763-780.

[42] P. D. Loewen, Limits of Fréchet normals in nonsmooth analysis, pp 178–188 in Optimization and Nonlinear Analysis, A. Ioffe, M. Marcus and S. Reich (editors), Longman, Essex, 1992.

[43] B. Mordukhovich, Maximum principle in an optimal time control problem with nonsmooth constraints, *J. Appl. Math. Mech.* **40** (1976), 960-969.

[44] B. Mordukhovich, Nonsmooth analysis with convex generalised subdifferentials and conjugate mappings, *Doklady Akad. Nauk BSSR* **28** (1984), 976-979.

[45] B. Mordukhovich, *Approximation Methods in Problems of Optimization and Control*, Nauka, Moscow 1988 (in Russian; English translation to be published by Wiley Interscience).

[46] B. Mordukhovich, Complete characterization of openness, metric regularity and Lipshitzian properties of multifunctions, *Trans. Amer. Math. Soc.* **340** (1993), 1-35.

[47] B. Mordukhovich and Yongheng Shao, Nonsmooth sequential analysis in Asplund spaces, *Trans. Amer. Math. Soc.* **384** (1996), 1235-1280.

[48] B. Mordukhovich and Yongheng Shao, Stability of set-valued mappings in infinite dimension: point criteria and applications, *SIAM J. Contro Optimization*, **35** (1997), 285-314.

[49] B. Mordukhovich and Yongheng Shao, Nonconvex differential calculus for infinite dimensional multifunction, *Set-Valued Analysis*, **4** (1996), 205-236.

[50] J.-P. Penot, Metric regularity, openness and Lipschitzean behavior of multifunctions, *Nonlinear Analysis. Theory, Methods, Appl.* **13** (1989), 629-643.

[51] J.-P. Penot, Personal communication.

[52] J. P. Penot, Compactness property, openness criteria and coderivatives, *Set-Valued Analysis* **6** (1998), 363-380.

[53] S.M. Robinson, Regularity and stability for convex multivalued functions, *Math. Oper. Res.* **1** (1976), 130-145.

[54] S.M. Robinson, Stability theory for systems of inequalities 2. Differentiable nonlinear systems, *SIAM J. Numer. Anal.* **132** (1976), 497-513.

[55] R.T. Rockafellar, *Convex Analysis*, Princeton Univ. Press 1970.

[56] R.T. Rockafellar, Directionally Lipschitzean functions and subdifferential calculus, *Proc. London Math. Soc.* **39** (1979), 331-355.

[57] R.T. Rockafellar, Extensions of subgradient calculus with applications to optimization, *Nonlinear Analysis. Theory, Methods, Appl* **9** (1985), 1553-1571.
[58] R.T. Rockafellar and R.B.J. Wets, *Variational Analysis*, Springer Verlag, 1998
[59] C. Ursescu, Multifunctions with closed convex graph, *Czekhoslovak Math.J.* **25** (1975), 430-441.
[60] Q.J. Zhu, The equivalence of several basic theorems for subdifferentials, *Set-Valued Analysis* **6** (1998), 171-185.
[61] A. Ioffe, Fuzzy principles and characterization of trustworthiness, *Set-valued Analysis* **6** (1998), 265-276.
[62] M. Lassonde, First order rules in non-smooth constrained optimization, to appear
[63] R. Henrion, thesis, A. Humboldt Univ., Berlin 1997.

DEPARTMENT OF MATHEMATICS, TECHNION – ISRAEL INST. OF TECHNOLOGY, HAIFA 32000, ISRAEL

E-mail address: ioffe@math.technion.ac.il (Alexander D. Ioffe)

Canadian Mathematical Society
Conference Proceedings
Volume **27**, 2000

Complete Dual Characterizations of Optimality for Convex Semidefinite Programming

Vaithilingam Jeyakumar and Michael J. Nealon

To Jonathan Borwein, on the occasion of his award of a Dhc

ABSTRACT. A convex semidefinite programming problem is a convex constrained optimization problem, where the constraints are linear matrix inequalities, for which the standard Lagrangian condition is sufficient for optimality. However, this condition requires constraint qualifications to completely characterize optimality. We present a necessary and sufficient condition for optimality without a constraint qualification. This is achieved by relaxing the standard Lagrangian condition to a new asymptotic form. A dual condition characterizing feasibility of a convex semidefinite program is also obtained. As applications of the optimality conditions we derive a version of Farkas' lemma for systems involving matrix inequalities and obtain a strong duality result for convex semidefinite programming problems. A numerical example is also given to illustrate the nature of the asymptotic optimality condition.

1. Introduction

Consider the convex semidefinite programming problem

$$(SDP) \quad \underset{x\in\mathbb{R}^m}{\text{minimize}} \quad f(x)$$

$$\text{subject to} \quad F_0 + \sum_{i=1}^{m} F_i x_i \succeq 0$$

where $f : \mathbb{R}^m \to \mathbb{R}$ is a convex function, and for $i = 0, \ldots, m$, $F_i \in S_n$, the space of symmetric $n \times n$ matrices and $\succeq$ denotes the Löwner partial order of S_n, that is, for $M, N \in S_n$, $M \succeq N$ if and only if $(M - N)$ is positive semidefinite. This problem is an extension of a linear programming problem, for which F_i's are diagonal matrices, and includes many important applications in systems and control, and combinatorial optimization (see [**2, 17, 21**]). Polynomial time interior-point algorithms are now routinely used for solving semidefinite programming problems

1991 *Mathematics Subject Classification.* Primary 65K10, 90C25, 90M45.

Key words and phrases. semidefinite convex programming, asymptotic optimality conditions, semidefinite Farkas' lemma, strong duality.

The research of the first author was partially supported by a grant from the Australian Research Council.

(see [**1, 5, 8, 17, 21**]). However, the standard Lagrangian for (SDP) requires constraint qualifications to avoid duality gaps and to completely characterize optimality. More recently much attention has been focussed on ensuring zero duality gaps and strong duality for linear semidefinite programming problems [**6, 18, 19**]. The purpose of this paper is to show that a simple asymptotic Lagrangian condition characterizing optimality for the convex semidefinite programming problems (SDP) holds without a constraint qualification.

Various characterizations of optimality without constraint qualifications have been given in the literature for the standard convex programming problems (see [**3, 4, 7, 12, 13, 15**]). Borwein and Wolkowicz [**3**] presented necessary and sufficient (non-asymptotic) conditions in terms of a modified Lagrangian function and the so-called minimal cones. This approach was used recently in [**19**] to obtain strong duality for semidefinite programming problems. Our approach is to derive an asymptotic Lagrangian condition which is both necessary and sufficient for optimality. This approach has recently been used to characterize optimality for both the standard convex minimization problems and the convex maximization problems (see [**7, 12, 13, 15**]).

We also derive a simple asymptotic dual condition characterizing feasibility of the convex semidefinite program. We present a numerical example to illustrate the case where a convex semidefinite program generally satisfies our asymptotic condition whereas the standard Lagrangian condition fails to hold. As applications we establish a version of Farkas' lemma for semidefinite matrix systems and derive a strong duality result for semidefinite programming problems.

2. A Feasibility Theorem

In this Section we establish a dual asymptotic condition characterizing feasibility of the semidefinite programming problem. We begin by fixing some definitions and preliminaries that will be used later in the paper.

We consider S_n as a vector space with the trace inner product $(M, N) := Tr[MN]$, where $Tr[.]$ refers to the trace operation. Let $S = \{M \in S_n \mid M \succeq 0\}$ be the closed convex cone of positive semidefinite $n \times n$ matrices. The feasible set D is defined by

$$D = \{x \in \mathbb{R}^m \mid F(x) \succeq 0\},$$

where $F(x) = F_0 + \sum_{i=1}^m F_i x_i$. The polar cone T^* of the set $T \subset S_n$ is given by

$$T^* = \{\theta \in S_n \mid (\theta, s) \geq 0, \forall s \in T\}.$$

Note that $S^* = S$ [**10**] and

$$M \in S \Leftrightarrow Tr[ZM] \geq 0, \forall Z \in S. \tag{2.1}$$

For $F_i \in S_n$, the linear operator $\hat{F} : \mathbb{R}^m \to S_n$ is defined by $\hat{F}(x) = \sum_{i=1}^m x_i F_i$. Then the adjoint operator $\hat{F}^* : S_n \to \mathbb{R}^m$ is given by

$$(\hat{F}^*(Z))_i = (Tr[F_i Z]).$$

Let

$$A := \bigcup_{(Z,\epsilon) \in S \times \mathbb{R}_+} \begin{pmatrix} \hat{F}^*(Z) \\ -Tr[ZF_0] - \epsilon \end{pmatrix}.$$

Clearly, A is a convex cone, but is not necessarily closed. We now see how a dual condition characterizing feasibility of the problem can be established in terms of A.

THEOREM 1 (Feasibility Theorem). *Let $F_i \in S_n$, for $i = 0, 1, \cdots, m$. Then the following statements are equivalent.*

(i) $\exists\, x \in \mathbb{R}^m, \quad F_0 + \sum_{i=1}^m F_i x_i \succeq 0$

(ii) $\begin{pmatrix} 0 \\ 1 \end{pmatrix} \notin cl\left(\bigcup_{(Z,\epsilon)\in S\times\mathbb{R}_+} \begin{pmatrix} \hat{F}^*(Z) \\ -Tr[ZF_0]-\epsilon \end{pmatrix}\right)$.

Proof.[(i)$\Longrightarrow$ (ii)]. Assume that $F(y) \succeq 0$ for some $y \in \mathbb{R}^m$. Suppose that $(0,1) \in cl(A)$. Then there exist sequences $\{(u_j, \alpha_j)\} \subset \mathbb{R}^m \times \mathbb{R}$ and $\{(Z_j, \epsilon_j)\} \subset S \times \mathbb{R}_+$ such that $u_j \to 0$, $\alpha_j \to 1$, $u_j = \hat{F}^*(Z_j)$ and $\alpha_j = -Tr[Z_j F_0] - \epsilon_j$. Hence, for each j,

$$\begin{aligned} Tr[Z_j F(y)] &= \hat{F}^*(Z_j)^T y + Tr[Z_j F_0] \\ &= u_j^T y - \alpha_j - \epsilon_j \\ &\le u_j^T y - \alpha_j. \end{aligned}$$

Since as $j \to \infty$, $u_j^T y - \alpha_j \to -1 < 0$, $\lim_{j\to\infty} Tr[Z_j F(y)] < 0$ and so there exists $Z_k \in S$ such that $Tr[Z_k F(y)] < 0$. This is a contradiction.

[(ii)$\Longrightarrow$ (i)]. Assume that (ii) holds. Since $cl(A)$ is a closed convex cone, it follows from a separation theorem that there exists (x, α) such that $(x, \alpha)^T(0,1) > 0$ and

$$u^T x + \gamma\alpha \le 0, \quad \forall (u, \gamma) \in \mathrm{cl}(A).$$

Then $\alpha > 0$ and let $\bar{x} = x/\alpha$, so $u^T\bar{x} + \gamma \le 0$, $\forall (u,\gamma) \in A$. Hence for each $Z \in S$, $\hat{F}^*(Z)^T\bar{x} - Tr[ZF_0] \le 0$. Thus, for each $Z \in S$, $Tr[Z(\sum_{i=1}^m F_i(-\bar{x})_i + F_0)] \ge 0$. Then $F(-\bar{x}) \succeq 0$ and so (i) holds. □

3. Complete Characterizations of Optimality

In this section we present a complete characterization of optimality without a constraint qualification. We state the necessary and sufficient optimality condition in terms of the convex subdifferential $\partial f(x_0)$ of f at the minimizer x_0 which is defined by

$$\partial f(x_0) := \{v \in \mathbb{R}^m \mid f(x) - f(x_0)) \ge v^T(x - x_0),\ \forall x \in \mathbb{R}^m\}.$$

THEOREM 2 (Characterization of Optimality). *Let x_0 be a feasible point of (SDP). Then x_0 is a minimizer of (SDP) if and only if there exists $v \in \partial f(x_0)$ such that*

$$\begin{pmatrix} v \\ v^T x_0 \end{pmatrix} \in cl\left(\bigcup_{(Z,\epsilon)\in S\times\mathbb{R}_+} \begin{pmatrix} \hat{F}^*(Z) \\ -Tr[ZF_0]-\epsilon \end{pmatrix}\right). \tag{3.1}$$

Proof. It follows from the standard result of convex optimization that the point x_0 is a minimizer of (SDP) if and only if there exists $v \in \partial f(x_0)$ such that

$$\forall x \in D, \quad v^T(x - x_0) \ge 0. \tag{3.2}$$

We shall now show that the condition (3.2) is equivalent to the optimality condition (3.1). Suppose that (3.1) holds. Then there exist sequences $\{Z_n\} \subset S$ and $\{\epsilon_n\} \subset \mathbb{R}_+$ such that

$$v = \lim_{n\to\infty} \hat{F}^*(Z_n) \text{ and } v^T x_0 = \lim_{n\to\infty} -Tr[Z_n F_0] - \epsilon_n.$$

Let $x \in D$. Then

$$\begin{aligned} v^T x - v^T x_0 &= \lim_{n\to\infty} \hat{F}^*(Z_n)^T x + Tr[Z_n F_0] + \epsilon_n \\ &= \lim_{n\to\infty} Tr[Z_n F] + \epsilon_n \\ &\geq 0, \end{aligned}$$

and so (3.2) holds.

Conversely, assume that (3.2) holds. Suppose, to the contrary, that

$$\begin{pmatrix} v \\ v^T x_0 \end{pmatrix} \notin \mathrm{cl}\left(\bigcup_{(Z,\epsilon)\in S\times\mathbb{R}_+} \begin{pmatrix} \hat{F}^*(Z) \\ -Tr[ZF_0] - \epsilon \end{pmatrix} \right).$$

Since x_0 is feasible to (SDP), it follows from the Feasibility Theorem 1 that $(0,1) \notin \mathrm{cl}(A)$. So, for each $\alpha \in [0,1]$, $\alpha(v, v^T x_0) + (1-\alpha)(0,1) \notin \mathrm{cl}(A)$. Otherwise, there exist $\alpha_0 \in (0,1)$ and sequences $\{Z_n\} \subset S$ and $\epsilon_n \subset \mathbb{R}_+$ such that

$$\alpha_0 v = \lim_{n\to\infty} \hat{F}^*(Z_n) \text{ and } \alpha_0 v^T x_0 + (1-\alpha_0) = \lim_{n\to\infty} Tr[Z_n F_0] - \epsilon_n.$$

So,

$$v = \lim_{n\to\infty} \hat{F}^*(\frac{Z_n}{\alpha_0}) \text{ and } v^T x_0 = \lim_{n\to\infty} Tr[\frac{Z_n}{\alpha_0} F_0] - \frac{\epsilon_n}{\alpha_0} - \frac{(1-\alpha_0)}{\alpha_0}.$$

This gives us that $(v, v^T x_0) \in \mathrm{cl}(A)$ which is a contradiction. Hence, $L \cap \mathrm{cl}(A) = \emptyset$, where $L := \{\alpha(v, v^T x_0) + (1-\alpha)(0,1) : \alpha \in [0,1]\}$ is the convex and compact line segment in $\mathbb{R}^m \times \mathbb{R}$. Then by a separation theorem **[9]** there exists $(\bar{x}, \beta) \in \mathbb{R}^m \times \mathbb{R}$ such that for each $\alpha \in [0,1]$,

$$[\alpha(v, v^T x_0) + (1-\alpha)(0,1)]^T (\bar{x}, \beta) < 0$$

and

$$u^T \bar{x} + \gamma\beta \geq 0, \;\; \forall (u,\gamma) \in \mathrm{cl}(A).$$

By letting $\alpha = 0$, we get $\beta < 0$. By letting $\alpha = 1$ we obtain the inequalities

$$v^T \bar{x} + (v^T x_0)\beta < 0$$

and

$$u^T \bar{x} + \gamma\beta \geq 0, \;\; \forall (u,\gamma) \in \mathrm{cl}(A).$$

Let $x' = \frac{-\bar{x}}{\beta}$. Then for each $(u,\gamma) \in \mathrm{cl}(A)$, $u^T x' - \gamma \geq 0$. So for each $Z \in S$,

$$Tr[Z(\sum_{i=1}^m F_i x'_i + F_0)] = \hat{F}^*(Z)^T x' + Tr[ZF_0] \geq 0.$$

This gives us $F(x') \succeq 0$. But

$$v^T \bar{x} + (v^T x_0)\beta = v^T(-\beta x') + (v^T x_0)\beta < 0.$$

So, $-v^T x' + v^T x_0 > 0$, since $\beta < 0$. This is a contradiction. □

We have here given a direct proof of Theorem 3.1 using a Separation Theorem. On the other hand, asymptotic optimality conditions for semidefinite programming problems may also be derived from a corresponding general result of an infinite dimensional convex programming problem. The interested reader is referred to [**7, 12, 15**] for asymptotic optimality conditions for convex programming problems. The following example illustrates the significance of our asymptotic optimality condition.

EXAMPLE 3. Consider the convex semidefinite programming problem

$$\begin{aligned} &\underset{x\in\mathbb{R}^2}{\text{minimize}} \quad x_1 + |x_2| \\ &\text{subject to} \quad \begin{pmatrix} 0 & x_1 & 0 \\ x_1 & x_2 & 0 \\ 0 & 0 & x_1+1 \end{pmatrix} \succeq 0. \end{aligned}$$

Then the feasible set $D = \{(x_1, x_2) \mid x_1 = 0,\ x_2 \geq 0\}$ and $x_0 = (0,0)$ is the minimizer for the problem. Note that the interior of D is empty and that

$$\lambda = \begin{pmatrix} \lambda_1 & \lambda_2 & \lambda_3 \\ \lambda_2 & \lambda_4 & \lambda_5 \\ \lambda_3 & \lambda_5 & \lambda_6 \end{pmatrix} \in S$$

satisfies $\lambda_1 \geq 0$, $\lambda_1\lambda_4 - \lambda_2^2 \geq 0$ and $\det(\lambda) \geq 0$. Then

$$\hat{F}^*(\lambda) = \begin{pmatrix} 2\lambda_2 + \lambda_6 \\ \lambda_4 \end{pmatrix},$$

and $Tr[\lambda F_0] = \lambda_6$. Now for the sequence $\{\lambda_j\} \subset S$ defined by

$$\lambda_j = \begin{pmatrix} \frac{j}{4} & \frac{1}{2} & 0 \\ \frac{1}{2} & \frac{1}{j} & 0 \\ 0 & 0 & 0 \end{pmatrix},$$

we see that $2\lambda_2 + \lambda_6 = 1$, $\lambda_4 = \frac{1}{j} \to 0$ as $j \to \infty$, and $\lambda_6 = 0$. Hence our asymptotic optimality condition is satisfied by $v = (1,0) \in \partial f(0) = \{(1,\alpha) : \alpha \in [-1,1]\}$. Thus,

$$\begin{pmatrix} v \\ v^T x_0 \end{pmatrix} = \begin{pmatrix} 1 \\ 0 \\ 0 \end{pmatrix} \in \text{cl}\left(\bigcup_{(\lambda,\epsilon)\in S\times\mathbb{R}_+} \begin{pmatrix} 2\lambda_2 + \lambda_6 \\ \lambda_4 \\ -\lambda_6 - \epsilon \end{pmatrix} \right).$$

Note however that the standard Lagrangian condition fails to hold here.

Let us now look at an application of Theorem 2 to derive an asymptotic form of the Farkas' lemma for positive semidefinite matrices. Recently, asymptotic Farkas' lemma for systems involving convex functions have been given in [**7, 11, 13, 15, 20**]. For related results for matrix inequalities see [**16**].

COROLLARY (Semidefinite Farkas Lemma). Let $F_i \in S_n$ for each $i = 1, \cdots, m$ and let $g : \mathbb{R}^m \to \mathbb{R}$ be a sublinear function. Then the following statements are equivalent

(i) $\sum_{i=1}^m F_i x_i \succeq 0 \Longrightarrow g(x) \geq 0$
(ii) $\partial g(0) \bigcap \text{cl}(\hat{F}^*(S)) \neq \emptyset$.

Proof. The implication (i) holds if and only if 0 is a minimizer of the convex semidefinite programming problem

$$\begin{aligned} &\underset{x\in\mathbb{R}^m}{\text{minimize}} \quad g(x) \\ &\text{subject to} \quad \sum_{i=1}^m F_i x_i \succeq 0. \end{aligned}$$

By Theorem 2 the attainment of minimum at 0 is equivalent to the dual condition that there exists $v \in \partial g(0)$ such that

$$v \in \mathrm{cl}\left(\bigcup_{Z \in S} \hat{F}^*(Z)\right) = \mathrm{cl}\left(\hat{F}^*(S)\right),$$

which is equivalent to (ii). □

A standard well known form of Farkas' lemma follows easily from Corollary 3.1. This result can also be deduced from the Farkas lemma for general linear systems (see [**11**]).

COROLLARY. Let $F_i \in S_n$ for each $i = 1, 2, \cdots, m$ and let $c \in \mathbb{R}^m$. Then the following statements are equivalent

(i) $\sum_{i=1}^m F_i x_i \succeq 0 \Longrightarrow c^T x \geq 0$

(ii) $c \in \mathrm{cl}(\hat{F}^*(S))$.

Proof. Let $g(x) = c^T x$. Then $\partial g(0) = \{c\}$ and so the conclusion follows from Corollary 3. □

Now consider the semidefinite linear programming problem

$$(SLP) \quad p^* := \inf_{x \in \mathbb{R}^m} \quad c^T x$$

$$\text{subject to} \quad F_0 + \sum_{i=1}^m F_i x_i \succeq 0.$$

We can easily characterize the optimal value of (SLP) using our semidefinite Farkas lemma.

COROLLARY. Let $\mu \in \mathbb{R}$. Let the feasible set D of (SLP) be non-empty. Then $\mu = p^*$ if and only if

$$\begin{pmatrix} c \\ \mu \end{pmatrix} \in \mathrm{cl}\left(\bigcup_{(Z,\epsilon) \in S \times \mathbb{R}_+} \begin{pmatrix} \hat{F}^*(Z) \\ -Tr[ZF_0] - \epsilon \end{pmatrix}\right). \tag{3.3}$$

Proof. Consider the homogenized problem of (SLP)

$$(HSLP) \quad \underset{(x,t) \in \mathbb{R}^m \times \mathbb{R}}{\text{minimize}} \quad c^T x - t\mu$$

$$\text{subject to} \quad tF_0 + \sum_{i=1}^m F_i x_i \succeq 0, \quad t \geq 0.$$

It is easy to see by the standard arguments that $\mu = p^*$ if and only if $(0, 0)$ is a minimizer of $(HSLP)$; thus, $\mu = p^*$ if and only if

$$G(x,t) \succeq_{S_n \times S_1} 0 \Longrightarrow (c, -\mu)^T (x,t) \geq 0, \tag{3.4}$$

where $\succeq_{S_n \times S_1}$ is the Löwner partial order on $S_n \times S_1$, that is, for $(M, m), (N, n) \in S_n \times S_1$, $(M, m) \succeq_{S_n \times S_1} (N, n)$ if and only if $M \succeq N$ and $m \geq n$, and $G : \mathbb{R}^m \times \mathbb{R} \to S_n \times S_1$ is given by

$$G(x,t) = (tF_0 + \sum_{i=1}^m F_i x_i, \ t)^T.$$

The adjoint $G^* : S_n \times S_1 \to \mathbb{R}^m \times \mathbb{R}$ is given by

$$G^*(Z, r) = (\hat{F}^*(Z), Tr[ZF_0] + r)^T.$$

It now follows from Corollary 3 that (3.4) holds if and only if

$$\begin{pmatrix} c \\ -\mu \end{pmatrix} \in \mathrm{cl}\left(\bigcup_{(Z,r)\in S\times\mathbb{R}_+} G^*(Z,r) \right),$$

and so, the conclusion holds. □

COROLLARY. The problem (SLP) attains its minimum at x_0 if and only if

$$\begin{pmatrix} c \\ c^T x_0 \end{pmatrix} \in \mathrm{cl}\left(\bigcup_{(Z,\epsilon)\in S\times\mathbb{R}_+} \begin{pmatrix} \hat{F}^*(Z) \\ -Tr[ZF_0]-\epsilon \end{pmatrix} \right).$$

Proof. The conclusion follows from Corollary 3 by taking $\mu = p^* = c^T x_0$. □

4. Strong Nonlinear Duality

In this Section we see how a strong duality result can be derived from our main Theorem 2.

Consider the dual problem to (SDP)

$$(4.1) \qquad \begin{array}{lll} (DSDP) & \underset{(x,Z)\in\mathbb{R}^m\times S_n}{\text{maximize}} & f(x) - Tr[Z(F_0 + \sum_{i=1}^m F_i x_i)] \\ & \text{subject to} & \hat{F}^*(Z) \in \partial f(x), \quad Z \succeq 0. \end{array}$$

Note that if $f(x) := c^T x$ is linear then the dual collapses to the following standard semidefinite linear programming dual

$$(4.2) \qquad \begin{array}{lll} (DLP) & \underset{Z\in S_n}{\text{maximize}} & -Tr[ZF_0] \\ & \text{subject to} & \hat{F}^*(Z) = c, \quad Z \succeq 0. \end{array}$$

For other forms of dual for semidefinite linear programming problems see [**6, 18, 19**]. Observe firstly the weak duality property that for each feasible x of (SDP) and (y, Z) of the dual program $(DSDP)$,

$$\begin{aligned} & f(x) - (f(y) - Tr[Z(F_0 + \sum_{i=1}^m F_i y_i)]) \\ & \geq F^*(Z)^T(x-y) + Tr[Z(F_0 + \sum_{i=1}^m F_i y_i)] \\ & = F^*(Z)^T x - Tr[Z\sum_{i=1}^m F_i y_i] + Tr[ZF_0] + Tr[Z\sum_{i=1}^m F_i y_i] \\ & = F^*(Z)^T x + Tr[ZF_0] = Tr[Z\sum_{i=1}^m F_i x_i] + Tr[ZF_0] \\ & \geq 0. \end{aligned}$$

Hence

$$f(x) \geq f(y) - Tr[Z(F_0 + \sum_{i=1}^m F_i y_i)].$$

In the following we deduce a strong duality theorem from Theorem 2 under a constraint qualification.

We say that the problem (SDP) satisfies a **constraint qualification** if the set

$$\bigcup_{(Z,\epsilon)\in S\times\mathbb{R}_+} \left(\hat{F}^*(Z),\ -Tr[ZF_0]-\epsilon\right)$$

is closed. Various conditions such as a generalized Slater condition ensure that this constraint qualification is satisfied (see [**14, 8**]).

THEOREM 4 (Strong Duality). *Assume that the problem (SDP) satisfies a constraint qualification. If x_0 is a minimizer of (SDP) then there exists $Z_0 \succeq 0$ such that (x_0, Z_0) is a maximizer of $(DSDP)$ and the optimal values are equal.*

Proof. Since

$$\bigcup_{(Z,\epsilon)\in S\times\mathbb{R}_+} \left(\hat{F}^*(Z),\ -Tr[ZF_0]-\epsilon\right)$$

is closed and x_0 is a minimizer, it follows from Theorem 2 that there exists $v \in \partial f(x_0)$ such that

$$\begin{pmatrix} v \\ v^T x_0 \end{pmatrix} \in \left(\bigcup_{(Z,\epsilon)\in S\times\mathbb{R}_+} \begin{pmatrix} \hat{F}^*(Z) \\ -Tr[ZF_0]-\epsilon \end{pmatrix} \right).$$

So, there exists $Z_0 \in S$ and $\epsilon_0 \geq 0$ such that

$$v = \hat{F}^*(Z_0) \quad \text{and} \quad v^T x_0 = -Tr[Z_0F_0] - \epsilon_0.$$

Thus, (x_0, Z_0) is feasible for $(DSDP)$. Moreover,

$$v^T x_0 = \hat{F}^*(Z_0)^T x_0 = -Tr[Z_0F_0] - \epsilon_0.$$

Hence $Tr[Z_0(F_0 + \sum_{i=1}^m F_i(x_0)_i)] \leq 0$. On the other hand, since $Z_0 \in S$ and $F_0 + \sum_{i=1}^m F_i(x_0)_i \succeq 0$, $Tr[Z_0(F_0 + \sum_{i=1}^m F_i(x_0)_i)] \geq 0$, and so, $Tr[Z_0(F_0 + \sum_{i=1}^m F_i(x_0)_i] = 0$. Now from the weak duality, for any feasible (y, Z) of the dual $(DSDP)$,

$$f(x_0) - Tr[Z(F_0 + \sum_{i=1}^m F_i(x_0)_i)] = f(x_0) \geq f(y) - Tr[Z(F_0 + \sum_{i=1}^m F_i y_i)].$$

Thus, (x_0, Z_0) is a maximizer of the $(DSDP)$ and the optimal values of both problems are equal. □

Acknowledgements. The authors are grateful to the referee for his/her comments and valuable suggestions which have contributed to the final preparation of the paper.

References

[1] Alizadeh, F., 'Interior point methods in semidefinite programming with applications to combinatorial optimization', *SIAM J. Optim.,* (1995),5, 13-51.

[2] Ben-Tal, A. and Nemirovskii, A., 'Robust truss topology design via semidefinite programming', *SIAM J. Optim.* (1997), 7(4), 991–1016.

[3] Borwein, J. M. and Wolkowicz, H., 'Characterizations of optimality without constraint qualification for the abstract convex program', *Math. Progr. Stud.*, (1982), 19, 77-100.

[4] Borwein, J. M. and Wolkowicz, H., 'Characterizations of optimality for the abstract convex program with finite dimensional range', *J. Austr. Math. Soc. Ser. A*, (1981), 30, 390-411.

[5] de Klerk, E., Roos, C. and Terlaky, T., 'On primal-dual path-following algorithms for semidefinite programming, New trends in mathematical programming', *Appl. Optim.*, Kluwer Acad. Publ.,(1998). 13, 137-157.

[6] de Klerk, E., Roos, C. and Terlaky, T., 'Initialization in semidefinite programming via self-dual skew-symmetric embedding', *Oper. Res. Lett.*, (1998), 20, 213-221.
[7] Glover, B.M., Jeyakumar, V. and Rubinov, A. M., 'Dual conditions characterizing optimality for convex multi-objective programs', *Math. Progr.*, (1999), 84, 201-217.
[8] Helmberg, C., Rendel, F., Vanderbei, R and Wolkowicz, H., 'An interior-point method for semidefinite programming', *SIAM J. Optim.*, (1996), 6, 342-361.
[9] Holmes, R. B., 'Geometric Functional Analysis', Springer-Verlag, 1975.
[10] Horn, R. and Johnson, C., 'Matrix Analysis', Cambridge University Press, England, 1985.
[11] Jeyakumar, V., 'Farkas' lemma and extensions', to appear in *Encyclopedia of Optimization*, Kluwer Academic Publishers.
[12] Jeyakumar, V., 'Asymptotic dual conditions characterizing optimality for infinite convex programs', *J. Optim. Theor. Appl.*, (1997), 93, 153-165.
[13] Jeyakumar, V. and Glover, B.M., 'Nonlinear extensions of Farkas' Lemma with applications to global optimization and least squares', *Math. Oper. Res.*, (1995), 20, 818-837.
[14] Jeyakumar, V. and Wolkowicz, H., 'Generalizations of Slater's constraint qualifications for infinite convex programs', *Math. Progr. Ser. B*, (1995), 57, 85-101.
[15] Jeyakumar, V. and Zaffaroni, A., 'Asymptotic Conditions for Weak and Proper Optimality in Infinite Dimensional Convex Vector Optimization', *Numer. Funct. Anal. and Optim.*, (1996) 17, 323-343.
[16] Lasserre, J.B., 'A New Farkas Lemma for Positive Semidefinite Matrices', *IEEE Trans. Auto. Cont.*, (1995), 40, 1131-1133.
[17] Nesterov, Y. and Nemirovskii, A., 'Interior-point methods in convex programming', *Studies in Applied Mathematics*, 13, SIAM Publications, 1994.
[18] Ramana, M. V., 'An exact duality theory for semidefinite programming and its complexity implications', *Math. Progr.*, (1997), 77, 129-162.
[19] Ramana, M. V., Tunçel, L. and Wolkowicz, H., 'Strong Duality for semidefinite programming', *SIAM J. Optim.*, (1997), 7, 644-662.
[20] Rubinov, A.M., Glover, B.M. and Jeyakumar, V., 'A general approach to dual characterizations of solvability of inequality systems with applications', *J. Convex Anal.*, (1995), 2(2), 309-344.
[21] Vandenberghe, L. and Boyd, S., 'Semidefinite Programming', *SIAM Rev.*, (1996), 38, 49-95.

School of Mathematics, University of New South Wales, Sydney 2052, Australia
E-mail address: `jeya@maths.unsw.edu.au (V. Jeyakumar)`

School of Mathematics, University of New South Wales, Sydney 2052, Australia
Current address: Operations Research Group Pty Ltd, North Sydney, NSW 2060, Australia
E-mail address: `Michael.Nealon@orgroup.com.au (M. Nealon)`

Canadian Mathematical Society
Conference Proceedings
Volume **27**, 2000

A Second-Welfare Theorem in Nonconvex Economies

Alejandro Jofré

Dédié à Jonathan Borwein

ABSTRACT. In this paper we prove an extension of the Second Welfare Theorem for nonconvex economies with an infinite number of goods, that is, when the preference set-valued maps or the production sets can be nonconvex. Firstly, it is proved a "viscous" version of this result without any compactness or epi-lipschitzianity assumption, secondly it is deduced the exact version introducing the compact epi-lipschitziany condition on a set. The key ingredient to prove this extension is a recent generalization of the Hahn-Banach Theorem proved recently on Banach spaces.

1. Introduction.

Many results in convex economic theory are proved by using a separation theorem or an extension of linear functional theorem. Both results are consequence of the Hahn-Banach Theorem which is one of the main results in Functional Analysis. For nonconvex sets the first results in this vein were proved by Mordukhovich [**24**], and Cornet and Rockafellar [**10**] in terms of normal cones in Asplud and finite dimensional spaces respectively. These results were extended by using the subdifferential of the distance function in finite and infinite Banach spaces in Jofré and Rivera [**16, 17**], and Borwein and Jofré [**5**].

Economics applications of these results were also given by J. Rivera and the author, namely, an extension of the Second Welfare Theorem in General Equilibrium Theory in $I\!R^n$. For infinite dimensional spaces in a nonconvex setting there are also some generalizations of the Welfare Theorem proved by using some tools of Variational Analysis such as Kuhn-Tucker conditions and intersection properties. These results prove the existence of a price in the Clarke or the approximate normal cone (see for example [**3, 2, 13, 20, 21, 22, 31**]).

In this abstract setting, an economy $\mathcal{E}$ with a finite numbers m and n of consumers indexed by i and producers indexed by j is defined as:

1991 *Mathematics Subject Classification.* Primary 47A15; Secondary 46A32, 47D20.
Key words and phrases. subgradients, nonconvex separation, General Equilibrium, Pareto Optimum, Welfare Theorem.
This work was partially supported by FONDAP-Matemáticas Aplicadas.

$$\mathcal{E} = ((X_i), (P_i), (Y_j), w),$$

where X_i is subset of a Banach space X corresponding to the consumption set of the consumer i, P_i is a set-valued map defined as $P_i : \prod_k X_k \mapsto X_i$, such that $P_i(x)$ represents those consumptions in X_i strictly preferred to x, Y_j is a subset of X determining the feasible productions associated to the producer j, and w is the total initial endowment in the economy to be exchanged. The second Welfare Theorem gives for each Pareto optimum point (social welfare) a decentralizing price, which means that for this price each consumer and producer has a maximizing behaviour on his utility function and profit respectively.

For infinite dimensional spaces X, that is, when the economic model contains an infinite number of goods the Banach spaces generally used are for example L^∞, l^∞ , L^2, $C[0,T]$, $C[0,\infty]$ or more general measure spaces. Thus, we work often on spaces with nonsmooth norms. This is one of the reasons to introduce in the first section an axiomatic notion of subdifferential on Banach spaces including limiting Fréchet, variational β-bornology or viscosity, approximate and other subdifferentials. The second section is devoted to recall an extension of the Hahn-Banach Theorem proved recently and some corollaries. In the third section it is proved the main result of this paper, that is, the Second Welfare Theorem for nonconvex nontransitive economies under very new weak assumptions or a "viscous" version of this result, which gives a decentralizing price belonging to the subdifferential of the distance function which is always bounded. In the last two sections we establish the "exact" version of the Welfare Theorem, that is, when we assume that one of the sets involved is compactly epi-lipschitzian. We finish this paper with an economical interpretation of these results in some particular Banach spaces.

2. A nonconvex separation property.

In what follows X will be a Banach space, X^* its topological dual, and B and B^* the respectively unit closed ball. We also denote $B(x,\epsilon)$ the closed ball with center at x and radius ϵ. In order to establish the main result we introduce an abstract subdifferential ∂, we remark that only properties on lipschitzian functions are required :

DEFINITION 1. By a subdifferential we mean any operator ∂ which possesses the following properties: For any $f, g : X \to I\!R$ locally Lipschitz functions and any point $x \in X$,

i) $\partial f(x) \subset X^*$.
ii) ∂f is the convex subdifferential whenever f is a convex function;
iii) $\partial f(x) = \partial g(x)$ if f and g coincide near x;
iv) if f is Lipschitz near a point x with constant l then $\|x^*\| \le l$ for all $x^* \in \partial f(x)$;
v) if x is a local minimum of $f+g$ then $0 \in \partial(f+g)(x) \subset \partial f(x) + \partial g(x)$;
vi) Given closed subsets Z_1, Z_2 in X, the function $h : Z_1 \times Z_2 \to I\!R$ defined by $h(z_1,z_2) := f(z_1) + g(z_2)$ satisfies $\partial h(z_1,z_2) = \partial f(z_1) \times \partial g(z_2)$.
vii) If $A : Y \to X$ is a bounded linear operator onto a Banach spaces Y, $\lambda > 0$ and $b \in X$ then $\partial(\lambda f(Ay+b)) \subset \lambda A^* \partial f(Ay+b)$

For a subset Z of X, we denote the boundary of Z by $bd(Z)$ and recall that the *distance function* to Z denoted d_Z is the function:

$$d_Z(x) := \inf_{z \in Z} \|x - z\|.$$

By using this function we can introduce the *normal cone* to a set Z at a point z as:

$$N_Z(z) := cl I\!R \ \ \partial d_Z(z) \tag{2.0.1}$$

We recall now the nonconvex separation property on Banach spaces proved by Borwein and Jofré [**5**]. This result will be a key ingredient to extend the Second Welfare Theorem for nonconvex sets on infinite dimensional spaces.

THEOREM 1. Let $Z_1, ..., Z_n$ be closed subsets of X. Suppose $(\overline{z}_1, \cdots, \overline{z}_n) \in Z_1 \times ... \times Z_n$ satisfies

$$\sum_{i=1}^{n} \overline{z}_i \in \text{bd}(\sum_{i=1}^{n} Z_i).$$

Then there exist $c > 0$ such that for any $\varepsilon > 0$ there exist $z_i \in Z_i \cap B(\overline{z}_i, \varepsilon)$ and $p^* \in X^*, p^* \neq 0$ such that

$$p^* \in \bigcap_{i=1}^{n} (\partial d_{Z_i}(z_i) + \varepsilon B^*) \quad \text{and } 0 < c \leq \|p^*\| \leq 1.$$

REMARK 1. ¿From the definition of normal cone we also have:

$$p^* \in \bigcap_{i=1}^{n} (N_{Z_i}(z_i) + \varepsilon B^*)$$

These viscous separation results become exact when we add a compactness-lipschitzianity condition on a set involved. In this sense, an important number of papers have been addressed to this question such as Borwein and Strowjas [**7**], Jourani and Thibault [**18, 19**], Loewen [**23**], Mordukhovich and Shao [**26, 27**], Penot [**30**] and Ioffe [**14**]. Among them a recent nice result have been proved by Ioffe [**14**] which shows how to transform weak*-converging sequences in strong ones when they belong to the normal cone. A. Ioffe shows that the compactly epi-lipschitzian early introduced by J. Borwein is exactly that condition. We say that a set Z is compactly epi-lipschitzian at z, if there exists a compact set $K \subset X, \epsilon > 0, \alpha > 0$, such that

$$Z \bigcap (z + \epsilon B) + tB \subset Z + tK, \quad \forall t \in [0, \alpha].$$

In $I\!R^n$ any closed subset is compactly epi-lipschitzian. We recall now Ioffe's result [**14**] as a lemma in which ∂ *denote any subdifferential included or equal to the approximate subdifferential.*

LEMMA 1. Let Z be a closed subset of X and $z \in Z$. The following properties are equivalent:

i) Z is compactly epi-lipschitzian at z;
ii) for any bounded net (z_α^*) such that $z_\alpha^* \in N_Z(z_\alpha)$ with $z_\alpha \in Z$ *and* $z_\alpha \to z$, we have that $z_\alpha^* \xrightarrow{w^*} 0$ implies $z_\alpha^* \xrightarrow{\|\|} 0$.

Proof. See Ioffe [**14**]. □

As a consequence of this result we obtain the following theorem which gives a *general "exact" separation property* in any Banach space.

THEOREM 2. Let $Z_1, \cdots, Z_n$ be closed subsets of X, and suppose $(\bar{z}_1, \cdots, \bar{z}_n) \in Z_1 \times \cdots \times Z_n$ is such that $\sum_{i=1}^{n} \bar{z}_i \in bd(\sum_{i=1}^{n} Z_i)$. If for some $i_0 \in \{1, \cdots, n\}$, Z_{i_0} is compactly epi-lipschitzian at $\bar{z}_{i_0}$, then there exist $p^* \neq 0$ such that

$$p^* \in \bigcap_{i=1}^{n} \partial d_{Z_i}(\bar{z}_i).$$

This result has some interesting consequences in several subjects in optimization and Variational Analysis. In what follows, we show two examples.

COROLLARY 1. Let Z be a closed subset of X, and suppose Z is compactly epi-lipschitzian at $z \in bd(Z)$ then, there exist $p^* \neq 0$ such that

$$p^* \in \partial d_Z(z) \subset N_Z(z).$$

In the case of convex sets we can establish the following *extension of the Hahn-Banach Theorem.*

COROLLARY 2. Let Z_1, Z_2 be closed convex subsets of X, and suppose $(z_1, z_2) \in Z_1 \times Z_2$ is such that $z_1 + z_2 \in \mathrm{bd}(Z_1 + Z_2)$. If Z_1 *or* Z_2 is compactly epi-Lipschitzian at z_1 *or* z_2 respectively then, there exists $p^* \neq 0$ such that

$$p^* \in \partial d_{Z_1}(z_1) \bigcap \partial d_{Z_2}(z_2).$$

In the convex case we will use in the sections below the recent characterization of the compactly epi-lipschitzian property proved by Borwein, Lucet and Mordukhovich [**6**] to obtain more explicit results.

PROPOSITION 1. We assume that Z is a closed convex subset of X then, this set is compactly epi-lipschitzian if and only if one of the following equivalent conditions is satisfied:

(i) There is convex compact set K such that $0 \in int(Z + K)$.
(ii) There is a finite dimensional space L such that $int(Z + L) \neq \emptyset$.
(iii) $L := span\ Z$ is a finite-codimensional closed subspace and the interior of Z in L is nonempty (relative interior).
(iv) There is a point $z \in Z$ such that the polar set $(Z - z)^0$ is w^*-locally compact.

Proof. See [**6**]. □

3. The Second Welfare Theorem in a general setting.

We start this section introducing some basic ingredients in the economic model. We denote by $m,\ n \in I\!N \backslash \{0\}$ the number of consumers and producers respectively, the *consumption set* for individual ith will be denoted as $X_i \subseteq X$ whereas $Y_j \subseteq X$ will represent the *production set* of the jth firm and $w \in X$ the total initial endowment for the economy. We assume subsets X_i and Y_j are all closed. The preferences of a consumer $i \in \{1, ..., m\}$ will be represented by a set-valued map:

$$P_i : \prod_k X_k \mapsto X_i,$$

Thus, for each bundle $x_i \in X_i$, $P_{i_0}(x_1, ..., x_m)$ $(clP_{i_0}(x_1, ..., x_m))$ represents those elements in X_{i_0} that are strictly preferred (resp. preferred) to $x_{i_0} \in X_{i_0}$. This notions generalize the preorder and utility concepts usually used on X_i. For example, if there is a preorder relation $\preceq_i$ on X_i or an utility function $\mu_i : X_i \to I\!R$, if $x_i^* \in X_i$ then the preference set-valued map P_i would be $P_i(x_1^*, ..., x_i^*, ..., x_m^*) := \{x_i \in X_i \mid x_i^* \prec_i x_i\}$ and $P_i := \{x_i \mid \mu_i(x_i^*) < \mu_i(x_i)\}$ respectively. Finally, in what follows we use the notation P_i^* instead of $P_i(x_1^*, ..., x_i^*, ..., x_m^*)$.

Now, we recall the notion of *nonconvex nontransitive economy.*

DEFINITION 2. A nonconvex nontransitive economy is defined as

$$\mathcal{E} = ((X_i), (P_i), (Y_j), w).$$

The extension of the Second Welfate Theorem requires the following hypothesis over the economy $\mathcal{E}$ which is weaker than the interiority assumptions used in a serie of works (Arrow and Debreu (1964), Arrow and Hahn (1971), Bonnisseau (1988), Brown (1991), Debreu (1954), Cornet (1986), Debreu (1959), Guesnerie (1975), etc). This hypothesis was already used in finite dimensional economies in Jofré and Rivera [**16, 17**] and it involves a constraint on the aggregated production and preference sets:

DEFINITION 3. The economy $\mathcal{E} = ((X_i), (P_i), (Y_j), w)$ satisfies the *Asymptotic Included Condition* (A.I.C.) at a point $((x_i^*), (y_j^*)) \in \prod X_i \times \prod Y_j$ if there exist $i_0 \in \{1, ..., m\}$, $\epsilon > 0$ and a sequence $h_k \to 0$ such that for $k \in I\!N$ (sufficiently large) we have that,

$$h_k + \sum_i clP_i^* \cap B(x_i^*, \epsilon) - \sum_j Y_j \cap B(y_i^*, \epsilon) \subseteq P_{i_0}^* + \sum_{i \neq i_0} clP_i^* - \sum_j Y_j,$$

We have mentioned in [**16, 17**] that a condition assuring (A.I.C.) in $I\!R^n$ is the epi-lipschitzianity of some P_i^* with $i \in \{1, ..., m\}$. At this point we recall that a subset $Z \subseteq X$ is epi-lipschitzian at a point $z \in Z$ if there exists $d \in X \setminus \{0\}$ and a couple of open neighborhoods N_z and N_d of z and d respectively, and $\lambda > 0$ such that

$$Z \cap N_z + [0, \lambda) N_d \subseteq Z.$$

It is easy to see that in infinite dimensional spaces the epi-lipschitzianity also become a sufficient conditions for A.I.C.. Other conditions implying A.I.C. are the following:

(a) There exists a sequence $h_k \to 0$ such that for some $i \in \{1, ..., m\}$ and k sufficiently large, $h_k + clP_i^* \subseteq P_i^*$.

(d) The free disposal hypothesis $Y_j - X_+ \subseteq Y_j$ for some j, when the Banach space X has $intX_+ \neq \emptyset$, which is stronger than a).

In order to establish the main result of this paper, we introduce the *Pareto optimum* notion for the economy $\mathcal{E}$. We begin with the definition of feasible allocation.

DEFINITION 4. We will say that $((x_i^*), (y_j^*)) \in \prod X_i \times \prod Y_j$ is a *feasible allocation* for the economy $\mathcal{E}$ if the following conditions are satisfied:

(a) for each $i = 1, ..., m$ and $j = 1, ..., n$, $x_i^* \in X_i$ and $y_j^* \in Y_j$.

(b) $\sum_i x_i^* - \sum_j y_j^* = w$.

DEFINITION 5. A feasible allocation $((x_i^*), (y_j^*)) \in \prod X_i \times \prod Y_j$ is a *Pareto optimum* for the economy $\mathcal{E}$ if there is no other feasible allocation $((x_i), (y_j)) \in \prod X_i \times \prod Y_j$ such that:

(a) for each $i \in \{1, ..., m\}$, $x_i \in clP_i(x_1^*, ..., x_m^*)$

(b) for some $i_0 \in \{1, ..., m\}$, $x_{i_0} \in P_{i_0}(x_1^*, ..., x_m^*, e^*)$.

We now establish the main theorem of this paper which is a viscous version of the Second Welfare Theorem.

THEOREM 3. Let $\mathcal{E} = ((X_i), (P_i), (Y_j), w)$ be an economy which satisfies the A.I.C at a Pareto optimum point $((x_i^*), (y_j^*)) \in \prod X_i \times \prod Y_j$. If for each $i \in \{1, ..., m\}$, $x_i^* \in clP_i^*$, then there exists $c > 0$ such that for any $\epsilon > 0$ there exist a price $p^* \in X^*$ with $\|p^*\| \geq c$, and $x_i \in clP_i^* \cap B(x_i^*, \epsilon)$, $y_j \in Y_j \cap B(y_i^*, \epsilon)$ for each i, j such that:

$$p^* \in \bigcap_j (\partial d_{Y_j}(y_j) + \epsilon B^*) \qquad - p^* \in \bigcap_i (\partial d_{clP_i^*}(x_i) + \epsilon B^*).$$

Proof. We will prove first that there exists $\epsilon_0 > 0$ such that

$$w \in bd\ (\sum_i clP_i^* \cap B(x_i^*, \epsilon_0) - \sum_j Y_j \cap B(y_i^*, \epsilon_0)),$$

where $w = \sum_i x_i^* + \sum_j y_j^*$. Indeed, otherwise for each $\epsilon > 0$, $w \in int\ (\sum_i clP_i^* \cap B(x_i^*, \epsilon) - \sum_j Y_j \cap B(y_i^*, \epsilon))$ because w belongs to this set, which allows us to deduce that for any sequence $h_k \to 0$,

$$w + h_k \in \sum_i clP_i^* \cap B(x_i^*, \epsilon) - \sum_j Y_j \cap B(y_i^*, \epsilon)$$

for k sufficiently large. Thus, from A.I.C we conclude that

$$w \in P_{i_0}^* + \sum_{i \neq i_0} clP_i^* - \sum_j Y_j,$$

which contradicts the Pareto optimum definition.

So that, $w \in bd\ (\sum_i clP_i^* \cap B(x_i^*, \epsilon_0) - \sum_j Y_j \cap B(y_i^*, \epsilon_0))$, for some ϵ_0. Thus, we can apply Theorem 1 to the sets $Z_i = clP_i^* \cap B(x_i^*, \epsilon)$ and $Z_j = -Y_j \cap B(y_j^*, \epsilon)$ for all $i = 1..m$ and $j = m+1, ..m+n$ respectively. Theorem 3 implies that for a given ϵ (enough small or less than ϵ_0) there exists $p \neq 0$, and $x_i \in clP_i^* \cap B(x_i^*, \epsilon)$, $y_j \in -Y_j \cap B(y_i^*, \epsilon)$ for each i, j such that

$$p^* \in \bigcap_j (\partial d_{-Y_j}(-y_j) + \epsilon B^*) \qquad - p^* \in \bigcap_i (\partial d_{clP_i^*}(x_i) + \epsilon B^*).$$

However, from the definition of subdifferential we know that $\partial d_{-Y_j}(-y_j^*) = -\partial d_{Y_j}(y_j^*)$ which allows us to conclude the proof. □

We point out that no interiority assumptions were made either over the consumption or the production sets. This result is new as well as the following corollaries including the convex and compactly epi-lispchitzian case. In the convex case we obtain an $\epsilon - |\cdot|$-maximizing profit behaviour for the producer over each production set and the $\epsilon - |\cdot|$-minimizing cost behaviour for the consumer over each preferred consumption set.

COROLLARY 3. Let $\mathcal{E} = ((X_i), (P_i), (Y_j), w)$ be an economy with that satisfies A.I.C at a Pareto optimum point $((x_i^*), (y_j^*)) \in \prod X_i \times \prod Y_j$. If for each $i \in \{1, ..., m\}$, $x_i^* \in clP_i^*$ and the sets clP_i^*, Y_j are convex for all i, j, then there exist $c > 0$ such that for any $\epsilon > 0$ there exist a price $p^* \in X^*$, $\|p^*\| \geq c$, and $x_i \in clP_i^* \cap B(x_i^*, \epsilon)$, $y_j \in Y_j \cap B(y_i^*, \epsilon)$ such that for each i, j:

$$< p^*, y_j' - y_j > \leq \epsilon \|y_j' - y_j\| \quad \text{for all} \quad y_j' \in Y_j$$

$$< p^*, x_i' - x_i > \geq -\epsilon \|x_i' - x_i\| \quad \text{for all} \quad x_i' \in clP_i^*$$

Proof. It is a consequence of the fact that for convex sets, the distance function is convex and then its subdifferential coincides with the convex analysis subdifferential. □

The points satisfying the last inequalities may be called *quasi-equilibrium.*

4. Exact version of the Second Welfare Theorem.

In this section we establish some consequences of the main theorem when we introduce a weak "compactness-epi-lispchitzianity" over one of the consumption or production sets. All these results are deduced by using Theorem 2 and their corollaries, that is, they correspond to the case ϵ going to zero in the above results. We recall at this point the Theorem 2 works where ∂ is any subdifferential included or equal to the approximate subdifferential. For this purpose, the lower bound of $\|p^*\|$ given by the main theorem become crucial.

PROPOSITION 2. Let $\mathcal{E} = ((X_i), (P_i), (Y_j), w)$ be an economy which satisfies the A.I.C at a Pareto optimum point $((x_i^*), (y_j^*)) \in \prod X_i \times \prod Y_j$. If for each $i \in \{1, ..., m\}$, $x_i^* \in clP_i^*$, and $clP_{i_0}^*$ or Y_{j_0} is compactly epi-lispchitz at $x_{i_0}^*$ or $y_{j_0}^*$ for some i_0 or j_0 respectively then, there exits a price $p^* \in X^*$ with $p^* \neq 0$ such that,

$$p^* \in \bigcap_j \partial d_{Y_j}(y_j^*) \qquad -p^* \in \bigcap_i \partial d_{clP_i^*}(x_i^*).$$

The following proposition show that the epi-lipschitzianity introduced by R.T. Rockafellar (see [**4**] for example) of one consumption or production set is enough to obtain an exact version of the Second Welfare Theorem.

PROPOSITION 3. Let $\mathcal{E} = ((X_i), (P_i), (Y_j), w)$ be an economy and $((x_i^*), (y_j^*)) \in \prod clP_i^* \times \prod Y_j$ a Pareto optimum point. If $clP_{i_0}^*$ is epi-lispchitzian at $x_{i_0}^*$ for some i_0 then, there exits a price $p^* \in X^*$ with $p^* \neq 0$ such that,

$$p^* \in \bigcap_j \partial d_{Y_j}(y_j^*) \qquad -p^* \in \bigcap_i \partial d_{clP_i^*}(x_i^*).$$

Proof. It follows readily from the fact that epi-lipschitzianity implies both the compactly epi-lipschitzian property and A.I.C. □

Now we will establish the main result for the convex case. We say that an economy $\mathcal{E} = ((X_i), (P_i), (Y_j), w)$ is *convex* if X_i, Y_j are convex for all i, j and P_i is convex-valued for all i.

PROPOSITION 4. Let $\mathcal{E}$ be a convex economy and $((x_i^*),(y_j^*)) \in \prod clP_i^* \times \prod Y_j$ a Pareto optimum point. If $clP_{i_0}^*$ or Y_{j_0} is compactly epi-lispchitz at $x_{i_0}^*$ or $y_{i_0}^*$ for some i_0 or j_0 respectively (or equivalently if it satisfies one the condition (i)-(iv) of Proposition 1) then, there exits a price $p^* \in X^*$ with $p^* \neq 0$ such that for all i, j:

$$x_i^* \in argmin\{< p^*, x_i > \ : x_i \in clP_i^*\}$$
$$y_j^* \in argmax\{< p^*, y_j > \ : y_j \in Y_j\}$$

Proof. First, we remark that for the convex sets clP_i^* and Y_j the compact epi-lipschitzianity characterized in Proposition 1 (iii) implies the (A.I.C.). Thus, we can apply Corollary 3 and the fact that when $clP_{i_0}^*$ and Y_j are convex we have a minimizing and maximizing behaviour for x_i^* and y_j^* respectively. □

REMARK 2. This result is an extension of the well-known Second-Welfare Theorem for the Arrow-Debreu convex model in infinite-dimensional spaces such as L^p with $1 \leq p \leq \infty$.

5. Limiting subgradient set and its economical interpretation

In the finite dimensional case (see Jofré and Rivera [**17**]), that is, when $X = I\!R^n$ it is possible to give an economical interpretation of the relations,

$$p^* \in \bigcap_j \partial d_{Y_j}(y_j^*) \qquad\qquad -p^* \in \bigcap_i \partial d_{clP_i^*}(x_i^*),$$

even for nonconvex economies by using the *Proximal normal formula* which gives the following representation of the subgradient to the distance function to for example the production subset Y (deleting the subscript j):

$$\partial d_Y(y^*) = \{\lim_k p_k^* \mid p_k^* \in PN_Y(y_k^*),\ y_k^* \to_Y y^*\},$$

where $PN_Y(y_k^*)$ is the set of *proximal normal vectors* to Y at y_k^* which is defined as follows: $p_k^* \in PN_Y(y_k^*)$ if and only if for some $t_k > 0$ the inequality

$$< p_k^*, y - y_k^* > \leq \frac{1}{2t_k}\|y - y_k^*\|^2$$

holds true for all $y \in Y$. This formula implies that $p^* \in \partial d_Y(y^*)$ if and only if there exist sequences $p_k^* \to p^*$ and $y_k^* \to y^*$, where $y_k^* \in Y$ is the solution of the following optimization problem,

$$\max_{y \in Y}[< p_k^*, y > - \frac{1}{2t_k}\|y - y_k^*\|^2].$$

An analogous minimizing behaviour can be proved for the consumers. However in the infinite dimensional case the Normal Proximal formula is only true in Hilbert Spaces. For this reason, instead of this formula we will use the *limiting subgradient representation* of a subdifferential or also the definition of a subdifferential itself in terms of limiting of more basic subgradients.

For example, when X is an Asplud space Mordukhovich and Shao [**25, 27, 26**] have developed a rich theory for limiting Fréchet subdifferential and normal cone which satisfy the conditions of Definition 1 and other nice properties. In this case, the conclusions of Proposition 1 can be written for each producer and consumer (removing the index i, j) as follows: there exist sequences $p_k \to p^*$ and $y_k \to y^*$, $x_k \to x^*$ where $y_k \in Y$ and $x_k \in clP^*$ such that,

$$< p_k, y - y_k > \leq \epsilon \|y - y_k\| \quad \text{for all} \quad y \in Y$$

$$< p_k, x - x_k > \geq -\epsilon \|x - x_k\| \quad \text{for all} \quad x \in clP^*.$$

So, in the general nonconvex setting it is possible to obtain for the consumer and producer an ϵ-quasi-equilibrium or $\epsilon - \| \cdot \|$ maximizing and minimizing behaviour analogously to the convex case showed in Proposition 3 for a sequence approaching a Pareto optimum point. The same formulae can be established for variational β-subdifferential for each β-bornology on Banach spaces with norm β-smooth. Thus, from a Pareto optimum point it is possible to deduce for nonconvex nontransitive economies a decentralizing price p^* producing on each consumer and producer ϵ-optimizing behaviours.

REMARK 3. After this paper was finished we received a copy of a paper by B. Mordukhovich on " The extremal principle and its applications to optimization and economics " [**28**] developing also some applications to the Welfare's Theorem in Asplund spaces by using the extremal principle and some ideas in [**16, 17**]. Some of these results are also extended to Banach spaces in our present work.

Acknowledgement. The author would like to thank Prof. M. Florenzano for helpful remarks improving the presentation of this paper.

References

[1] K.J. Arrow, An extension of the basic theorem of classical welfare economics. Proceedings of the Second Berkeley Symposium, University of California Press (1951).
[2] J.M. Bonnisseau, Habilitation à diriger des recherches. Thèorie de l'Equilibre Général et ensembles de production non-convexes, U. Paris 1 (1990).
[3] J.M. Bonnisseau, B. Cornet, Valuation equilibrium and Pareto optimum in nonconvex economies. *Journal of Mathematical Economics.* Vol. 17 (1988), 293-308.
[4] J. Borwein, Epi-Lipschitz like sets in Banach spaces: theorems and examples, *Nonlinear Analysis. Theory, Methods, Appl.* 11 (1987), 1207-1217.
[5] J. Borwein and A. Jofré, A nonconvex separation property in Banach spaces. *J. of Operational Research and Applied Mathematics,* 48 (1997), 169-180.
[6] J. Borwein, Y. Lucet and B. Mordukhovich, Compactly epi-lipschitzian convex sets and functions in normed spaces. To appear in *J. Convex Analysis.*
[7] J. Borwein and H.Strojwas, Tangential approximations, *Nonlinear Analysis TMA* **9** (1985), 1347-1366.
[8] D. J. Brown, Equilibrium analysis with nonconvex technologies, Handbook of mathematical economics, Vol IV, chapter 36. W. Hildenbrand and H. Sonnenschein. eds. , North-Holland, Amsterdam (1991).
[9] B. Cornet, The second welfare theorem in nonconvex economies. Preprint. CERMSEM Université Paris 1 (1986).
[10] B. Cornet, R.T. Rockafellar, Separation theorems and supporting price theorems for nonconvex sets. Preprint. Verbal comunication (1989).
[11] G. Debreu, Valuation equilibrium and Pareto optimum. *Proceedings of the National Academy of Sciences, U.S.A.*,40 (1954), 588-542.
[12] G. Debreu, Theory of value, an axiomatic analysis of economic equilibrium , John Wiley, New York (1959).
[13] R. Guesnerie, Pareto optimality in nonconvex economics. *Econometrica.* Vol. 43 (1975), 1-29.
[14] A. D. Ioffe, Codirectional compactness, metric regularity and subdifferential calculus. Submitted (1997).
[15] A.D. Ioffe, Approximate subdifferentials and applications. I: The finite dimensional theory. *Transactions of the Amer. Math. Soc.*, 281 (1984), 389-416.
[16] A. Jofré, J. Rivera, A nonconvex separation property and applications. To appear in *Math. Progr.*.

[17] A. Jofré, J. Rivera, The second Welfare Theorem for a nonconvex nontransitive economy. Submitted to *J. of Math. Econ.*(1998).

[18] A. Jourani, Compactly epi-lipschitzian sets and A-subdifferential in WT-spaces. *Optimization* , 34 (1995), 1-17.

[19] A. Jourani and L. Thibault, Coderivatives of multivalued mappings, locally compact cones and metric regularity, preprint (1995).

[20] Ali M. Kahn, R. Vohra, An extension of the second welfare theorem to economics with nonconvexities and public goods. *Quarterly Journal of Economics*, 102 (1987), 223-241.

[21] Ali M. Kahn, R. Vohra, Pareto optimal allocations of nonconvex economics in locally convex spaces, *Nonlinear Analysis*, 12 (1988), 934-950.

[22] Ali M. Kahn, Mordukhovich's normal cone and the foundations of welfare economics. To appear(1999).

[23] Ph. Loewen, Limits of Fréchet normals in Nonsmooth Analysis, Optimization and Nonlinear Analysis, A. Ioffe, M. Marcus and S. Reich (editors), Longman, Essex (1992), 178-188.

[24] B. Mordukhovich, Metric approximations and necessary optimality conditions for general classes of non-smooth extremal problems.*Dokl. Akad. Nauk. SSR*, 254 (1980), 1072-1076.

[25] B. Mordukhovich, Generalized differential calculus for nonsmooth and set-valued mapping,*J. Math. Anal. Appl.*, 183 (1994),250-288.

[26] B. Mordukhovich and Yongheng Shao, Nonsmooth sequential analysis in Asplund spaces, *Trans. Amer. Math. Soc.*, 384 (1996), 1235-1280.

[27] B. Mordukhovich and Yongheng Shao, Nonconvex differential calculus for infinite dimensional multifunction,*Set-Valued Analysis*,**4** (1996), 205-236.

[28] B. Mordukhovich, The extremal principle and its applications to optimization and economics. Submitted to "Progress in Optimization: Contribution from Australasia (A. M. Rubinov, ed.), Kluwer. Research Report No. 8 (1999), Dept. of Math., Wayne State University.

[29] V. Pareto, Manual d'économie politique. Lousanne, Rouge (1909).

[30] J. P. Penot, Compactness property, openness criteria and coderivatives, preprint 1995.

[31] M. Quinzii, Rendements croissants et equilibre général. Thesis, Université de Paris II (1986).

[32] R.T. Rockafellar and R.J.B. Wets, *Variational Analysis*, Springer Verlag, (1997).

DEPARTAMENTO DE INGENIERÍA MATEMÁTICA, UNIVERSIDAD DE CHILE, CASILLA 170, CORREO 3, SANTIAGO, CHILE

E-mail address: `ajofre@dim.uchile.cl`

Canadian Mathematical Society
Conference Proceedings
Volume **27**, 2000

Auxiliary problem principle and the approximation of variational inequalities with non-symmetric multi-valued operators

A. Kaplan and R. Tichatschke

Dédié à Jonathan Borwein
à l'occasion de la remise de son titre de Docteur Honoris Causa

ABSTRACT. An extension of the auxiliary problem principle to variational inequalities with non-symmetric multi-valued operators in Hilbert spaces is studied. This extension supposes that the operator of the variational inequality is splitted into the sum of a maximal monotone operator $\mathcal{Q}$ and a single-valued operator $\mathcal{F}$, and $\mathcal{F}$ is linked with a sequence $\{\mathcal{L}^k\}$ of non-symmetric components of auxiliary operators by a kind of pseudo DUNN property.
The current auxiliary problem is constructed by fixing $\mathcal{F}$ at the previous iterate, whereas $\mathcal{Q}$ (or its single-valued approximation $\mathcal{Q}^k$) is considered at a variable point. Using auxiliary operators of the form $\mathcal{L}^k + \chi_k \nabla h$, with $\chi_k > 0$, the standard for the auxiliary problem principle assumption of the strong convexity of the function h is weakened by exploiting mutual properties of $\mathcal{Q}$ and h.
Convergence of the general scheme is analyzed allowing that the auxiliary problems are solved approximately, and some applications are sketched briefly.

1. Introduction

The *auxiliary problem principle* (APP), originally introduced by COHEN [**3**], [**4**] as a general framework to analyze optimization algorithms of gradient and subgradient types as well as decomposition algorithms, was extended later to different numerical methods for solving variational inequalities. In the majority of the papers dedicated to such extensions, the $(k+1)$-th auxiliary problem is constructed by applying the operator of the variational inequality (here called main operator) to the k-th iterate. The idea to take an additive component of this operator at a variable point leads to a scheme which appears to be a generalization of proximal-like methods, too.

1991 *Mathematics Subject Classification.* Primary 90C48, 90C25, 90C26; Secondary 49J40, 65K05.

This research is supported by The Rheinland/Pfalz Foundation for Innovation, grant no. 8312-386261/307.

In order to illustrate this, let us consider the problem

$$(1.1) \qquad find \quad x \in X: \quad 0 \in \Psi(x),$$

with Ψ a given multi-valued maximal monotone operator in a Hilbert space X. The APP is taken in the form

$$(1.2) \qquad find \quad x^{k+1} \in X: \quad 0 \in \frac{1}{\epsilon_k}\left[\Omega(x^{k+1}) - \Omega(x^k)\right] + \Psi(x^k),$$

with $\Omega : X \to X$ an auxiliary operator, $\epsilon_k > 0$, and this iterative scheme can be modified as follows:

$$(1.3) \qquad find \quad x^{k+1} \in X: \quad 0 \in \frac{1}{\epsilon_k}\left[\Omega(x^{k+1}) - \Omega(x^k)\right] + \Psi(x^{k+1}),$$

or

$$(1.4) \qquad find \quad x^{k+1} \in X: \quad 0 \in \frac{1}{\epsilon}\left[\Omega(x^{k+1}) - \Omega(x^k)\right] + \Psi_1(x^k) + \Psi_2(x^{k+1}),$$

where $\epsilon > 0$ and Ψ is decomposed into a sum of a single-valued monotone operator Ψ_1 and a maximal monotone operator Ψ_2.
Then, in case $\Omega = \mathcal{I}$ ($\mathcal{I}$ - identity operator), the following well-known methods arise:

– inclusion (1.2) leads to

$$x^{k+1} \in x^k - \epsilon_k \Psi(x^k),$$

which is an analogon of the subgradient method;

– on using (1.3) we obtain

$$x^{k+1} = (\mathcal{I} + \epsilon_k \Psi)^{-1}(x^k),$$

the proximal point method;

– (1.4) produces

$$(1.5) \qquad x^{k+1} = (\mathcal{I} + \epsilon\Psi_2)^{-1}(1 - \epsilon\Psi_1)(x^k),$$

the splitting algorithm, suggested by LIONS/MERCIER [**17**] (see also GABAY [**9**]) and PASSTY [**23**].

Obviously, the latter algorithm can be represented as

$$z^k = x^k - \epsilon\Psi_1(x^k), \qquad x^{k+1} = (\mathcal{I} + \epsilon\Psi_2)^{-1}(z^k),$$

where x^{k+1} is calculated from z^k by means of the proximal mapping. TSENG [**30**], [**31**] has used method (1.5) (with a variable ϵ) as a basic process to investigate convergence of several known but also new splitting methods for variational inequalities with separability properties as well as for related convex optimization problems with linear constraints and for linear complementarity problems.

In the present paper the APP is studied for variational inequalities of the type

$$(P) \quad find \quad x^* \in K: \quad \langle \mathcal{F}(x^*) + \mathcal{Q}(x^*), x - x^* \rangle \geq 0 \quad \forall x \in K,$$

with K a convex closed subset of a Hilbert space $(X, \|\cdot\|)$, $\mathcal{F}$ a single-valued operator from X into the dual space X', $D(\mathcal{F}) \supset K$, and $\mathcal{Q} : X \to 2^{X'}$ a maximal monotone (in general, non-symmetric and multi-valued) operator; $\langle \cdot, \cdot \rangle$ denotes the duality pairing between X and X'. In the case of monotone operators, the notion "symmetric" means that the operator is the subdifferential of a convex functional. Auxiliary problems of the form

$$
\begin{aligned}
(P^k) \qquad & find\ \bar{x}^{k+1} \in K: \\
& \langle \mathcal{F}(x^k) + \mathcal{Q}^k(\bar{x}^{k+1}) + \mathcal{L}^k(\bar{x}^{k+1}) - \mathcal{L}^k(x^k) \\
& + \chi_k(\nabla h(\bar{x}^{k+1}) - \nabla h(x^k)), x - \bar{x}^{k+1}\rangle \geq 0 \quad \forall x \in K,
\end{aligned}
$$

are considered, with $h : K \to I\!R$ a convex Gáteaux-differentiable functional, $\mathcal{Q}^k : X \to X'$ a monotone operator approximating $\mathcal{Q}$, $\mathcal{L}^k : X \to X'$ a monotone operator, $D(\mathcal{L}^k) \supset K$, x^k an approximate solution of the previous problem and χ_k a positive scalar. In this case, $\mathcal{L}^k + \chi_k \nabla h$ corresponds to the customary notion of an auxiliary operator.
To make the APP numerically tractable we investigate the iteration process

$$
\begin{aligned}
(P^k_\epsilon) \qquad & find\ x^{k+1} \in K: \\
& \langle \mathcal{F}(x^k) + \mathcal{Q}^k(x^{k+1}) + \mathcal{L}^k(x^{k+1}) - \mathcal{L}^k(x^k) \\
& + \chi_k(\nabla h(x^{k+1}) - \nabla h(x^k)), x - x^{k+1}\rangle \geq -\epsilon_k \|x - x^{k+1}\| \quad \forall x \in K,
\end{aligned}
$$

where $\{\epsilon_k\}$ is a given non-negative sequence, $\lim_{k\to\infty} \epsilon_k = 0$ and $x^1 \in K$ is an arbitrary starting point. We call this process the *proximal auxiliary problem method* (PAP-method). It is easy to see, that replacing $\mathcal{Q}^k$ by $\mathcal{F} + \mathcal{Q}^k$, $\mathcal{F}$ by $\mathbf{0}$ and taking $\mathcal{L}^k \equiv \mathbf{0}$, proximal-like methods can be derived from (P^k_ϵ).
For the exact conditions which are supposed to be valid for the both problems (P) and (P^k) see the Assumptions 1 and 2 below. These conditions allow the solution set of (P) to be infinite-dimensional and/or unbounded, i.e. we deal with essentially ill-posed problems.

Applying for the sources of the PAP-method, we begin with a version of the APP for convex optimization problems, where a linearization of the Gáteaux-differentiable term J_1 of an objective functional $J = J_1 + J_2$ is used. This was already studied in [**3**]. Taking this partial linearization, the term $\langle \nabla J_1(x^k), \cdot - x^k\rangle + J_2(\cdot)$ is inserted into the objective functional of the $(k+1)$-th auxiliary problem. Such an approach is of special interest for constructing decomposition methods. In fact, if the problem $\min\{J_2(x) :\ x \in K\}$ splits up into independent subproblems, then the mentioned linearization permits to provide the same splitting in the framework of the APP for the original problem $\min\{J(x) :\ x \in K\}$, i.e. the corresponding auxiliary problems can be splitted up, too.
The general scheme oriented to decomposition methods for variational inequalities of the form

$$
(1.6) \qquad find\quad x^* \in X: \quad \langle \Psi(x^*), x - x^*\rangle + f(x) - f(x^*) \geq 0, \quad \forall x \in X,
$$

where Ψ is a single-valued monotone operator and the functional f is convex, lower semicontinuous (lsc) and additive w.r.t. a Cartesian factorization of the space X, has been developed in the paper of MAKLER-SCHEIMBERG / NGUYEN / STRODIOT [**18**]. Here the APP is combined with the approximation of the functional f on the basis of the concept of the MOSCO-convergence. An extension of this scheme (without accentuating decomposition methods), described by SALMON / NGUYEN / STRODIOT [**28**], is connected with a relaxation of the monotonicity condition for Ψ and with the use of a wider class of auxiliary operators (as distinct from [**18**], these operators may be non-symmetric). The auxiliary problems in [**28**] can be written as

$$\begin{aligned} &find\ \ x^{k+1} \in K: \\ &\quad \langle \Psi(x^k) + \mathcal{L}^k(x^{k+1}) - \mathcal{L}^k(x^k) + \chi_k(\nabla h^k(x^{k+1}) - \nabla h^k(x^k)), x - x^{k+1}\rangle \\ &\quad + f^k(x) - f^k(x^{k+1}) \geq 0 \ \ \forall x \in K, \end{aligned}$$

and the conditions concerning $\mathcal{L}^k$, h^k and f^k made there permit to cover a lot of earlier versions of the APP and special algorithms.

Our PAP-method may be considered as a generalization of the method studied by ZHU/MARCOTTE ([**32**], Sect. 4.2) in the case that $X = I\!R^n$ and $\mathcal{F} + \mathcal{Q}$ is a continuous operator: the auxiliary problems in [**32**] correspond to (P^k) with $\mathcal{Q}^k = \mathcal{Q}$, $\mathcal{L}^k = \mathcal{L}$, $\chi_k = \chi$, under stronger assumptions w.r.t. h and $\mathcal{F} - \mathcal{L}$.
The paper of RENAUD/COHEN [**25**] should also be mentioned, where for Problem (1.1) the APP is studied in the form

$$(1.7) \qquad find\ \ x^{k+1} \in X: \quad 0 \in \frac{1}{\epsilon}\left[\Xi(x^{k+1}) - \Xi(x^k)\right] + \Psi(x^{k+1}),$$

with a single-valued (in general, non-symmetric) auxiliary operator $\Xi = \Xi_1 + \epsilon\Xi_2$. Here, Ξ_1 is supposed to be the gradient of a strongly convex functional, and observing the following relation between Ψ and Ξ_2:

$$\begin{aligned} \exists\ \gamma_0 > 0: \quad &\langle \psi(x) - \psi(y), x - y\rangle \geq \gamma_0\|\Xi_2(x) - \Xi_2(y)\|^2, \\ &\forall \psi(x) \in \Psi(x),\ \psi(y) \in \Psi(y),\ \ \forall x, y \in dom\Psi, \end{aligned}$$

the operator Ξ_2 may be hemicontinuous only. Note that the auxiliary problem (1.4) is equivalent to (1.7) setting $\Xi = \Omega - \epsilon\Psi_1$.
In [**25**] the general convergence results for method (1.7) have been also adapted to prove convergence of a new algorithm for solving saddle-point problems with convex-concave functions on the product of convex sets. Depending on the decomposition of the related maximal monotone operator, the ARROW - HURWICZ algorithm and the proximal point method can be obtained as particular cases.

The PAP-method studied in the present paper has the following peculiarities:

- as distinct from [**28**], [**18**], the operator $\mathcal{Q}$ is supposed to be not necessarily the subdifferential of a convex functional (note that the variational inequality (1.6) corresponds to Problem (P) with $\mathcal{Q} = \partial f$);
- as distinct from [**25**], the main operator $\mathcal{F} + \mathcal{Q}$ is not necessarily monotone, an approximation of $\mathcal{Q}$ is included, and the auxiliary operator may vary after each step.

Besides, we weaken the standard (for the APP) assumption on strong convexity of the auxiliary function h in the Problems (P^k): h is supposed to be convex and the operators $\mathcal{Q}^k + \nabla h$ have to be strongly monotone with a common modulus for all k. This permits to extend the APP-scheme to proximal-like methods with weak regularization and regularization on a subspace [**11**], as well as to methods using Bregman functions (see references in [**7**]), if the chosen Bregman function provides the mentioned property for $\mathcal{Q}^k + \nabla h$.
Note that the conditions joining the main and auxiliary operators are not completely comparable in the schemes suggested here and in [**18**], [**25**], [**28**].

The paper is organized as follows: In Section 2 we start with the full description of the problem under consideration and discuss some assumptions. The conditions w.r.t. the successive approximation of the problem and the convergence analysis of the PAP-method are described in Section 3, and Section 4 contains some applications.

2. Proximal Auxiliary Problem method

We consider the variational inequality (P) under the following basic assumptions.

ASSUMPTION 1. (i) *$K \subset X$ is a non-empty, convex closed set;*

(ii) *$\mathcal{Q} : X \to 2^{X'}$ is a maximal monotone operator, $D(\mathcal{Q}) \cap K$ is a convex set, and*

$$\mathcal{Q}_K : y \to \begin{cases} \mathcal{Q}(y) & \text{if } y \in K \\ \emptyset & \text{otherwise} \end{cases}$$

is locally hemi-bounded at each point of $D(\mathcal{Q}) \cap K$;

(iii) *the operator $\mathcal{Q} + \mathcal{N}_K$ is maximal monotone, where*

$$\mathcal{N}_K : y \to \begin{cases} \{z \in X' : \langle z, y - x \rangle \geq 0 \ \forall x \in K\} & \text{if } y \in K \\ \emptyset & \text{otherwise} \end{cases}$$

is the normality operator for K;

(iv) *$\mathcal{F} : X \to X'$ is a single-valued compact on K operator, or $\mathcal{F}$ is a single-valued, monotone, hemicontinuous and bounded on X operator;*

(v) *given a family $\{\mathcal{L}_y\}, \mathcal{L}_y : X \to X'$, of monotone on K operators parameterized by $y \in K$; if*

$$\langle \mathcal{F}(x) + q(x), y - x \rangle \geq 0$$

holds for some $x \in D(\mathcal{Q}) \cap K$, $y \in K$ and some $q(x) \in \mathcal{Q}(x)$, then

$$\langle \mathcal{F}(y) - \mathcal{L}_y(y) + \mathcal{L}_y(x) + q(x), y - x \rangle \geq \gamma \|\mathcal{F}(y) - \mathcal{L}_y(y) - \mathcal{F}(x) + \mathcal{L}_y(x)\|^2_{X'}$$

is valid, with $\gamma > 0$ independent of x, y;

(vi) *Problem (P) is solvable.*

Referring in the sequel to the separate conditions described in the Assumptions 1 and 2 (below), we write (1-*i*), (1-*ii*),... and (2-*i*), (2-*ii*),..., respectively.

Let us discuss some mentioned notions and conditions.

- By definition, an element $x^* \in K$ is a solution of Problem (P) if the inequality
$$\langle \mathcal{F}(x^*) + q^*(x^*), x - x^* \rangle \geq 0 \ \ \forall x \in K$$
is valid for some $q^*(x^*) \in \mathcal{Q}(x^*)$.
In the sequel, the notation $q^*(\cdot)$ taken to diverse elements of the solution set X^* has the same meaning.
- Local hemi-boundedness of an operator $\mathcal{M}$ at a point $x^0 \in D(\mathcal{M})$ means: for each $x \in D(\mathcal{M})$, $x \neq x^0$, there exists a number $t_0(x^0, x) > 0$ such that $x^0 + t(x - x^0) \in D(\mathcal{M})$ holds for $0 \leq t \leq t_0(x^0, x)$ and the set
$$\bigcup_{0 < t \leq t_0(x^0, x)} \mathcal{M}(x^0 + t(x - x^0)) \text{ is bounded in } X'.$$

Here we use a weakened notion of local hemi-boundedness: the standard notion supposes boundedness of $\cup_{0\le t\le t_0(x^0,x)}\mathcal{M}(x^0+t(x-x^0))$. The simple example $\mathcal{M}=\mathcal{N}_C$, where $C=\{x\in I\!R^2:\ x_1^2+x_2^2\le 1\}$, shows that this relaxation may be very essential.

- With $\mathcal{Q}$ a maximal monotone operator and K a convex closed set, the operator $\mathcal{Q}+\mathcal{N}_K$ is maximal monotone if, for instance, $intD(\mathcal{Q})\cap K\neq\emptyset$ or $\mathcal{Q}$ is locally bounded[1] at some $x\in K\cap clD(\mathcal{Q})$ (see [**26**]).
- Assumption (1-v) is not so strange as it seems at the first glance. For instance, (1-v) is certainly fulfilled if the operators $\mathcal{F}-\mathcal{L}_y$ possess the DUNN property (are co-coercive) with a common modulus $\gamma>0$:

$$\begin{aligned}&\langle\mathcal{F}(x)-\mathcal{L}_y(x)-\mathcal{F}(v)+\mathcal{L}_y(v),x-v\rangle\\&\quad\ge\gamma\|\mathcal{F}(x)-\mathcal{L}_y(x)-\mathcal{F}(v)+\mathcal{L}_y(v)\|^2_{X'}\ \forall x,v\in K,\ \forall y\in K,\end{aligned}$$

which is a rather standard hypothesis for the APP.
On the other hand, the simple example

$$X=I\!R^1,\ \mathcal{F}:x\to x^2,\ \mathcal{Q}:x\to x+4,\ K=[-1,1]\ \text{ and } \mathcal{L}_y\equiv\mathbf{0}$$

illustrates the situation that (1-v) is fulfilled, although the operator $\mathcal{F}-\mathcal{L}_y$ does not possess the DUNN property and even not the following weaker pseudo DUNN property:

$$\exists\gamma_1>0:\quad\langle\mathcal{F}(v)-\mathcal{L}_y(v),v-x\rangle\ge\gamma_1\|\mathcal{F}(v)-\mathcal{L}_y(v)-\mathcal{F}(x)+\mathcal{L}_y(x)\|^2_{X'}$$

holds, whenever $x,\ v\in K$ and

$$\langle\mathcal{F}(x)-\mathcal{L}_y(x),v-x\rangle\ge 0.$$

It is worth to be noted that the operator $\mathcal{F}+\mathcal{Q}$ is not monotone.

Now, the method suggested reads as follows:

Proximal Auxiliary Problem method (PAP-method)

Starting with $x^1\in K$, the sequence $\{x^k\}$ is defined by solving successively the auxiliary problems (P^k_ϵ), $k=1,2,...$, where $\mathcal{L}^k=\mathcal{L}_y|_{y=x^k}$.

3. Convergence analysis

We study the convergence of the PAP-method using Assumption 1 and the following conditions w.r.t. the data of the auxiliary problems (P^k).

ASSUMPTION 2. (i) *The operators $\mathcal{L}_y$, $y\in K$, are monotone and Lipschitz continuous on K, with a common Lipschitz constant $l_{\mathcal{L}}$;*

(ii) *h is a convex functional on K and the mapping ∇h is Lipschitz continuous on K, with a Lipschitz constant l_h;*

(iii) *for each k, $\mathcal{Q}^k$ is a single-valued operator, $K\cap D(\mathcal{Q}^k)\neq\emptyset$ and*

$$\langle\mathcal{Q}^k(x)-\mathcal{Q}^k(y),x-y\rangle\ge\langle\mathcal{B}(x-y),x-y\rangle\quad\forall x,y\in K\cap D(\mathcal{Q}^k),$$

where $\mathcal{B}:X\to X'$ is a given linear continuous and monotone operator with the symmetry property $\langle\mathcal{B}x,y\rangle=\langle\mathcal{B}y,x\rangle$;

[1] Local boundedness of $\mathcal{Q}$ at x means that $\mathcal{Q}$ carries some neighborhood of x into a bounded set.

(iv) *with given constants* $\tilde{\chi} > 0,\ m > 0$, *the inequality*

$$\frac{1}{2}\tilde{\chi}\langle \mathcal{B}(x-y), x-y\rangle + h(x) - h(y) - \langle \nabla h(y), x-y\rangle \geq m\|x-y\|^2$$

is valid for all $x, y \in K$;

(v) *the regularization parameters satisfy* $0 < \underline{\chi} \leq \chi_k \leq \chi_{k+1} \leq \bar{\chi} < \infty \quad \forall k$;

(vi) *for all* k *and* $y \in K$, *the operators* $\mathcal{Q}^k + \mathcal{L}_y + \mathcal{N}_K + \chi_k \nabla h$ *are maximal monotone;*

(vii) *for each* $w \in D(\mathcal{Q}) \cap K$, *there exists a sequence* $\{w^k\}$, $w^k \in D(\mathcal{Q}^k) \cap K$, *such that*

$$\lim_{k\to\infty} \|w^k - w\| = 0, \qquad \lim_{k\to\infty} \|\mathcal{Q}^k(w^k) - q(w)\|_{X'} = 0,$$

with $q(w) \in \mathcal{Q}(w)$ *(in general,* $q(w)$ *depends on* $\{w^k\}$*);*

(viii) *for some solution* x^* *of Problem* (P) *there exist a constant* $\alpha > 1$ *and a sequence* $\{w^k\}$, $w^k \in D(\mathcal{Q}^k) \cap K$, *such that*

$$\lim_{k\to\infty} k^\alpha \|w^k - x^*\| = 0, \qquad \lim_{k\to\infty} k^\alpha \|\mathcal{Q}^k(w^k) - q^*(x^*)\|_{X'} = 0.$$

Under (1-*i*), (2-*i*), (2-*ii*), Assumption (2-*vi*) is valid if each operator $\mathcal{Q}^k$ is maximal monotone and locally bounded at some $x \in clD(\mathcal{Q}^k) \cap K$, or (regarding also (2-*iii*)) if $D(\mathcal{Q}^k) \supset K$ and $\mathcal{Q}^k$ is hemicontinuous on K (this follows from the Theorems 1 and 3 in [**26**]). The other conditions ensuring (2-*vi*) can be derived from the results about the maximality of sums of two monotone operators in [**26**] and [**1**].

The assumptions (2-*vii*), (2-*viii*), concerning the successive approximation of the operator $\mathcal{Q}$, are closely related to MOSCO's conditions [**19**] for the approximation of variational inequalities by using the BROWDER-TICHONOV regularization.

If $\chi_k \equiv \chi \in [\underline{\chi}, \bar{\chi}]$ and $\mathcal{G}$ is a symmetric monotone operator such that $\mathcal{L}_y - \chi\mathcal{G}$ (taken as $\mathcal{L}_y$) satisfies (2-*i*), we have a chance to weaken the Assumptions (1-*v*), (2-*iv*) considering $\mathcal{L}_y - \chi\mathcal{G}$ as $\mathcal{L}_y$ and $h + g$ as h, where $\mathcal{G} = \nabla g$.

In [**12**] and subsection 4.3 below it is shown how the function h can be chosen for special variational inequalities in mathematical physics in order to fulfill (2-*iii*) and (2-*iv*).

Now, we consider an example explaining the choice of operators Q^k in accordance with the conditions (2-*vii*) and (2-*viii*). A similar technique can be used to approximate a "friction" operator in some problems of elasticity or a "dissipation" operator in the Bingham problem, etc.

EXAMPLE 1. *Let* $V = I\!R^2, \mathcal{F} \equiv \mathbf{0},\ \mathcal{Q} : y \to \binom{z(y_1)+y_2}{-y_1}$, *where*

$$z(y_1) = \begin{cases} -1 & if \quad y_1 < -1, \\ [-1, 0] & if \quad y_1 = -1, \\ 0 & if \quad y_1 \in (-1, 1), \\ [0, 1] & if \quad y_1 = 1, \\ 1 & if \quad y_1 > 1, \end{cases}$$

and we consider two cases:

$$K = K_1 = \{y \in I\!R^2 : \ |y_1| \leq 2\}$$

and

$$K = K_2 = \{y \in I\!R^2 : y_1 \geq 1,\ y_2 \leq 0\}.$$

The operator $\mathcal{Q}$ is maximal monotone, because

$$\mathcal{Q}(y) = \{\partial_{y_1} f(y_1, y_2), \partial_{y_2}(-f(y_1, y_2))\},$$

where

$$f(y_1, y_2) = \max\{0, -y_1 - 1\} + \max\{0, y_1 - 1\} + y_1 y_2$$

is a convex-concave function. But, the operator $\mathcal{Q}$ is neither strictly monotone (take $y^1 = (-\frac{1}{2}, 1)$, $y^2 = (\frac{1}{2}, 1)$), nor symmetric.
The operators $\mathcal{Q}^k$ are chosen as follows:

$$\mathcal{Q}^k : y \to \begin{pmatrix} z^k(y_1) + y_2 \\ -y_1 \end{pmatrix},$$

with

$$z^k(y_1) = \frac{1}{2}\left[\frac{1 + y_1}{\sqrt{(1+y_1)^2 + k^{-6}}} + \frac{y_1 - 1}{\sqrt{(y_1 - 1)^2 + k^{-6}}}\right].$$

Obviously,

$$\mathcal{Q}^k(y) = \{\nabla_{y_1} f^k(y_1, y_2), -\nabla_{y_2} f^k(y_1, y_2)\},$$

in accordance with the convex-concave function

$$f^k(y) = \frac{1}{2}\left(\sqrt{(1+y_1)^2 + k^{-6}} + \sqrt{(y_1 - 1)^2 + k^{-6}}\right) + y_1 y_2.$$

Hence, $\mathcal{Q}^k$ is maximal monotone, too.
Taking $w = (w_1, w_2)$, $w^k = (w_1^k, w_2)$, any $q(w) \in \mathcal{Q}(w)$ has the form

$$q(w) = \begin{pmatrix} \xi(w_1) + w_2 \\ -w_1 \end{pmatrix},$$

where $\xi(w_1) \in z(w_1)$, thus

$$\|\mathcal{Q}^k(w^k) - q(w)\|^2 = \left(z^k(w_1) - \xi(w_1)\right)^2 + \left(w_1^k - w_1\right)^2.$$

It is not difficult to verify that the following choice of w_1^k ensures

$$\lim_{k\to\infty} k^{\frac{3}{2}} \|\mathcal{Q}^k(w^k) - q(w)\| = 0$$

with a given w and arbitrary $q(w) \in \mathcal{Q}(w)$

- *if $w_1 \neq \pm 1$, then $w_1^k = w_1$;*
- *if $w_1 = -1$ and $\xi(w_1) = -1$ or $w_1 = 1$ and $\xi(w_1) = 0$, then $w_1^k = w_1 - k^{-2}$;*
- *if $w_1 = 1$ and $\xi(w_1) = 1$ or $w_1 = -1$ and $\xi(w_1) = 0$, then $w_1^k = w_1 + k^{-2}$;*
- *if $w_1 = -1$ and $\xi(w_1) \in (-1, 0)$, then $w_1^k = -1 + ck^{-3}$ with* $c = \frac{1+2\xi(w_1)}{\sqrt{1-(1+2\xi(w_1))^2}}$;
- *if $w_1 = 1$ and $\xi(w_1) \in (0, 1)$, then $w_1^k = 1 - ck^{-3}$ with $c = \frac{1-2\xi(w_1)}{\sqrt{1-(1-2\xi(w_1))^2}}$.*

In case $K = K_1$ or $K = K_2$, $w \in \{y \in K : y_1 > 1\}$, then the choice of w^k described above satisfies $w^k \in K$, at least for sufficiently large k.
But in case $K = K_2$ and $w \in \{y \in K : y_1 = 1\}$, the rules given above not always guarantee that $w^k \in K$, even for large k. Nevertheless, $w^k \in K$ holds if $\xi(w_1) = 1$, i.e. if $q(w) \equiv \tilde{q}(w) = \binom{1+w_2}{-1}$. It is sufficient to guarantee (2-vii), (2-viii), because the unique solution of Problem (P) (with $\mathcal{F}, \mathcal{Q}$ and $K = K_2$ given in this example) is $x^* = (1, 0)$ and $q^*(x^*) = \binom{1}{-1}$, i.e. $q^*(x^*) = \tilde{q}(w)|_{w=x^*}$.

We start the convergence analysis with some preliminary results. The symbol $\rightharpoonup$ denotes weak convergence.

LEMMA 1. *Let $C \subset X$ be a convex closed set, the operators $\mathcal{A}_0 : X \to 2^{X'}$, $\mathcal{A}_0 + \mathcal{N}_C$ be maximal monotone and $D(\mathcal{A}_0) \cap C$ be a convex set. Moreover, assume that the operator*

$$\mathcal{A}_C : v \to \begin{cases} \mathcal{A}_0(v) & \text{if } v \in C \\ \emptyset & \text{otherwise} \end{cases}$$

is locally hemi-bounded at each point $v \in D(\mathcal{A}_0) \cap C$ and that, for some $u \in D(\mathcal{A}_0) \cap C$ and each $v \in D(\mathcal{A}_0) \cap C$, there exists $\eta(v) \in \mathcal{A}_0(v)$ satisfying

$$\langle \eta(v), v - u \rangle \geq 0. \tag{3.1}$$

Then, with some $\eta \in \mathcal{A}_0(u)$, the inequality

$$\langle \eta, v - u \rangle \geq 0 \tag{3.2}$$

holds for all $v \in C$.

PROOF. In view of the maximal monotonicity of $\mathcal{A}_0$ and $\mathcal{A}_0 + \mathcal{N}_C$, the operators $\mathcal{A} : v \to \mathcal{A}_0(v) + \mathcal{I}(v-u)$ and $\mathcal{A}_1 = \mathcal{A} + \mathcal{N}_C$ (with $\mathcal{I} : X \to X'$ a canonical injection) are also maximal monotone. Moreover, they are strongly monotone. Therefore, there exists $w \in D(\mathcal{A}_0) \cap C$ such that $0 \in \mathcal{A}(w) + \mathcal{N}_C(w)$, and due to the definition of the normality operator, this yields

$$\langle \eta(w), v - w \rangle \geq 0 \quad \forall v \in C, \tag{3.3}$$

with some $\eta(w) \in \mathcal{A}(w)$.
If $w = u$, then, of course, $\eta(w) \in \mathcal{A}_0(w)$, hence, the conclusion of the lemma is valid. Otherwise, we use the relation

$$\langle \bar{\eta}(v), v - u \rangle \geq 0 \quad \forall v \in D(\mathcal{A}_0) \cap C, \tag{3.4}$$

which follows from (3.1) taking $\bar{\eta}(v) = \eta(v) + \mathcal{I}(v-u) \in \mathcal{A}(v)$.
Let $w_\lambda = u + \lambda(w - u)$ for $\lambda \in (0, 1]$. Obviously, $w_\lambda \in D(\mathcal{A}_0) \cap C$, and according to (3.4) there exists $\bar{\eta}(w_\lambda) \in \mathcal{A}(w_\lambda)$ ensuring

$$\langle \bar{\eta}(w_\lambda), w - u \rangle \geq 0.$$

Because the operator $\mathcal{A}_C$ is locally hemi-bounded at u, the set $\{\bar{\eta}(w_\lambda) : \lambda \in (0, \lambda_0]\}$ is bounded in V' for a sufficiently small $\lambda_0 > 0$. Hence, if λ tends to 0 in an appropriate manner, the corresponding sequence $\{\bar{\eta}(w_\lambda)\}$ converges weakly in V' to some $\bar{\eta}$. Taking into account that $\lim_{\lambda \to 0} \|w_\lambda - u\| = 0$ and that $\mathcal{A}$ is maximal monotone, one can conclude that $\bar{\eta} \in \mathcal{A}(u)$ and

$$0 \leq \lim \langle \bar{\eta}(w_\lambda), w - u \rangle = \langle \bar{\eta}, w - u \rangle.$$

Combining this inequality and inequality (3.3) given with $v = u$, we obtain

$$\langle \bar{\eta} - \eta(w), u - w \rangle \leq 0,$$

but that contradicts the strong monotonicity of $\mathcal{A}$. □

REMARK 1. *The Assumptions (2-ii) - (2-v) provide strong monotonicity of the operator $\mathcal{Q}^k + \chi_k \nabla h$ on $K \cap D(\mathcal{Q}^k)$, and together with (1-i), (2-i), this yields strong monotonicity of $\mathcal{Q}^k + \mathcal{L}^k + \mathcal{N}_K + \chi_k \nabla h$. Moreover, according to (2-vi), the operator $\mathcal{Q}^k + \mathcal{L}^k + \mathcal{N}_K + \chi_k \nabla h$ is maximal monotone. Hence, for each k, Problem (P^k) is uniquely solvable.*

With $x^* \in X^*$, $q^*(x^*)$ as in (2-*viii*), define

$$
\begin{aligned}
\Gamma^k(x^*, x) = {} & \tilde{\chi}\langle \mathcal{B}(x - x^*), x - x^* \rangle \\
& + h(x^*) - h(x) - \langle \nabla h(x), x^* - x \rangle \\
& + \frac{1}{\chi_k} \langle \mathcal{F}(x^*) + q^*(x^*), x - x^* \rangle.
\end{aligned} \tag{3.5}
$$

For $x \in K$, under (2-*iv*), (2-*v*) it holds

$$
\Gamma^k(x^*, x) \geq m\|x^* - x\|^2 \quad \text{and} \quad \Gamma^{k+1}(x^*, x) \leq \Gamma^k(x^*, x). \tag{3.6}
$$

The sequence $\{\Gamma^k\}$ plays the role of a Ljapunov function in the further analysis.

LEMMA 2. *Let the Assumptions (1-i), (1-v), (1-vi) and (2-i) - (2-vi), (2-viii) be fulfilled. Moreover, let* $\sum_{k=1}^{\infty} \epsilon_k < \infty$ *and*

$$
\frac{1}{4\gamma m} < \underline{\chi}, \quad 2\tilde{\chi}\bar{\chi} < 1. \tag{3.7}
$$

Then the sequence $\{x^k\}$*, generated by the PAP-method, is bounded,* $\lim_{k\to\infty} \|x^{k+1} - x^k\| = 0$ *and the sequence* $\{\Gamma^k(x^*, x^k)\}$ *converges.*

This statement is proved in the Appendix by modifying the proof of Theorem 2.1 in [**28**].

LEMMA 3. *Let the Assumptions (1-i) - (1-iv), (1-vi) and (2-i) - (2-iii), (2-vii) be fulfilled, and* $0 < \chi_k \leq \bar{\chi}$ *holds for all* k.
Moreover, let the sequence $\{x^k\}$ *generated by the PAP-method be bounded, and* $\lim_{k\to\infty} \|x^{k+1} - x^k\| = 0$.
Then each weak limit point of $\{x^k\}$ *is a solution of Problem* (P).

PROOF. Let $\bar{x}$ be an arbitrary weak limit point of $\{x^k\}$ and let $\{x^k\}_{k\in\mathfrak{K}}$ converge weakly to $\bar{x}$. Since $\lim_{k\to\infty} \|x^{k+1} - x^k\| = 0$, one gets $x^{k+1} \rightharpoonup \bar{x}$ if $k \in \mathfrak{K}$, $k \to \infty$.
According to (2-*vii*), for each $y \in D(\mathcal{Q}) \cap K$ one can choose a sequence $\{y^k\}$, $y^k \in D(\mathcal{Q}^k) \cap K$ such that $\lim_{k\to\infty} \|y^k - y\| = 0$, and

$$
\lim_{k\to\infty} \|\mathcal{Q}^k(y^k) - q(y)\|_{X'} = 0 \tag{3.8}
$$

holds true with some $q(y) \in \mathcal{Q}(y)$. By definition of x^{k+1},

$$
\begin{aligned}
&\langle \mathcal{F}(x^k) + \mathcal{Q}^k(x^{k+1}) + \mathcal{L}^k(x^{k+1}) - \mathcal{L}^k(x^k) \\
&\quad \chi_k \left(\nabla h(x^{k+1}) - \nabla h(x^k)\right), y^k - x^{k+1} \rangle \geq -\epsilon_k \|y^k - x^{k+1}\|
\end{aligned}
$$

is valid, and the monotonicity of $\mathcal{Q}^k$ (see (2-*iii*)) leads to

$$
\begin{aligned}
&\langle \mathcal{F}(x^k) + \mathcal{Q}^k(y^k) + \mathcal{L}^k(x^{k+1}) - \mathcal{L}^k(x^k) \\
&\quad \chi_k \left(\nabla h(x^{k+1}) - \nabla h(x^k)\right), y^k - x^{k+1} \rangle \geq -\epsilon_k \|y^k - x^{k+1}\|.
\end{aligned} \tag{3.9}
$$

In case $\mathcal{F}$ is a compact operator on K, passing to the limit for $k \in \mathfrak{K}$ in (3.9), from

$$
\lim_{k\to\infty} \|y^k - y\| = 0, \quad x^{k+1} \rightharpoonup \bar{x} \ (k \in \mathfrak{K}), \quad \lim_{k\to\infty} \|x^{k+1} - x^k\| = 0,
$$

as well as $\lim_{k\to\infty} \epsilon_k = 0$, $0 < \chi_k \leq \bar{\chi}$, inequality (3.8) and the conditions (2-*i*), (2-*ii*), we obtain

$$
\langle \mathcal{F}(\bar{x}) + q(y), y - \bar{x} \rangle \geq 0.
$$

Moreover, due to (1-*ii*), (1-*iii*), the operators $\mathcal{Q}_0 : y \to \mathcal{F}(\bar{x}) + \mathcal{Q}(y)$ and $\mathcal{Q}_0 + \mathcal{N}_K$ are maximal monotone, and the operator

$$\mathcal{Q}_K : y \to \begin{cases} \mathcal{Q}_0(y) & if\ v \in K \\ \emptyset & otherwise \end{cases}$$

is locally hemi-bounded at each point of K. Thus, the application of Lemma 1 with $u = \bar{x}$, $C = K$ and $\mathcal{A}_0 = \mathcal{Q}_0$, yields

$$\langle \mathcal{F}(\bar{x}) + q(\bar{x}), y - \bar{x} \rangle \geq 0 \quad \forall y \in K,$$

where $q(\bar{x}) \in \mathcal{Q}(\bar{x})$. Hence, $\bar{x}$ is a solution of (P).
Now, let us suppose, instead of the compactness of $\mathcal{F}$, that the operator $\mathcal{F}$ is monotone, hemicontinuous and bounded on X. Then

$$\langle \mathcal{F}(x^k), y^k - x^{k+1} \rangle \leq \langle \mathcal{F}(y), y - x^k \rangle + \langle \mathcal{F}(x^k), y^k - y \rangle + \langle \mathcal{F}(x^k), x^k - x^{k+1} \rangle$$

is valid, and inserting this inequality in (3.9), with $k \to \infty, k \in \mathfrak{K}$, one gets

$$\langle \mathcal{F}(y) + q(y), y - \bar{x} \rangle \geq 0 \quad \forall y \in D(\mathcal{Q}) \cap K.$$

In view of (1-*iii*) and Theorem 2 in [**2**] the operators $\mathcal{F} + \mathcal{Q}$ and $\mathcal{F} + \mathcal{Q} + \mathcal{N}_K$ are maximal monotone in this case, and regarding (1-*ii*) the operator

$$y \to \begin{cases} \mathcal{F}(y) + \mathcal{Q}(y) & if\ y \in K \\ \emptyset & otherwise \end{cases}$$

is locally hemi-bounded at each point of K.
Thus, from Lemma 1 given with $u = \bar{x}$, $C = K$ and $\mathcal{A}_0 = \mathcal{F} + \mathcal{Q}$, we also conclude that $\bar{x}$ solves Problem (P). □

THEOREM 1. *Let the Assumptions 1 and 2 and condition (3.7) be fulfilled and let* $\sum_{k=1}^{\infty} \epsilon_k < \infty$. *Then the following conclusions are true:*

(i) *Problem* (P^k) *is uniquely solvable for each k, the sequence* $\{x^k\}$ *generated by the PAP-method is bounded, and each weak limit point of* $\{x^k\}$ *is a solution of Problem* (P)*;*

(ii) *if, in addition, (2-viii) is valid for all* $x \in X^*$ *(with common* $\alpha > 1$*) and*

$$z^k \rightharpoonup z \text{ in } X,\ z^k \in K \implies \nabla h(z^k) \rightharpoonup \nabla h(z) \text{ in } X', \tag{3.10}$$

then the whole sequence $\{x^k\}$ *converges weakly to a solution* x^* *of Problem* (P)*;*

(iii) *if, moreover,*

$$\lim_{k\to\infty} h(x^k) = h(x^*) \tag{3.11}$$

holds with x^* *as in (ii), then* $\{x^k\}$ *converges strongly to* x^*.

PROOF. Conclusion (i) follows immediately from the Lemmata 2, 3 and Remark 1. To prove (ii), suppose that $\{x^k\}_{k\in\mathfrak{K}_1}$, $\{x^k\}_{k\in\mathfrak{K}_2}$ are two subsequences converging weakly to $\bar{x}$, $\tilde{x}$, respectively. Then, according to (i), $\bar{x}$ and $\tilde{x}$ belong to X^*, and because (2-*viii*) is valid for each $x \in X^*$, Lemma 2 ensures that the sequences $\{\Gamma^k(\bar{x}, x^k)\}_{k\in\mathbb{N}}$, $\{\Gamma^k(\tilde{x}, x^k)\}_{k\in\mathbb{N}}$ are convergent.

By definition of Γ^k and with regard to $\bar{x} \in X^*$, the symmetry property (in (2-*iii*)) of the operator $\mathcal{B}$ and (2-*ii*), (2-*iv*), we obtain for $x \in K$

$$\begin{aligned}
&\Gamma^k(\bar{x},x) - \Gamma^k(\tilde{x},x) \\
&\quad = (h(\bar{x}) - h(\tilde{x}) - \langle \nabla h(\tilde{x}), \bar{x} - \tilde{x}\rangle) + \langle \nabla h(\tilde{x}) - \nabla h(x), \bar{x} - \tilde{x}\rangle \\
&\quad + \frac{1}{\chi_k}\langle \mathcal{F}(\bar{x}) + q^*(\bar{x}), \tilde{x} - \bar{x}\rangle \\
&\quad + \frac{1}{\chi_k}\langle \mathcal{F}(\bar{x}) + q^*(\bar{x}) - \mathcal{F}(\tilde{x}) - q^*(\tilde{x}), x - \tilde{x}\rangle \\
&\quad + \tilde{\chi}\langle \mathcal{B}(\bar{x} - \tilde{x}), \bar{x} - \tilde{x}\rangle + 2\tilde{\chi}\langle \mathcal{B}(\bar{x} - \tilde{x}), \tilde{x} - x\rangle \\
&\quad \geq m\|\bar{x} - \tilde{x}\|^2 + \langle \nabla h(\tilde{x}) - \nabla h(x), \bar{x} - \tilde{x}\rangle \\
&\quad + \frac{1}{\chi_k}\langle \mathcal{F}(\bar{x}) + q^*(\bar{x}) - \mathcal{F}(\tilde{x}) - q^*(\tilde{x}), x - \tilde{x}\rangle \\
&\quad + 2\tilde{\chi}\langle \mathcal{B}(\bar{x} - \tilde{x}), \tilde{x} - x\rangle.
\end{aligned} \tag{3.12}$$

Inserting $x = x^k$ in (3.12) and passing to the limit for $k \in \mathfrak{K}_2$, one can conclude from (3.10) and (3.12) that

$$\bar{\gamma} - \tilde{\gamma} \geq m\|\bar{x} - \tilde{x}\|^2,$$

where $\bar{\gamma} = \lim_{k\to\infty} \Gamma^k(\bar{x}, x^k)$, $\tilde{\gamma} = \lim_{k\to\infty} \Gamma^k(\tilde{x}, x^k)$.
Obviously, in the same way the "symmetric" inequality

$$\tilde{\gamma} - \bar{\gamma} \geq m\|\bar{x} - \tilde{x}\|^2$$

can be concluded, and therefore $\bar{x} = \tilde{x}$ is valid, proving the uniqueness of the weak limit point for $\{x^k\}$.
Denoting this limit point by x^*, now we assume additionally that relation (3.11) is fulfilled. With $\{w^k\}$ chosen according to (2-*viii*), Assumption (2-*iii*) yields

$$\begin{aligned}
&\langle \mathcal{B}(x^{k+1} - x^*), x^{k+1} - x^*\rangle \\
&\quad = \langle \mathcal{B}(x^{k+1} - w^k), x^{k+1} - w^k\rangle - \langle \mathcal{B}(x^* - w^k), x^{k+1} - x^*\rangle \\
&\quad - \langle \mathcal{B}(x^{k+1} - w^k), x^* - w^k\rangle \\
&\quad \leq \langle \mathcal{Q}^k(x^{k+1}) - \mathcal{Q}^k(w^k), x^{k+1} - w^k\rangle - \langle \mathcal{B}(x^{k+1} - x^*), x^* - w^k\rangle \\
&\quad - \langle \mathcal{B}(x^{k+1} - w^k), x^* - w^k\rangle.
\end{aligned} \tag{3.13}$$

To estimate the term $\langle \mathcal{Q}^k(x^{k+1}), x^{k+1} - w^k\rangle$, we use Problem (P^k_ϵ). Together with (3.13) this gives

$$\begin{aligned}
&\langle \mathcal{B}(x^{k+1} - x^*), x^{k+1} - x^*\rangle \\
&\quad \leq \langle \mathcal{Q}^k(w^k) - q^*(x^*), w^k - x^{k+1}\rangle \\
&\quad + \langle q^*(x^*), w^k - x^*\rangle + \langle q^*(x^*), x^* - x^{k+1}\rangle \\
&\quad + \chi_k\langle \nabla h(x^{k+1}) - \nabla h(x^k), w^k - x^{k+1}\rangle \\
&\quad + \langle \mathcal{F}(x^k) - \mathcal{F}(x^*), w^k - x^{k+1}\rangle \\
&\quad + \langle \mathcal{L}^k(x^{k+1}) - \mathcal{L}^k(x^k), w^k - x^{k+1}\rangle + \langle \mathcal{F}(x^*), w^k - x^*\rangle \\
&\quad + \langle \mathcal{F}(x^*), x^* - x^{k+1}\rangle + \langle \mathcal{B}x^* + \mathcal{B}w^k - 2\mathcal{B}x^{k+1}, x^* - w^k\rangle \\
&\quad + \epsilon_k\|w^k - x^{k+1}\|.
\end{aligned} \tag{3.14}$$

If $\mathcal{F}$ is a compact operator, then using (2-*i*) - (2-*iii*), (2-*viii*) and taking into account the boundedness of $\{w^k\}$, $x^k \rightharpoonup x^*$, $\|x^k - x^{k+1}\| \to 0$ and $\lim_{k\to\infty} \epsilon_k = 0$, the relation

$$\lim_{k\to\infty} \langle \mathcal{B}(x^{k+1} - x^*), x^{k+1} - x^* \rangle = 0$$

can be deduced from (3.14). Together with (3.11) and (2-*iv*) this ensures conclusion (iii).

In case $\mathcal{F}$ is a monotone, hemicontinuous and bounded operator on X the modification in the proof is connected only with the estimation of the term $\langle \mathcal{F}(x^k) - \mathcal{F}(x^*), w^k - x^{k+1} \rangle$. Due to the monotonicity of $\mathcal{F}$, we obtain

$$\begin{aligned} &\langle \mathcal{F}(x^k) - \mathcal{F}(x^*), w^k - x^{k+1} \rangle \\ &\quad = \langle \mathcal{F}(x^k) - \mathcal{F}(x^*), w^k - x^* + x^k - x^{k+1} \rangle + \langle \mathcal{F}(x^k) - \mathcal{F}(x^*), x^* - x^k \rangle \\ &\quad \le \langle \mathcal{F}(x^k) - \mathcal{F}(x^*), w^k - x^* + x^k - x^{k+1} \rangle \end{aligned}$$

and the boundedness of $\mathcal{F}$ together with (2-*viii*) and $\|x^k - x^{k+1}\| \to 0$ ensure

$$\overline{lim}_{k\to\infty} \langle \mathcal{F}(x^k) - \mathcal{F}(x^*), w^k - x^{k+1} \rangle \le 0.$$

□

REMARK 2. *Obviously, the conditions (3.7), used in Lemma 2 and also in Theorem 1, are compatible if and only if* $2\gamma m \ge \tilde{\chi}$. *But, they are certainly compatible, for instance, if the regularizing functional* h *is strongly convex. Then, assuming* m *is the modulus of the strong convexity of* h, *in (2-iv) an arbitrary small* $\tilde{\chi} > 0$ *is appropriate. Also, the inequality* $2m\gamma \ge \tilde{\chi}$ *can be always satisfied if* $\mathcal{F}$ *is a monotone operator and we deal with proximal-like methods, which correspond formally to the PAP-method by setting* $\mathcal{Q}^k := \mathcal{Q}^k + \mathcal{F}$, $\mathcal{F} := \mathbf{0}$, $\mathcal{L}^k := \mathbf{0}$. *In this case, Assumption (1-v) is valid for arbitrary large* γ. *Using conditions which are similar to (2-iii), (2-iv), proximal-like methods with* weak regularization *and* regularization on a subspace *have been developed in* [**11**], [**12**], [**14**] *for solving problems in elasticity theory and optimal control, and in* [**13**] *for abstract variational inequalities with non-symmetric, multi-valued and monotone operators.*

In general, one should choose the constants $\tilde{\chi}$ *and* m *such that (2-iv) is satisfied and the ratio* $\frac{\tilde{\chi}}{m}$ *is as small as possible.*

If the Assumptions 1, (2-i), (2-vii) and (2-viii) (for all $x \in X^*$*) are fulfilled, the operators* $\mathcal{Q}^k + \mathcal{L}_y + \mathcal{N}_K$ *are maximal monotone and* $\mathcal{Q}^k$ *are strongly monotone on* K *with a common modulus* m_1, *such that* $2\gamma m_1 \ge 1$, *then the iterates of the PAP-method applied with* $h \equiv 0$ *converge strongly to a solution of Problem (P).*

To adapt this analysis to the case when the auxiliary problems do not include an approximation of the operator $\mathcal{Q}$ (see, for instance [**6**], [**7**] [**25**]), one has to modify Problem (P_ϵ^k) and Assumption 2 as follows:

- the point $x^{k+1} \in K$ is defined such that

$$\begin{aligned} &\langle \mathcal{F}(x^k) + q^{k+1} + \mathcal{L}^k(x^{k+1}) - \mathcal{L}^k(x^k) \\ &\quad + \chi_k \left(\nabla h(x^{k+1}) - \nabla h(x^k)\right), x - x^{k+1} \rangle \\ &\quad \ge -\epsilon_k \|x - x^{k+1}\| \quad \forall x \in K \end{aligned} \tag{3.15}$$

  is valid with some $q^{k+1} \in \mathcal{Q}(x^{k+1})$;
- in (2-*iii*) the operator $\mathcal{Q}^k$ is replaced by $\mathcal{Q}$, and the inequality

$$\langle q(x) - q(y), x - y \rangle \ge \langle \mathcal{B}(x - y), x - y \rangle$$

  has to be fulfilled for all $x, y \in D(\mathcal{Q}) \cap K$ and all $q(x) \in \mathcal{Q}(x), q(y) \in \mathcal{Q}(y)$;
- in (2-*vi*) $\mathcal{Q}^k$ is replaced by $\mathcal{Q}$;

- the conditions (2-*vii*) and (2-*viii*) are skipped.

THEOREM 2. *The Lemmata 2,3, and Theorem 1 remain true under the modifications described above.*

The proof of Lemma 2 remains true in this case if we take $w^k = x^*$ and replace $\mathcal{Q}^k(w^k)$ by $q^*(x^*)$ and $\mathcal{Q}^k(x^{k+1})$ by q^{k+1}.
In the proof of Lemma 3 one has only to take $y^k \equiv y$ and to substitute an arbitrary $q(y) \in \mathcal{Q}(y)$ for $\mathcal{Q}^k(y^k)$. The proof of Theorem 1 needs no alterations.

REMARK 3. *If $x^{k+1} \in K$ and a residual vector $r^k \in X'$ satisfy*

$$\begin{aligned} &\mathcal{F}(x^k) + \mathcal{Q}^k(x^{k+1}) + \mathcal{L}^k(x^{k+1}) - \mathcal{L}^k(x^k) \\ &\chi_k\left(\nabla h(x^{k+1}) - \nabla h(x^k)\right) + \mathcal{N}_K(x^{k+1}) \ni r^k, \end{aligned} \tag{3.16}$$

with $\|r^k\|_{X'} \le \epsilon_k$, then the direct use of the definition of the normal cone shows that x^{k+1} is a solution of Problem (P_ϵ^k). An analogous conclusion can be made concerning the modified PAP-method (3.15).
*This permits to apply the convergence analysis above to methods using a criterion of type (3.16) ([**27**], [**7**]).*

4. Applications

We start with some observations concerning the application of the PAP-method to different types of variational inequalities.
For the variational inequality

$$(\tilde{P}) \quad find\ x^* \in X: \quad \langle \mathcal{F}(x^*) + \mathcal{Q}(x^*), x - x^* \rangle + f(x) - f(x^*) \ge 0, \quad \forall x \in X,$$

where $f : X \to \bar{I\!R} \equiv I\!R \cup \{+\infty\}$ is a convex lsc functional and the operators $\mathcal{Q} + \partial f$ (instead of $\mathcal{Q}$) and $\mathcal{F}$ satisfy Assumption 1 with $K = dom f$, the studied scheme can be directly applied if we construct the auxiliary problems (P_ϵ^k) being caused from the equivalent formulation for $(\tilde{P})$:

$$find\ x^* \in dom f: \quad \langle \mathcal{F}(x^*) + \mathcal{Q}(x^*) + \partial f(x^*), x - x^* \rangle \ge 0 \ \ \forall x \in dom f.$$

However, one can handle more convenient requirements for the approximation of f (see [**28**]) if the auxiliary problems have the form

$$\begin{aligned} (\tilde{P}_\epsilon^k) \quad & find\ x^{k+1} \in X: \\ & \langle \mathcal{F}(x^k) + \mathcal{Q}^k(x^{k+1}) + \mathcal{L}^k(x^{k+1}) - \mathcal{L}^k(x^k) \\ & \quad + \chi_k\left(\nabla h(x^{k+1}) - \nabla h(x^k)\right), x - x^{k+1} \rangle \\ & \quad + f^k(x) - f^k(x^{k+1}) \ge -\epsilon_k \|x - x^{k+1}\| \ \ \forall x \in X, \end{aligned}$$

where $f^k : X \to \bar{I\!R}$ is a convex lsc functional. In this case, compiling the convergence analysis in [**28**] and in this paper, Theorem 1 can be proved under the following modifications concerning the Assumptions 1 and 2:

- K is replaced by $dom f$ and $\mathcal{N}_K$ by ∂f in the Assumptions 1, (2-*i*), (2-*ii*) and (2-*iv*);
- (1-*v*)': for $y \in dom f$, $\mathcal{L}_y$ is monotone on $dom f$, and

$$\begin{aligned} &\langle \mathcal{F}(y) - \mathcal{L}_y(y) + \mathcal{L}_y(x) + q(x), y - x \rangle + f(y) - f(x) \\ &\quad \ge \gamma \|\mathcal{F}(y) - \mathcal{L}_y(y) - \mathcal{F}(x) + \mathcal{L}_y(x)\|_{X'}^2 \quad (\gamma > 0 - \text{ const. }) \end{aligned}$$

holds true whenever $x \in D(\mathcal{Q}) \cap dom f$, $y \in dom f$ and

$$\langle \mathcal{F}(x) + q(x), y - x\rangle + f(y) - f(x) \geq 0 \quad \text{with some } q(x) \in \mathcal{Q}(x);$$

- (2-*iii*)': $\langle \mathcal{Q}^k(x) - \mathcal{Q}^k(y), x - y\rangle + f^k(x) - f^k(y) - \langle g^k(y), x - y\rangle \geq \langle \mathcal{B}(x - y), x - y\rangle$, $\forall x, y \in D(\mathcal{Q}^k) \cap D(\partial f^k)$, $\forall g^k(y) \in \partial f^k(y)$, with $\mathcal{B}$ as in (2-*iii*);
- (2-*vi*)': for all k and $y \in dom f$, the operators $\mathcal{Q}^k + \mathcal{L}_y + \partial f^k + \chi_k \nabla h$ are maximal monotone;
- (2-*vii*)': $f^k \geq f$ and for each $w \in D(\mathcal{Q}) \cap dom f$ there exists a sequence $\{w^k\}$, $w^k \in D(\mathcal{Q}^k) \cap dom f^k$, such that

$$\lim_{k\to\infty} \|w^k - w\| = 0, \ \lim_{k\to\infty} f^k(w^k) = f(w), \ \lim_{k\to\infty} \|\mathcal{Q}^k(w^k) - q(w)\|_{X'} = 0,$$

with $q(w) \in \mathcal{Q}(w)$;
- (2-*viii*)': for some solution x^* of Problem $(\tilde{P})$ there exist a constant $\alpha > 1$ and a sequence $\{w^k\}$, $w^k \in D(\mathcal{Q}^k) \cap dom f^k$, such that

$$\lim_{k\to\infty} k^\alpha \|w^k - x^*\| = 0, \quad \lim_{k\to\infty} k^\alpha \|\mathcal{Q}^k(w^k) - q^*(w^*)\|_{X'} = 0$$

and $\lim_{k\to\infty} k^\alpha \max\{f^k(w^k) - f(x^*),\ 0\} = 0$.

REMARK 4. *Of course, the conditions (2-vii), (2-viii) and especially (2-vii)', (2-viii)' are not intended for a wide class of monotone operators, because real possibilities to satisfy these conditions are connected with individual properties of $\mathcal{Q}$ or mutual properties of $\mathcal{Q}$ and f. The general way based on the Moreau-Yosida approximation is very expensive (however, the use of this approximation within a special algorithm of the APP* [**8**] *and the proximal method* [**20**] *seems to be promising). A couple of problems in mathematical physics (the problem of linear elasticity with friction, Bingham's problem, etc.) takes the form of a variational inequality $(\tilde{P})$ with a single-valued operator $\mathcal{Q}$, and a uniform approximation of f by a sequence of differentiable functionals is possible (see* [**10**], [**21**], [**13**]*). In this case, there are no serious difficulties in satisfying (2-vii)', (2-viii)'.*

The form $(\tilde{P}^k_\epsilon)$ of auxiliary problems is also convenient to include in the APP a successive approximation of the set K (in (P)) or of the space X (in $(\tilde{P})$) by a family $\{K^k\}$ of convex sets or $\{X^k\}$ of subspaces, respectively. Then one can take $f^k = \delta(\cdot|K^k)$ and $f^k = f + \delta(\cdot|X^k)$, respectively, with $\delta(\cdot|G)$ the indicator function of G. A simultaneous approximation of f and X is also possible (see subsection 4.3 below).

The extension of the Theorems 1 and 2 to the methods developed for inclusion (1.1) can be carried out using the relationship

$$0 \in \Psi(x) \quad \Longleftrightarrow \quad \langle \Psi(x), y - x\rangle \geq 0 \ \forall y \in K,$$

where K is an arbitrary convex closed set containing $D(\Psi)$.

For instance, method (1.5) applied to inclusion (1.1) can be rewritten in the form

$$\text{(4.1)} \qquad \textit{find } x^{k+1} \in X: \quad 0 \in \frac{1}{\epsilon}\left(x^{k+1} - x^k\right) + \Psi_1(x^k) + \Psi_2(x^{k+1}),$$

and since Ψ_2 is supposed to be a maximal monotone operator, the operator

$$\Psi^k : x \to \frac{1}{\epsilon}\left(x - x^k\right) + \Psi_1(x^k) + \Psi_2(x),$$

is maximal monotone, too, and $D(\Psi^k) = D(\Psi_2)$. Hence, (4.1) is equivalent to

$$find\ x^{k+1} \in K :$$

$$\langle \Psi_1(x^k) + \frac{1}{\epsilon}\left(x^{k+1} - x^k\right) + \Psi_2(x^{k+1}), x - x^{k+1}\rangle \geq 0\ \forall x \in K, \tag{4.2}$$

with $K \supset D(\Psi_2)$ an arbitrary convex closed set, and (4.2) is a partial case of Problem (3.15) where

$$\mathcal{F} = \Psi_1,\ \mathcal{Q} = \Psi_2,\ \mathcal{L}^k = \mathbf{0},\ \chi_k = \frac{1}{\epsilon},\ h : x \to \frac{1}{2}\|x\|^2 \text{ and } \epsilon_k = 0.$$

Analogously, taking $\Psi = \Psi_1 + \Psi_2$, the auxiliary problem (1.7) may be rewritten in the form (3.15) with

$$\mathcal{F} = \Psi_1,\ \mathcal{Q} = \Psi_2,\ \nabla h = \Xi_1,\ \mathcal{L}^k = \Xi_2 + \Psi_1,\ \chi_k = \frac{1}{\epsilon},\ \text{ and } \epsilon_k = 0$$

and $K \supset D(\Psi_2)$ a convex closed set. Of course, the decomposition $\Psi = \Psi_1 + \Psi_2$ has to be performed maintaining the conditions for $\mathcal{F}$ and $\mathcal{Q}$ in Assumption 1.

On the basis of Remark 3, Theorem 2 can be extended also to the inexact version of method (1.7):

$$find\quad x^{k+1} \in X : \quad \Psi(x^{k+1}) \ni \frac{1}{\epsilon}\left[\Xi(x^{k+1}) - \Xi(x^k)\right] + r^k,$$

with residual vectors r^k such that $\sum_{k=1}^{\infty} \|r^k\|_{X'} < \infty$. In [**25**] convergence has been studied for the exact method only.

4.1. Decomposition. Due to the splitting of the main operator in (P) into a sum $\mathcal{F} + \mathcal{Q}$, certain decomposition properties are already inherent in the auxiliary problem (P^k) as well as in (3.15), where x^{k+1} is calculated with $\mathcal{F}$ fixed at the point x^k. Such a splitting can be caused by several reasons, in particular,

- usually, applications of the APP in its traditional way assume that the operator in the variational inequality is single-valued, whereas - applying proximal methods - the operator is supposed to be monotone, but not necessary single-valued (concerning relaxations of these conditions see [**4**] Sect. 3, [**15**], [**29**]). The class of problems admitting the use of the PAP-method is wider, including variational inequalities with an operator $\mathcal{F} + \mathcal{Q}$, whose "geometrical" properties are defined by Assumptions (1-*ii*), (1-*v*);
- for some variational inequalities, the problems (P^k_ϵ) or (3.15) can be solved easier than the auxiliary problems arising in proximal methods. This fact was the motivation for LIONS-MERCIER's method (1.5) for Problem (1.1), although the conditions assumed for the operator Ψ in [**17**], [**9**] permit a straightforward use of the proximal point method, too.

It is of special interest when, under an appropriate splitting of the operator, the application of the PAP-method leads to a decomposition of the auxiliary problem in the space X.

Let X be a Cartesian product of the Hilbert spaces X_1 and X_2 with their duals X_1' and X_2', respectively. In this subsection, subscript i, $i = 1, 2$, indicates that a

point or a set belongs to X_i or that an operator acts from X_i into X_i'.
Assume that $K = K_1 \times K_2$ and the operator $\mathcal{Q}$ possesses the following separability property:

$$\langle \mathcal{Q}(x), y\rangle = \langle \mathcal{Q}_1(x_1), y_1\rangle + \langle \mathcal{Q}_2(x_2), y_2\rangle$$

for $x = (x_1, x_2)$, $y = (y_1, y_2)$. In this case, it is natural that the approximating operators $\mathcal{Q}^k$ (if an approximation of $\mathcal{Q}$ is needed) have the same separability property. Then, choosing the auxiliary operators $\mathcal{L}^k : x \to \left(\mathcal{L}_1^k(x_1), \mathcal{L}_2^k(x_2)\right)$ and the function $h(x) = h_1(x_1) + h_2(x_2)$, Problem (P_ϵ^k) with

$$-\epsilon_k \left(\|x_1 - x_1^{k+1}\|_{X_1} + \|x_2 - x_2^{k+1}\|_{X_2}\right)$$

in the right-side part of the inequality can be decomposed into the pair of variational inequalities $(i = 1, 2)$:

$$\begin{aligned} &\textit{find } x_i^{k+1} \in K_i : \\ &\quad \langle \mathcal{F}_i(x_i^k) + \mathcal{Q}_i^k(x_i^{k+1}) + \mathcal{L}_i^k(x_i^{k+1}) - \mathcal{L}_i^k(x_i^k) \\ &\quad + \chi_k(\nabla h_i(x_i^{k+1}) - \nabla h_i(x_i^k)), x_i - x_i^{k+1}\rangle \geq -\epsilon_k \|x_i - x_i^{k+1}\|_{X_i} \quad \forall x_i \in K_i, \end{aligned}$$

where $\mathcal{F}_i : X_i \to X_i'$ is the composition of $\mathcal{F}$ and the canonical projection onto X_i'.
A related method for Problem $(\tilde{P})$ with $\mathcal{Q} \equiv \mathbf{0}$ and $\mathcal{F}$ a monotone operator was investigated in [**18**].
For the variational inequality (P) with $X = I\!R^n$, $\mathcal{Q} = \mathbf{0}$, $K = K_1 \times K_2$ and $\mathcal{F}$ a monotone affine operator of a special structure, TSENG [**30**] developed a decomposition algorithm where at each step two minimization problems with separable, strongly convex, quadratic functions and feasible sets K_1 and K_2, respectively, have to be solved one after the other. This algorithm was derived from the *asymmetric projection method*:

$$\text{(4.3)} \qquad \textit{find } x^{k+1} \in K : \quad \langle \mathcal{F}(x^k) + \mathcal{D}(x^{k+1} - x^k), x - x^{k+1}\rangle \geq 0 \ \ \forall x \in K,$$

by the use of an appropriately chosen positive definite (non-symmetric) matrix $\mathcal{D}$. Obviously, the auxiliary variational inequality (4.3) can be considered as a partial case of Problem (P^k) with $\mathcal{Q}^k \equiv \mathbf{0}$, $\chi_k \equiv \chi$, $\mathcal{L}^k \equiv \mathcal{D} - \chi\mathcal{I}$ and $h(x) = \frac{1}{2}\|x\|^2$. It is worth to note that the auxiliary operator corresponding to TSENG's decomposition algorithm does not possess the mentioned separability property, in general.

4.2. Linear approximation methods. These methods have been studied extensively, mainly for finite - dimensional problems (P) with $\mathcal{Q} \equiv \mathbf{0}$, and we refer to the paper [**22**] and the bibliography therein. The operator $\mathcal{F} : I\!R^n \to I\!R$ is approximated at the point x^k by $\mathcal{F}(x^k) + \mathcal{A}(x^k)(x - x^k)$, where $\mathcal{A}(x^k)$ is an $n \times n$-matrix, and $x^{k+1} \in K$ is defined by solving the variational inequality

$$\text{(4.4)} \qquad \langle \mathcal{F}(x^k) + \mathcal{A}(x^k)(x^{k+1} - x^k), x - x^{k+1}\rangle \geq 0 \ \ \forall x \in K.$$

Depending on the choice of $\mathcal{A}(x^k)$ well-known methods can be derived:

- choosing $\mathcal{A}(x^k) = \nabla\mathcal{F}(x^k)$, $\mathcal{A}(x^k) = \frac{1}{2}\left[\nabla\mathcal{F}(x^k) + \nabla\mathcal{F}(x^k)^T\right]$ or $\mathcal{A}(x^k)$ is an approximation of $\nabla\mathcal{F}(x^k)$, the Newton-, symmetrized Newton- or quasi Newton method are obtained, respectively;
- the cases $\mathcal{A}(x^k) = \mathcal{L}(x^k) + \frac{1}{\omega}\mathcal{D}(x^k)$ or $\mathcal{A}(x^k) = \mathcal{U}(x^k) + \frac{1}{\omega}\mathcal{D}(x^k)$ ($\mathcal{L}(x^k), \mathcal{U}(x^k)$ and $\mathcal{D}(x^k)$ are strictly lower triangular, strictly upper triangular and diagonal parts of $\nabla\mathcal{F}(x^k)$, respectively, and $0 < \omega < 2$), correspond to SOR-methods;

- in case $\mathcal{A}(x^k) = \mathcal{M}$, where $\mathcal{M}$ is a symmetric positive-definite matrix, a projection method is designed, etc.

If there exists a symmetric positive definite matrix $\mathcal{S}$ such that $\mathcal{A}(x) - \mathcal{S}$ is positive semi-definite for all $x \in K$, then Problem (4.4) can be transformed into the auxiliary problem (P^k) of the PAP-method with $\mathcal{Q}^k \equiv \mathbf{0}$, $\mathcal{L}^k : x \to (\mathcal{A}(x^k) - \mathcal{S})x$, $h(x) = \frac{1}{2}x^T\mathcal{S}x$ and $\chi_k \equiv 1$.
For the general variational inequality (P) with $X = \mathbb{R}^n$, the methods considered can be extended using the same $\mathcal{L}^k$, h and χ_k and an appropriate approximation $\mathcal{Q}^k$ of $\mathcal{Q}$ as in Assumption 2. Moreover, taking into account the mutual properties of $\mathcal{Q}^k$ and h, the requirements for h may be weakened according to (2-*ii*), (2-*iii*), (2-*iv*).
Of course, in both cases the choice of $\mathcal{S}$ has to satisfy Assumption (1-v).

4.3. On the use of weak regularization. Weak regularization in proximal methods means that the regularizing functional h is not strongly convex on $(X, \|\cdot\|)$, but strongly convex on X endowed by a weaker norm than $\|\cdot\|$ (see [**13**] - [**12**]). Taking into account the convergence results for the classical proximal point method in [**27**], an application of weak regularization is justified if the mutual properties of the operator of the variational inequality and the regularizing functional provide weak convergence of the iterates in the space $(X, \|\cdot\|)$. We illustrate this approach using the well-known model of linear elasticity with friction (see [**5**], Sect.3 and [**21**], Sect.5).
Let $\Omega \subset \mathbb{R}^2$ be a bounded domain with a Lipschitz-continuous boundary Γ, $g \in L_\infty(\Gamma)$ be a given non-negative function ($mes_\Gamma\ supp\ g > 0$), $a_{mnpq} \in L_\infty(\Omega)$ $(m, n, p, q = 1, 2)$ be given functions with symmetry property

$$a_{mnpq} = a_{nmpq} = a_{pqmn}. \tag{4.5}$$

Moreover, it is supposed that there exists a positive constant c_0 such that

$$a_{mnpq}(\xi)\sigma_{mn}\sigma_{pq} \geq c_0\sigma_{mn}\sigma_{mn} \text{ a.e. on } \Omega \tag{4.6}$$

holds for all symmetric matrices $[\sigma_{mn}]_{m,n=1,2}$ (here we follow EINSTEIN's summation convention, i.e. the summation is performed over terms with repeating indices).
Let $X = [H^1(\Omega)]^2$. We consider the variational inequality

$$find\ \ x \in X: \quad \langle \mathcal{A}x - l, y - x\rangle + j(y) - j(x) \geq 0 \ \ \forall y \in X, \tag{4.7}$$

where $l \in X'$ is a given linear functional,

$$\begin{aligned} \langle \mathcal{A}x, y\rangle &= \int_\Omega a_{mnpq}\epsilon_{mn}(x)\epsilon_{pq}(y)d\Omega, \\ j(x) &= \int_\Gamma g|x_t|d\Gamma, \end{aligned}$$

with $\epsilon_{mn}(x) = \frac{1}{2}\left(\frac{\partial x_m}{\partial \xi_n} + \frac{\partial x_n}{\partial \xi_m}\right)$ the component of the strain tensor and x_t the tangential (w.r.t. Γ) component of x.
The linear operator $\mathcal{A}$ is not strongly (and even not strictly) monotone, its kernel has the structure

$$\{z = (z_1, z_2) : \ z_1 = a_1 - b\xi_2, \ z_2 = a_2 + b\xi_1\}$$

with arbitrary $a_1, a_2, b \in \mathbb{R}$. However, the second KORN inequality (see [**21**]) and(4.5), (4.6) ensure that there exists a positive constant c_1 such that

$$\langle \mathcal{A}x, x\rangle + \|x\|^2_{[L_2(\Omega)]^2} \geq c_1 \|x\|^2_X,$$

and hence, taking $h(x) = \|x\|^2_{[L_2(\Omega)]^2}$, the operator $\mathcal{A}x + \nabla h(x)$ (with $\nabla h : X \to X'$) is strongly monotone in X, although h is not strongly convex w.r.t. H^1-norm, but only w.r.t. L_2-norm.
The proximal method considered in [**12**] couples weak regularization with successive discretization of the original problem (by means of the finite element method on a sequence of subspaces $\{X_k\}$) and a successive smooth approximation of the non-differential functional j. This reads as follows:

$$\begin{aligned} & find \quad x^{k+1} \in X : \\ & \langle \mathcal{A}x^{k+1} - l + \chi_k \left(\nabla h(x^{k+1}) - \nabla h(x^k)\right), y - x^{k+1}\rangle \\ & + j^k(y) - j^k(x^{k+1}) \geq -\epsilon_k \|y - x^{k+1}\| \quad \forall y \in X, \end{aligned} \tag{4.8}$$

where

$$j^k(x) = \begin{cases} \int_\Gamma g\sqrt{x_t^2 + s_k}\, d\Gamma & if\ x \in X_k \\ +\infty & otherwise \end{cases}$$

and for the smoothing parameter it holds: $s_k > 0$, $\lim_{k\to\infty} s_k = 0$.
Obviously, this method is a special case of the PAP-method for a variational inequality in the form $(\tilde{P})$ with auxiliary problems $(\tilde{P}^k_\epsilon)$, in which

$$\mathcal{F} \equiv \mathbf{0}, \ \mathcal{Q}^k \equiv \mathcal{Q} : x \to \mathcal{A}x - l, \ \mathcal{L}^k \equiv \mathbf{0} \ \text{ and } \ f^k = j^k.$$

If a solution of (4.7) belongs to $[H^2(\Omega)]^2$, then a suitable choice of $\{X_k\}$ and $\{s_k\}$, as in [**12**], Sect. 5.2, permits to satisfy the modified Assumptions 1 and 2 at the beginning of this section, moreover, (2-*viii*) can be guaranteed for any solution of (4.7).

Using Theorem 1 for the iteration process (4.8) it is worth to note that relation (3.11) follows from the other conditions of this theorem (due to the compactness of the embedding of $[H^1(\Omega)]^2$ into $[L_2(\Omega)]^2$). This situation is typical for the application of proximal methods with weak regularization to variational inequalities in mathematical physics, and of course, it means that weak convergence of the iterates x^k ensures automatically that $\{x^k\}$ converges strongly in $(X, \|\cdot\|)$.
For the contact problem of two elastic bodies, a similar interpretation of the in [**12**] described proximal methods with weak regularization and regularization on a subspace can be easily performed if the application of a finite element method ensures the inclusion $K^k \subset K$ for the approximated sets K^k.

REMARK 5. *Replacing in (4.8) $\mathcal{A}x^{k+1}$ by $\mathcal{A}x^k$, we obtain a perturbed variant of the prox-gradient method for variational inequality (4.7), which is also a particular case of the PAP-method with $\mathcal{F} : x \to \mathcal{A}x - l$, $\mathcal{Q}^k \equiv \mathcal{Q} = \mathbf{0}$, $\mathcal{L}^k \equiv \mathbf{0}$ and $f^k = j^k$ in Problem $(\tilde{P}^k_\epsilon)$. The operator $\mathcal{A}$, being symmetric, linear and maximal monotone, possesses the* DUNN *property, and this ensures the validity of Assumption (1-v)$'$. However, in order to apply the convergence analysis above, here we have to take a strongly convex (in X) functional h. For the straightforward study of the prox-gradient method see* [**16**].

5. Appendix

Proof of Lemma 2

In the sequel, we make use of the following inequalities, which are valid for arbitrary $a, b, x \in X,\ p \in X'$ and $\epsilon > 0$:

$$\langle p, a\rangle \le \frac{1}{2\epsilon}\|p\|^2_{X'} + \frac{\epsilon}{2}\|a\|^2, \tag{A.1}$$

$$\begin{aligned}&\langle \mathcal{B}(a-b), a-b\rangle \\ &\quad \le (1+\epsilon)\langle \mathcal{B}(a-x), a-x\rangle + \frac{1+\epsilon}{\epsilon}\langle \mathcal{B}(b-x), b-x\rangle.\end{aligned} \tag{A.2}$$

In order to estimate $\Gamma^{k+1}(x^*, x^{k+1}) - \Gamma^k(x^*, x^k)$, where Γ^k is defined by (3.5), we obtain from (3.6) that

$$\Gamma^{k+1}(x^*, x^{k+1}) - \Gamma^k(x^*, x^k) \le \Gamma^k(x^*, x^{k+1}) - \Gamma^k(x^*, x^k)$$

(the existence of x^* and x^k for all k is guaranteed by the Assumptions (1-vi) and (2-vi)). The right-hand side of this inequality can be splitted up as follows:

$$\Gamma^k(x^*, x^{k+1}) - \Gamma^k(x^*, x^k) \equiv s_1 + s_2 + s_3 + s_4,$$

with

$$\begin{aligned}
\text{(A.3)}\ s_1 &= h(x^k) - h(x^{k+1}) + \langle \nabla h(x^k), x^{k+1} - x^k\rangle,\\
s_2 &= \langle \nabla h(x^k) - \nabla h(x^{k+1}), w^k - x^{k+1}\rangle + \frac{1}{\chi_k}\langle \mathcal{F}(x^*) + q^*(x^*), x^{k+1} - x^k\rangle,\\
s_3 &= \langle \nabla h(x^k) - \nabla h(x^{k+1}), x^* - w^k\rangle
\end{aligned}$$

and

$$s_4 = \tilde{\chi}\langle \mathcal{B}(x^{k+1} - x^*), x^{k+1} - x^*\rangle - \tilde{\chi}\langle \mathcal{B}(x^k - x^*), x^k - x^*\rangle. \tag{A.4}$$

We suppose that the sequence $\{w^k\}$ satisfies (2-$viii$). Then (2-ii) and (A.1) yield

$$s_3 \le \frac{\tau}{2}\|x^{k+1} - x^k\|^2 + \frac{1}{2\tau} l_h^2 \|x^* - w^k\|^2, \tag{A.5}$$

with an arbitrary $\tau > 0$.
Setting $x = w^k$ in Problem (P^k_ϵ), we obtain

$$\begin{aligned}
s_2 \le\ &\frac{1}{\chi_k}\langle \mathcal{F}(x^k) + \mathcal{Q}^k(x^{k+1}) + \mathcal{L}^k(x^{k+1})\\
&-\mathcal{L}^k(x^k), w^k - x^{k+1}\rangle + \frac{1}{\chi_k}\langle \mathcal{F}(x^*) + q^*(x^*), x^{k+1} - x^k\rangle\\
&+\frac{\epsilon_k}{\chi_k}\|w^k - x^{k+1}\|,
\end{aligned}$$

and due to (2-*iii*),

$$
\begin{aligned}
\chi_k s_2 \le\ & \langle \mathcal{F}(x^k) + \mathcal{Q}^k(w^k) + \mathcal{L}^k(x^{k+1}) - \mathcal{L}^k(x^k), w^k - x^{k+1}\rangle \\
& - \langle \mathcal{B}(w^k - x^{k+1}), w^k - x^{k+1}\rangle + \langle \mathcal{F}(x^*) + q^*(x^*), x^{k+1} - x^k\rangle \\
& + \epsilon_k \|w^k - x^{k+1}\| \\
=\ & -\langle \mathcal{B}(w^k - x^{k+1}), w^k - x^{k+1}\rangle + \langle \mathcal{F}(x^k) + q^*(x^*), x^* - x^k\rangle \\
& + \langle \mathcal{Q}^k(w^k) - q^*(x^*), x^* - x^{k+1}\rangle + \langle \mathcal{F}(x^k) + q^*(x^*), x^k - x^{k+1}\rangle \\
& + \langle \mathcal{Q}^k(w^k) - q^*(x^*), w^k - x^*\rangle + \langle q^*(x^*), w^k - x^*\rangle \\
& + \langle \mathcal{F}(x^k), w^k - x^*\rangle + \langle \mathcal{L}^k(x^{k+1}) - \mathcal{L}^k(x^k), w^k - x^{k+1}\rangle \\
& + \langle \mathcal{F}(x^*) + q^*(x^*), x^{k+1} - x^k\rangle + \epsilon_k \|w^k - x^{k+1}\|.
\end{aligned}
$$

But, by definition of x^*, $q^*(x^*)$,

$$\langle \mathcal{F}(x^*) + q^*(x^*), x^k - x^*\rangle \ge 0,$$

and with (1-*v*) one can continue:

$$
\begin{aligned}
\chi_k s_2 \le\ & -\langle \mathcal{B}(w^k - x^{k+1}), w^k - x^{k+1}\rangle \\
& - \gamma \|\mathcal{F}(x^k) - \mathcal{L}^k(x^k) - \mathcal{F}(x^*) + \mathcal{L}^k(x^*)\|^2_{X'} \\
& + \langle \mathcal{Q}^k(w^k) - q^*(x^*), x^* - x^{k+1}\rangle + \langle \mathcal{Q}^k(w^k) - q^*(x^*), w^k - x^*\rangle \\
& + \langle q^*(x^*), w^k - x^*\rangle + \langle \mathcal{F}(x^k), w^k - x^*\rangle \\
& + \langle \mathcal{L}^k(x^{k+1}) - \mathcal{L}^k(x^k), w^k - x^{k+1}\rangle + \langle \mathcal{L}^k(x^*) - \mathcal{L}^k(x^k), x^k - x^*\rangle \\
& + \langle \mathcal{F}(x^k) - \mathcal{F}(x^*), x^k - x^{k+1}\rangle + \epsilon_k \|w^k - x^{k+1}\| \\
=\ & -\gamma \|\mathcal{F}(x^k) - \mathcal{L}^k(x^k) - \mathcal{F}(x^*) + \mathcal{L}^k(x^*)\|^2_{X'} - \langle \mathcal{B}(w^k - x^{k+1}), w^k - x^{k+1}\rangle \\
& + \langle \mathcal{F}(x^k) - \mathcal{L}^k(x^k) - \mathcal{F}(x^*) + \mathcal{L}^k(x^*), x^k - x^{k+1}\rangle \\
& + \langle \mathcal{F}(x^k) - \mathcal{L}^k(x^k) - \mathcal{F}(x^*) + \mathcal{L}^k(x^*), w^k - x^*\rangle \\
& + \langle \mathcal{L}^k(x^*) - \mathcal{L}^k(x^{k+1}), x^{k+1} - w^k\rangle \\
& + \langle \mathcal{F}(x^*) + q^*(x^*), w^k - x^*\rangle + \langle \mathcal{Q}^k(w^k) - q^*(x^*), x^* - x^{k+1}\rangle \\
& + \langle \mathcal{Q}^k(w^k) - q^*(x^*), w^k - x^*\rangle + \epsilon_k \|w^k - x^{k+1}\|.
\end{aligned}
\tag{A.6}
$$

Due to the monotonicity of $\mathcal{L}^k$,

$$\langle \mathcal{L}^k(x^*) - \mathcal{L}^k(x^{k+1}), x^{k+1} - w^k\rangle \le \langle \mathcal{L}^k(x^{k+1}) - \mathcal{L}^k(x^*), w^k - x^*\rangle$$

is valid, and using the inequalities (A.1), (A.2) together with (2-*i*) in order to estimate the right-hand side of (A.6) as well as the term

$\langle \mathcal{L}^k(x^{k+1}) - \mathcal{L}^k(x^*), w^k - x^* \rangle$, we obtain

$$\begin{aligned}
\chi_k s_2 \leq & \frac{\mu + \eta - 2\gamma}{2} \|\mathcal{F}(x^k) - \mathcal{L}^k(x^k) - \mathcal{F}(x^*) + \mathcal{L}^k(x^*)\|^2_{X'} \\
& + \frac{1}{2\mu}\|x^k - x^{k+1}\|^2 + \frac{1}{2\eta}\|w^k - x^*\|^2 \\
& - \frac{1}{1+\epsilon}\langle \mathcal{B}(x^{k+1} - x^*), x^{k+1} - x^* \rangle + \frac{1}{\epsilon}\langle \mathcal{B}(w^k - x^*), w^k - x^* \rangle \\
& + \|\mathcal{F}(x^*) + q^*(x^*)\|_{X'}\|w^k - x^*\| + \frac{l_{\mathcal{L}}}{\theta_k}\|w^k - x^*\|^2 \\
& + \epsilon_k\|w^k - x^*\| + \epsilon_k\left(\frac{1}{4c} + c\|x^{k+1} - x^*\|^2\right) \\
& + l_{\mathcal{L}}\theta_k\|x^{k+1} - x^*\|^2 + \left(\frac{1}{2l_{\mathcal{L}}\theta_k} + \frac{\theta_k}{2l_{\mathcal{L}}}\right)\|\mathcal{Q}^k(w^k) - q^*(x^*)\|^2_{X'}, \qquad \text{(A.7)}
\end{aligned}$$

with arbitrary positive μ, η, θ_k, c and ϵ.
Choosing

$$\epsilon = \frac{1}{2\tilde{\chi}\bar{\chi}} - 1, \quad \mu \in \left(\frac{1}{2\underline{\chi}m}, \ 2\gamma\right), \quad \tau = m - \frac{1}{2\underline{\chi}\mu}, \quad \eta \in (0, \ 2\gamma - \mu), \qquad \text{(A.8)}$$

one can conclude from (2-iv), (2-v), (3.7) and (A.2) that

$$\begin{aligned}
& \left(\tilde{\chi} - \frac{1}{\chi_k}\frac{1}{1+\epsilon}\right)\langle \mathcal{B}(x^{k+1} - x^*), x^{k+1} - x^* \rangle - \tilde{\chi}\langle \mathcal{B}(x^k - x^*), x^k - x^* \rangle \\
& \quad - h(x^{k+1}) + h(x^k) + \langle \nabla h(x^k), x^{k+1} - x^k \rangle + \left(\frac{1}{2\chi_k\mu} + \frac{\tau}{2}\right)\|x^{k+1} - x^k\|^2 \\
& \leq -\tilde{\chi}\left[\langle \mathcal{B}(x^{k+1} - x^*), x^{k+1} - x^* \rangle + \langle \mathcal{B}(x^k - x^*), x^k - x^* \rangle\right] \\
& \quad - h(x^{k+1}) + h(x^k) + \langle \nabla h(x^k), x^{k+1} - x^k \rangle \\
& \quad + \left(\frac{1}{2\chi_k\mu} + \frac{m}{2} - \frac{1}{4\underline{\chi}\mu}\right)\|x^{k+1} - x^k\|^2 \\
& \leq -\frac{\tilde{\chi}}{2}\langle \mathcal{B}(x^{k+1} - x^k), x^{k+1} - x^k \rangle - h(x^{k+1}) + h(x^k) + \langle \nabla h(x^k), x^{k+1} - x^k \rangle \\
& \quad + \left(\frac{m}{2} + \frac{1}{4\underline{\chi}\mu}\right)\|x^{k+1} - x^k\|^2 \leq \left(-\frac{m}{2} + \frac{1}{4\underline{\chi}\mu}\right)\|x^{k+1} - x^k\|^2, \qquad \text{(A.9)}
\end{aligned}$$

and that $-\frac{m}{2}+\frac{1}{4\underline{\chi}\mu}<0$. Now, we sum up (A.3), (A.4) and the estimates for s_2, s_3 in (A.5), (A.7), and insert (A.8), (A.9) in this sum. This yields

$$\begin{aligned}\Gamma^{k+1}(x^*,x^{k+1})-\Gamma^k(x^*,x^k)&\le\left(-\frac{m}{2}+\frac{1}{4\underline{\chi}\mu}\right)\|x^{k+1}-x^k\|^2\\&+\frac{1}{\chi_k}\frac{\mu+\eta-2\gamma}{2}\|\mathcal{F}(x^k)-\mathcal{L}^k(x^k)-\mathcal{F}(x^*)+\mathcal{L}^k(x^*)\|^2_{X'}\\&+\frac{1}{\chi_k}\left[\left(\frac{1}{2\eta}+\frac{l_{\mathcal{L}}}{\theta_k}+\frac{l_h^2}{2\tau}\right)\|x^*-w^k\|^2+l_{\mathcal{L}}\theta_k\|x^{k+1}-x^*\|^2\right.\\&+\left(\frac{1}{2l_{\mathcal{L}}\theta_k}+\frac{\theta_k}{2l_{\mathcal{L}}}\right)\|\mathcal{Q}^k(w^k)-q^*(x^*)\|^2_{X'}\\&\left.+\frac{1}{\epsilon}\langle\mathcal{B}(w^k-x^*),w^k-x^*\rangle+\|\mathcal{F}(x^*)+q^*(x^*)\|_{X'}\|w^k-x^*\|\right]\\&+\frac{\epsilon_k}{\chi_k}\|w^k-x^*\|+\frac{\epsilon_k}{\chi_k}\left(\frac{1}{4c}+c\|x^{k+1}-x^*\|^2\right).\end{aligned}\tag{A.10}$$

But, from the first inequality in (3.6) one gets

$$\|x^{k+1}-x^*\|^2\le\frac{1}{m}\Gamma^{k+1}(x^*,x^{k+1})$$

and (A.10) leads to

$$\begin{aligned}&\left(1-\frac{l_{\mathcal{L}}\theta_k}{\underline{\chi}m}-\frac{c\epsilon_k}{\underline{\chi}m}\right)\Gamma^{k+1}(x^*,x^{k+1})\\&\le\Gamma^k(x^*,x^k)+\frac{1}{\underline{\chi}}\left[\left(\frac{1}{2\eta}+\frac{l_{\mathcal{L}}}{\theta_k}+\frac{l_h^2}{2\tau}\right)\|w^k-x^*\|^2\right.\\&+\left(\frac{1}{2l_{\mathcal{L}}\theta_k}+\frac{\theta_k}{2l_{\mathcal{L}}}\right)\|\mathcal{Q}^k(w^k)-q^*(x^*)\|^2_{X'}\\&\left.+\frac{1}{\epsilon}\langle\mathcal{B}(w^k-x^*),w^k-x^*\rangle+\|\mathcal{F}(x^*)+q^*(x^*)\|_{X'}\|w^k-x^*\|\right]\\&+\left(-\frac{m}{2}+\frac{1}{4\underline{\chi}\mu}\right)\|x^{k+1}-x^k\|^2+\frac{\epsilon_k}{\underline{\chi}}\|w^k-x^*\|+\frac{\epsilon_k}{4c\underline{\chi}}.\end{aligned}\tag{A.11}$$

Choose $c\in\left(0,\frac{\underline{\chi}m}{2\sup\epsilon_k+1}\right)$ and set $\theta_k=\theta k^{-\alpha}$, where $\theta\in\left(0,\frac{\underline{\chi}m}{2l_{\mathcal{L}}}\right)$ and $\alpha>1$ is the same as in (2-*viii*). Then we obtain with $c_0=\frac{l_{\mathcal{L}}\theta}{\underline{\chi}m}$, $c_1=\frac{c}{\underline{\chi}m}$ that $1-c_0k^{-\alpha}-c_1\epsilon_k>0$

and

$$
\begin{aligned}
\Gamma^{k+1}(x^*, x^{k+1}) \le & \frac{1}{1 - c_0 k^{-\alpha} - c_1 \epsilon_k} \Gamma^k(x^*, x^k) \\
& + \frac{1}{\underline{\chi}(1/2 - c_0)} \left[\left(\frac{1}{2\eta} + \frac{l_{\mathcal{L}}}{\theta} k^\alpha + \frac{l_h^2}{2\tau} \right) \|w^k - x^*\|^2 \right. \\
& + \left(\frac{k^\alpha}{2 l_{\mathcal{L}} \theta} + \frac{\theta}{2 l_{\mathcal{L}}} \right) \|\mathcal{Q}^k(w^k) - q^*(x^*)\|_{X'}^2 \\
& \left. + \frac{1}{\epsilon} \langle \mathcal{B}(w^k - x^*), w^k - x^* \rangle + \|\mathcal{F}(x^*) + q^*(x^*)\|_{X'} \|w^k - x^*\| \right] \\
& + \left(-\frac{m}{2} + \frac{1}{4\underline{\chi}\mu} \right) \|x^{k+1} - x^k\|^2 \\
& + \frac{\epsilon_k}{\underline{\chi}} \|w^k - x^*\| + \frac{\epsilon_k}{4c\underline{\chi}}, \quad k = 1, 2, \ldots \qquad \text{(A.12)}
\end{aligned}
$$

Now, in view of

$$
\frac{1}{1 - c_0 k^{-\alpha} - c_1 \epsilon_k} \le 1 + \frac{c_0 k^{-\alpha} + c_1 \epsilon_k}{1/2 - c_0}, \quad \sum_{k=1}^\infty \epsilon_k < \infty, \quad -\frac{m}{2} + \frac{1}{4\underline{\chi}\mu} < 0
$$

and Assumption (2-*viii*), convergence of the sequence $\{\Gamma^k(x^*, x^k)\}$ follows from Lemma 2.2.2 in [**24**], and the first inequality in (3.6) ensures boundedness of the sequence $\{x^k\}$. Passing to the limit in (A.12), one can immediately conclude that $\lim_{k \to \infty} \|x^{k+1} - x^k\| = 0$. □

References

[1] H. Attouch, J. B. Baillon, and M. Thera. Sum of maximal monotone operators revisited: The variational sum. *J. of Convex Analysis*, 1:1–29, 1994.

[2] F. E. Browder. Nonlinear maximal monotone operators in banach spaces. *Math. Annalen*, 175:89–113, 1968.

[3] G. Cohen. Auxiliary problem principle and decomposition of optimization problems. *JOTA*, 32:277–305, 1980.

[4] G. Cohen. Auxiliary problem principle extended to variational inequalities. *JOTA*, 59:325–333, 1988.

[5] G. Duvaut and J.-L. Lions. *Les Inéquations en Mécanique et en Physique*. Dunod, Paris, 1972.

[6] J. Eckstein. Nonlinear proximal point algorithms using bregman functions, with application to convex programming. *Math. Oper. Res.*, 18:202–226, 1993.

[7] J. Eckstein. Approximate iterations in bregman-function-based proximal algorithms. *Math. Programming*, 83:113–123, 1998.

[8] N. El-Farouq and G. Cohen. Progressive regularization of variational inequalities and decomposition algorithms. *JOTA*, 97:407–433, 1998.

[9] D. Gabay. Applications of the method of multipliers to variational inequalities. In M. Fortin and R. Glowinski, editors, *Augmented Lagrangian Methods: Applications to the Numerical Solution of Boundary-Value Problems*, pages 299–331, North-Holland, Amsterdam, 1983.

[10] R. Glowinski. *Numerical Methods for Nonlinear Variational Problems*. Springer, New York - Berlin - Heidelberg - Tokyo, 1984.

[11] A. Kaplan and R. Tichatschke. *Stable Methods for Ill-Posed Variational Problems - Prox-Regularization of Elliptical Variational Inequalities and Semi-Infinite Optimization Problems*. Akademie Verlag Berlin, 1994.

[12] A. Kaplan and R. Tichatschke. Prox-regularization and solution of ill-posed elliptic variational inequalities. *Applications of Mathematics*, 42:111–145, 1997.

[13] A. Kaplan and R. Tichatschke. Proximal methods for variational inequalities with set-valued monotone operators. In R. Gilbert and P.D. Panagiotopoulos, editors, *From Convexity to Nonconvexity*. Kluwer Akad. Publishers, 1998. to appear.
[14] A. Kaplan and R. Tichatschke. Proximal penalty methods for ill-posed parabolic optimal control problems. In W. Desch, F. Kappel, and K. Kunisch, editors, *Control and Estimation of Distributed Parameter Systems*, volume 126, pages 169–182, ISNM, Birkhäuser, 1998.
[15] A. Kaplan and R. Tichatschke. Proximal point methods in nonconvex optimization. *J. of Global Optim.*, 4:389–406, 1998.
[16] B. Lemaire. Coupling optimization methods and variational convergence. *ISNM*, 84:163–179, 1988.
[17] P.-L. Lions and B. Mercier. Splitting algorithm for the sum of two nonlinear operators. *SIAM J. Numer. Anal.*, 16:964–979, 1979.
[18] S. Makler-Scheimberg, V. H. Nguyen, and J. J. Strodiot. Family of perturbation methods for variational inequalities. *JOTA*, 89:423–452, 1996.
[19] U. Mosco. Convergence of convex sets and of solutions of variational inequalities. *Advances in Mathematics*, 3:510–585, 1969.
[20] K. Mouallif and P. Tossings. Une méthode de pénalisation exponentielle associée à une régularisation proximale. *Bull. Soc. Roy. Sc. de Liège*, 56:181–192, 1987.
[21] P. D. Panagiotopoulos. *Inequality Problems in Mechanics and Applications. Convex and Nonconvex Energy Functionals.* Birkhäuser-Verlag, Boston - Basel - Stuttgart, 1985.
[22] J. S. Pang and D. Chan. Iterative methods for variational and complementarity problems. *Math. Programming*, 24:284–313, 1982.
[23] G. B. Passty. Ergodic convergence to a zero of the sum of monotone operators in hilbert space. *J. Math. Anal. Appl.*, 72:383–390, 1979.
[24] B. T. Polyak. *Introduction to Optimization.* Optimization Software, Inc. Publ. Division, New York, 1987.
[25] A. Renaud and G. Cohen. An extension of the auxiliary problem principle to nonsymmetric auxiliary operators. *ESAIM: Control Optimization and Calculus of Variations*, 2:281–306, 1997.
[26] R. T. Rockafellar. On the maximality of sums of nonlinear monotone operators. *Trans. Amer. Math. Soc.*, 149:75–88, 1970.
[27] R. T. Rockafellar. Monotone operators and the proximal point algorithm. *SIAM J. Control and Optimization*, 14:877–898, 1976.
[28] G. Salmon, V. H. Nguyen, and J. J. Strodiot. Coupling the auxiliary problem principle and the epiconvergence theory for solving general variational inequalities. *JOTA*, 104, 2000. (to appear).
[29] J. E. Spingarn. Submonotone mappings and the proximal point algorithm. *Numer. Funct. Anal. and Optimiz.*, 4:123–150, 1981-1982.
[30] P. Tseng. Further applications of a splitting algorithm to decomposition in variational inequalities and convex programming. *Math. Programming*, 48:249–263, 1990.
[31] P. Tseng. Applications of a splitting algorithm to decomposition in convex programming and variational inequalities. *SIAM J. Control Optim.*, 29:119–138, 1991.
[32] D. L. Zhu and P. Marcotte. Co-coercivity and its role in the convergence of iterative schemes for solving variational inequalities. *SIAM J. Optimization*, 6:714–726, 1996.

DEPARTMENT OF MATHEMATICS, UNIVERSITY OF TRIER, D-54286 TRIER, GERMANY
E-mail address: `tichat@uni-trier.de`

Canadian Mathematical Society
Conference Proceedings
Volume **27**, 2000

Dynamic Emission Tomography – Regularization and Inversion

Jean Maeght, Dominikus Noll, and Stephen Boyd

To Jonathan Borwein, on the occasion of his award of a Dhc

ABSTRACT. The problem of emission tomography, inverting the attenuated Radon transform, is moderately ill-posed if the unknown emission source is static. Here we consider the case where the emission source is dynamic due to the movement of molecules after metabolism. Inversion now requires a model of the underlying dynamics and presents a seriously ill-posed inverse problem. We present a regularization approach and a numerical experiment to validate our method.

1. Problem Setting

Single Photon Emission Computed Tomography (SPECT) is a non-invasive diagnostic technology which is used to show blood flow in the heart muscle, extent of damage in stroke patients, presence and degree of malignancy of tumors, and much else. SPECT images the function of the body through a *tracer*, a biochemical molecule labelled with radioactivity. The radioactive material is incorporated by the patient and metabolized by the organ under investigation. The emissions are recorded by a SPECT camera rotating around the patient, and a 3D visualization is created from the 2D projection data via an inverse mathematical method.

The traditional inverse method in tomography, filtered back-projection (FBP), may no longer be used if the unknown radiating source is dynamic, i.e., changes significantly during the scan. However, representing *dynamic* phenomena is the truly interesting case, since it allows to assess the movements of molecules during biodistribution, metabolism and washout, and consequently gives a much deeper insight into the function of the body. Recently, a first step towards *dynamic* SPECT reconstruction has been achieved through a mathematical approach replacing FBP [**22, 19, 23, 13, 1**]. This approach will be further developed here.

Key words and phrases. Dynamic emission tomography, dynamic attenuated Radon transform, compartment model, exponential fit, augmented Lagrangian SQP method, photon transport equation.

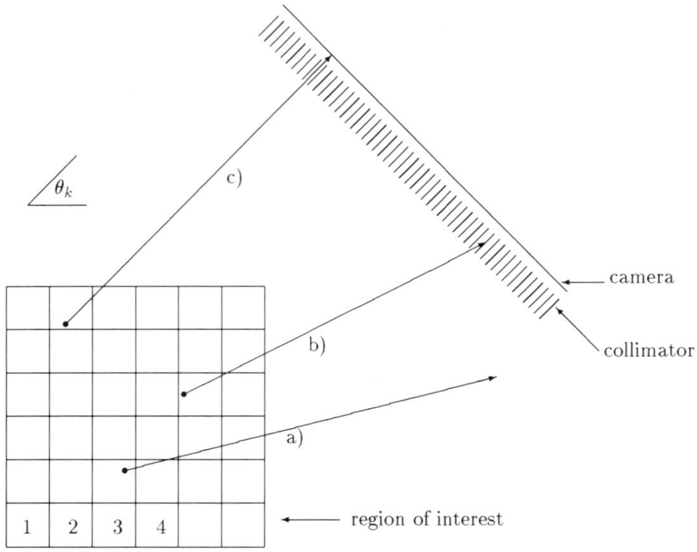

Figure 1. Photons radiating from the region of interest. a) misses the camera, b) is absorbed by the collimator, c) passes the collimator and hits the camera.

Similar to the static case, the reconstruction of the unknown 3D dynamic source may be decoupled into a series of 2D slices if Compton scatter is ignored in the transport model. The details of this process are recalled in Appendix 1. In each slice, the unknown dynamic source $f(x,t)$, representing the radio activity at position x and time t, is then related to the projection data $d(t,s,\omega)$ through the *dynamic attenuated Radon transform*

$$(1)\quad \mathcal{R}[\mu, f(\cdot,t)](s,\omega) = \int_{-\infty}^{\infty} f(s\omega^\perp + \tau\omega, t)\, e^{-\int_\tau^\infty \mu(s\omega^\perp + \tau'\omega)\, d\tau'} d\tau = d(t,s,\omega).$$

Here $\omega = (\cos\theta, \sin\theta)$ and $\omega^\perp = (-\sin\theta, \cos\theta)$, and the integrals in (1) are line integrals along the lines $L(s,\omega) = \{s\omega^\perp + \tau\omega : \tau \in R\}$. The term $d(t,s,\omega)$ represents the number of counts recorded at time t by a sensor positioned on the line $L(s,\omega)$. Figure 1 illustrates the situation in the case of a parallel hole collimator which at a given angular position admits only photons traveling perpendicular to the camera surface. It is the effect of collimation which allows attributing a direction ω of propagation to a count detected at time t and position s on the camera.

Notice that $\mu(x)$, the unknown attenuation map, which essentially represents the density of the tissue at position x, is not the primordial goal of the SPECT reconstruction. Knowledge of $\mu(x)$ is required, however, in order to reconstruct the unknown emission source, $f(x,t)$, and the basic problem in SPECT, therefore, is to solve (1) for the unknown source and attenuation map. In practice, two types of approaches are employed, leading to different mathematical problems.

A first possibility is to do a transmission scan (CT) in parallel with the SPECT acquisition, which allows to calculate μ beforehand. This reduces the complexity of the problem, but requires technology not always available.

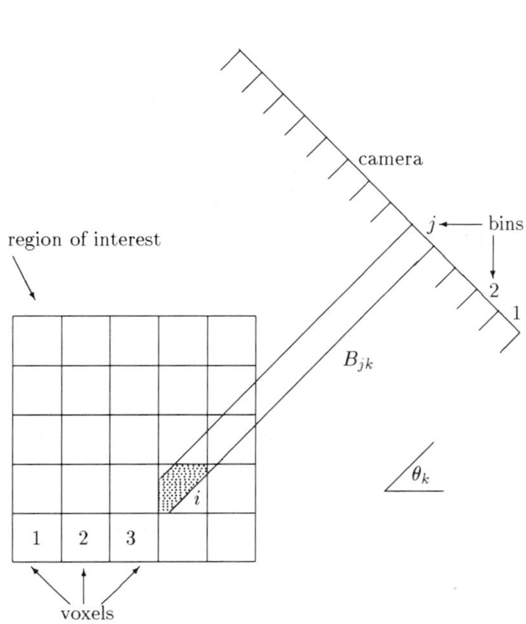

Figure 2. The coefficient c_{ijk} is the relative volume of voxel i lying within the beam B_{jk} connecting i and the receptor bin j during the camera position θ_k.

The second line of attack is to estimate μ entirely by mathematical means, using the emission data $d(t, s, \omega)$ only. This may be done either by simultaneously solving (1) for both unknowns, [**12**], a difficult task even in the static case, or by estimating μ beforehand via Helgason's consistency formula, an approach followed in [**26, 28**], and more recently in [**37**]. More elementary methods assume for instance a constant attenuation over the contour of the body if the latter can be recovered, as is often the case, from the emission data. This is usually satisfactory in brain or renal SPECT, but less satisfactory in myocardiac imaging.

We will eventually assume that the attenuation map μ is known, estimated e.g. by any of the above techniques. This renders the integral equation (1) linear in f. In the static case, inverting (1) is a well-studied moderately ill-posed linear inverse problem. The state of the art is to attack it by iterative techniques based on versions of the EM-algorithm [**36, 6**].

In the dynamic case, the problem remains intrinsically more difficult even with known μ, as we proceed to indicate. Considering exemplarily the case of a rotating single head camera, schematically displayed in Figure 1, it is crucial to observe that the camera head, rotating e.g. over 180^0, will acquire *inconsistent projection data.* By that we mean that since a considerable time (usually of the order of 20 minutes) will have elapsed between the initial and final positions of the camera, the unknown source will have changed significantly. The acquired projections will therefore correspond to entirely different dynamic states of the source, which means that the integral equation (1) is largely under determined. Formally, this may be seen by the fact that we have to reconstruct a 3D function $f(x, t)$ on the basis of data given on a the 2D submanifold

$$\mathcal{M} = \{(t, s, \omega) : \omega = (\cos\theta, \sin\theta),\ \theta = 2\pi t/T,\ 0 \le t \le T\},$$

of the full 3D data manifold $[0,T] \times R \times S^1$ (T the time of the scan). In order to address this fact, we have to provide additional information to reconstruct the unknown dynamic image $f(x,t)$.

2. Tracer Dynamics

The missing information is based on a suitable model of the chemical reaction the tracer molecule undergoes during metabolism. As this is highly dependent on the situation, we proceed to give a rather general description as it applies to our situation.

Based on theoretical knowledge of the underlying reaction, it is usually possible to identify a number of different chemical states, called *compartments*, the tracer molecule will assume. The compartment model generally represents a simplified version of the complex reaction. With $c_\nu(t)$ the concentration of tracer in the νth compartment, the reactive process is represented by a dynamical system

$$c'(t) = Kc(t) + c_B(t) \tag{2}$$

with a system matrix $K = (k_{ij})$ involving exchange rates k_{ij} between compartments, and a driving term c_B representing the supply from the blood pool. As radio active events originating from any one of the chemical states c_ν, $\nu = 1, \ldots, P$, could not possibly be distinguished from each others, the emission source is of the form $f(t) = \sum_{\nu=1}^{P} c_\nu(t)$, which means that we cannot assess the tracer concentrations c_ν directly.

So far we have represented the traditional point of view of compartmental modeling which seeks a *single* dynamic model $f(t)$ for the organ as a whole. This point of view changes with tomography, where we wish to account for spatially varying behavior. Assuming that the chemical reaction remains in principal the same over the entire organ, we allow the exchange rates to vary locally. This leads to a *localized compartment model* of the form $c'(x,t) = K(x)c(x,t) + c_B(x,t)$ involving localized exchange rates $k_{ij}(x)$ and leading to a source term of the form

$$f(x,t) = \sum_{\nu=1}^{P} c_\nu(x,t).$$

The inverse problem of dynamic emission tomography, in a rather general form, is then the following:

Find $f(x,t)$, $\mu(x)$ and exchange rates $k_{ij}(x)$ such that

$$(3) \qquad (\mathcal{P}) \quad \begin{aligned} &(i) && \mathcal{R}[\mu, f(\cdot,t)](s,\omega) = d(t,s,\omega) \text{ for } (t,s,\omega) \in \mathcal{M} \\ &(ii) && c'(x,t) = K(x)c(x,t) + c_B(x,t) \\ &(iii) && f(x,t) = \sum_{\nu=1}^{P} c_\nu(x,t) \end{aligned}$$

are satisfied.

This model may in practice be completed by prior information about the exchange rates such as bounds or physiologically meaningful ranges.

The shape of the input function, an impulse, is qualitatively known and may conveniently be modeled as $c_B(t) = Ae^{-\mu t} - Ae^{-\nu t}$ with $A > 0$ and $\mu > \nu > 0$, if $t = 0$ corresponds to the moment of injection of the pharmaceutical. If a considerable time between the injection of the drug and the initialization of the scan has elapsed, the source term may generally have a simpler form and is often even assumed constant. Assuming that the system matrices $K(x)$ are diagonalizable, which in practice is always the case, the source term will have the functional form

$$f(x,t) = \sum_{\nu=1}^{P+2} a_\nu(x)\, e^{-\lambda_\nu(x)t}. \tag{4}$$

Recovering the exchange rates $k_{ij}(x)$ from (4) is in general impossible, but the parameters $\lambda_i(x)$, $a_j(x)$ still carry useful diagnostic information.

Dynamics of existing radio tracers for SPECT use scarcely more than two compartments (cf. [**16**]), even though more complex models with up to five compartments have been proposed for some positron emitters (cf. [**34**]). During the following, we shall therefore restrict our analysis to a two compartment model with constant input function c_B. This is already of sufficient generality for a number of applications, including for instance a myocardiac viability study presented by Limber et al. [**19**], or a renal dynamic SPECT scan discussed in [**13**]. The functional form of the unknown emission source, accordingly, is a biexponential

$$f(x,t) = a_1 e^{-\lambda_1(x)t} + a_2 e^{-\lambda_2(x)t} + a_3(x), \tag{5}$$

and we may substitute (5) for the equations (ii), (iii) of $(\mathcal{P})$. The dynamic SPECT inversion may now be transformed into a nonlinear optimization problem:

$$(NL) \quad \begin{array}{ll} \text{minimize} & \|\mathcal{R}[\mu]f - d\| \\ \text{subject to} & f(x,t) = a_1(x)e^{-\lambda_1(x)t} + a_2(x)e^{-\lambda_2(x)t} + a_3(x). \\ & \lambda_1(x) \geq 0,\ \lambda_2(x) \geq 0. \end{array}$$

This approach is the starting point for our algorithmic discussion. Before continuing this line, we discuss a possible discretization of $(\mathcal{P})$.

3. Discretization

We assume that the camera cross-section has been divided into M bins $j = 1,\ldots,M$ of size Δs, the jth bin corresponding to an interval $[s_j - \Delta s/2, s_j + \Delta s/2]$. Suppose the camera takes S angular positions $\omega_k = (\cos\theta_k, \sin\theta_k)$, $k = 1,\ldots,S$, with $\theta_k = (k-1)\Delta\theta$, and remains in the kth position during the time interval $[t_k - \Delta t/2, t_k + \Delta t/2]$. Then the datum acquired in the jth camera bin during the kth stop is

$$y_{jk} = \int_{t_k-\Delta t/2}^{t_k+\Delta t/2} \int_{s_j-\Delta s/2}^{s_j+\Delta s/2} d(t,s,\omega_k)\, ds\, dt.$$

Integrating (1) accordingly gives the condensed form of equation (1):

$$\int_{t_k-\Delta t/2}^{t_k+\Delta t/2} \int_{s_j-\Delta s/2}^{s_j+\Delta s/2} \int_{L(s,\omega_k)} f(x,t)\, e^{-\int_\tau^\infty \mu(s\omega^\perp + \tau'\omega)\, d\tau'} d\tau\, ds\, dt = y_{jk}.$$

We proceed to discretize the unknown image into N pixels P_i, $i = 1,\ldots,N$ of size $\Delta x \times \Delta x$, and assume the activity $f(x,t)$ constant on a pixel at every instant t, i.e., $f(x,t) = f(x_i,t)$ for the barycenter x_i of the ith pixel. Then the unknown variables are the activities originating from these pixels and cumulated over the stops

$$f_{ik} := \int_{t_k-\Delta t/2}^{t_k+\Delta t/2} \int_{P_i} f(x,t)\, dx\, dt = |P_i| \int_{t_k-\Delta t/2}^{t_k+\Delta t/2} f(x_i,t)\, dt. \tag{6}$$

Assuming the attenuation map $\mu(x)$ constant on each pixel, and using (6), the discretized form of (1) becomes

$$\sum_{i=1}^{N} R_{ijk}[\mu]\, f_{ik} = y_{jk} \tag{7}$$

where the coefficients $R_{ijk}[\mu]$ are

$$R_{ijk}[\mu] = \int_{t_k-\Delta t/2}^{t_k+\Delta t/2} \int_{s_j-\Delta s/2}^{s_j+\Delta s/2} \int_{L(s,\omega_k)\cap P_i} e^{-\int_\tau^\infty \mu(s\omega^\perp + \tau'\omega)\, d\tau'}\, d\tau\, ds\, dt. \tag{8}$$

During the sequel, (7) will be abbreviated as $R[\mu]f = y$, the discretized version of (1). It is instructive to see the meaning of the coefficients in the unattenuated case $\mu = 0$. Here $R_{ijk}[0]$ is the surface of that part of the ith pixel lying in the strip connecting to the jth camera bin during stop k (see Figure 2).

Observe that, unless $\mu = 0$, (8) has to be calculated numerically, which by itself requires a discretization. We mention that our approach is not the only possible, and various other discretizations for $\mathcal{R}$ have been proposed, including wavelet bases [**40**], or C^0-elements [**12**].

Let us proceed to discretize (5). Using (6) we obtain the estimate

$$f_{ik} = \big(a_1(x_i)\, e^{-\lambda_1(x_i)t_k} + a_2(x_i)\, e^{-\lambda_2(x_i)t_k} + a_3(x_i)\big)(\Delta t \Delta x^2 + o(\Delta t \Delta x^2))$$

if x_i is the barycenter of the ith pixel. Setting $\lambda_{1i} = \lambda_1(x_i)\Delta t$, $\lambda_{2i} = \lambda_2(x_i)\Delta t$ and $a_{\nu i} = a_\nu(x_i)\Delta t \Delta x^2$, we are led to replace the constraint (5) by its discretized version

$$f_{ik} = a_{1i}e^{-\lambda_{1i}k} + a_{2i}e^{-\lambda_{2i}k} + a_{3i}. \tag{9}$$

4. Prony's Method Extended

The projection data are in practice always corrupted by noise, and it is necessary to solve the discretized problem (7), (9) in the least squares sense. Moreover, since the problem is ill-posed, it is wise to include a regularizing term $I[f]$ into the objective function. Assuming that μ is known suggests the following approach via nonlinear programming:

$$(NL) \qquad \begin{array}{ll} \text{minimize} & \ell(f) = \|R[\mu]f - y\|^2 + I[f] \\ \text{subject to} & f_{ik} = a_{1i}e^{-\lambda_{1i}k} + a_{2i}e^{-\lambda_{2i}k} + a_{3i} \\ & \lambda_{1i} \geq 0,, \lambda_{2i} \geq 0 \end{array}$$

Possible choices for $I[f]$ will be discussed in Section 8. The problem has been analyzed by Borwein and Sun [**3**] in the case $I[f] = 0$ and found well-behaved if the matrix $R[\mu]$ has maximal rank and MS is larger than $5N$, the number of unknown variables in (NL). While the rank condition is often satisfied, in particular if every camera bin can see at least one pixel at each stop, the second condition is in practice only satisfied for SPECT devices with a large number of receptors, and does not hold for the usual rotating camera systems. As an example, consider the case of a camera cross section divided into $M = 64$ bins, the camera taking $S = 64$ stops, and divide the region of interest into 64×64 pixels. Then $5N > MS$ for a single head camera, and even $5N > 3MS$ when a triple head camera is used.

And in fact, our implementation, based on the limited memory BFGS approach of Nocedal et al. [**21, 5**] shows that (NL) is a difficult problem subject to instabilities in particular when the eigenvalues tend to coalesce, $\lambda_{1i} \approx \lambda_{2i}$. Reconstructing a *typical slice* above may take between 1 and 2 hours CPU, which is too slow for convenient clinical applications.

In order to circumvent the numerical difficulties of (NL), we consider a generalization of Prony's method for fitting exponentials (cf. [**31**]). The key idea is the simple fact that a curve f_k of biexponential form (9) satisfies a difference equation

$$f_k = \alpha_1 f_{k-2} + \alpha_2 f_{k-1} + \alpha_3, \tag{10}$$

where the five parameters $(a_1, a_2, a_3, \lambda_1, \lambda_2)$ in (9) correspond to the five pieces of information $\alpha_1, \alpha_2, \alpha_3$ along with the initial values f_0, f_1 in (10). The links are the following. If f_k has the form (9), then

$$\alpha_1 = -e^{-(\lambda_1+\lambda_2)},\ \alpha_2 = e^{-\lambda_1} + e^{-\lambda_2},\ \alpha_3 = (1-\alpha_1-\alpha_2)a_3, \tag{11}$$

along with the obvious expressions for f_0, f_1. Conversely, if f_k satisfies a difference equation (10) and if we impose conditions to avoid the possible oscillatory solutions, we retrieve the form (9) through

$$a_1 + a_2 = f_0 - \frac{\alpha_3}{1-\alpha_1-\alpha_2},\ a_1u_1 + a_2u_2 = f_1 - \frac{\alpha_3}{1-\alpha_1-\alpha_2} \tag{12}$$

where u_i are the (real) roots of $x^2 - \alpha_2 x - \alpha_1 = 0$, $u_1 = e^{-\lambda_1}$, and $u_2 = e^{-\lambda_2}$.

In dynamic emission tomography we are mainly interested in two cases: (i) washout only, where the dynamic profiles are decaying, and (ii) uptake and washout, where activity in the organ initially increases and after reaching the peak activity, decays (cf. [**4, 10, 15, 18, 19, 30, 34**]). The corresponding information may be expressed in terms of the difference equation:

Lemma 1. *The following statements are equivalent:*

1. $f_k = a_1 e^{-\lambda_1 k} + a_2 e^{-\lambda_2 k} + a_3$ *with* $\lambda_1 \neq \lambda_2$ *is monotonically decreasing and nonnegative on* $k = 0, 1, \ldots$
2. f_k *satisfies the difference equation* $f_k = \alpha_1 f_{k-2} + \alpha_2 f_{k-1} + \alpha_3$, *along with the constraints*

$$-1 < \alpha_1 < 0,\ 0 < \alpha_2 < 2,\ \alpha_2^2 + 4\alpha_1 > 0,\ \alpha_1 + \alpha_2 < 1 \text{ and } a_1 \geq 0,\ a_2 \geq 0,$$

where a_i *are expressed in terms of the parameters* α_i, f_j *via (12).* □

Similarly, the increasing-decreasing case is covered by the following

Lemma 2. *The following are equivalent:*

1. f_k *of the form (9) is increasing-decreasing and nonnegative on the range* $k = 0, 1, \ldots$ *with* $\lambda_1 \neq \lambda_2$.
2. f_k *satisfies the difference equation (10) along with the constraints*

$$-1 < \alpha_1 < 0,\ 0 < \alpha_2 < 2,\ \alpha_2^2 + 4\alpha_1 > 0,\ \alpha_1 + \alpha_2 < 1 \text{ and } a_1 + a_2 \geq 0,\ a_2 \geq 0,$$

where the a_i *are expressed in terms of* α_i, f_j *via (12).* □

Let us now show in which way the idea of representing exponentials by difference equations may be exploited algorithmically. We start with the case (i) where we expect decaying activity. We replace program (NL) by the following

$$(NL_1)\qquad \begin{array}{ll} \text{minimize} & \ell(f,\alpha) = \|R[\mu]f - y\|^2 + I[f,\alpha] \\ \text{subject to} & f_{i1} \geq f_{i2} \geq \cdots \geq f_{iS} \geq 0, \\ & f_{ik} - \alpha_{1i} f_{i,k-2} - \alpha_{2i} f_{i,k-1} - \alpha_{3i} = 0, \\ & \text{for}\quad i = 1, \ldots, N, \quad k = 3, \ldots, S. \end{array}$$

with unknown variables $f = (f_{ik})$ and $\alpha = (\alpha_{\nu i})$, and where the regularizing term $I[f,\alpha]$ may involve both variables. The linear inequality constraints, which in tandem with the difference equation will yield a decaying positive exponential (9) will be referred to as *shape constraints.* These constraints form a cocktail of the conditions we have isolated in Lemma 1, which are easier to handle than the nonlinear constraints on α in condition 2 of Lemma 1. As we shall see, the shape constraints turn out extremely useful during the initial period of the minimization, but become more and more redundant as the iteration approaches a local minimum.

Before discussing this model any further, let us catch up with more general dynamic profiles. The most interesting case is where we expect increasing-decreasing curves. Here the shape constraints are more delicate to calibrate, since we do not, in general, know the exact position of the peak activity in each dixel (dixel: short for dynamic pixel). We propose the program

$$(NL_2)\qquad \begin{array}{ll} \text{minimize} & \ell(f,\alpha) = \|R(\mu)f - y\|^2 + I[f,\alpha] \\ \text{subject to} & 0 \le f_{i1} \le f_{i2} \le \cdots \le f_{i,m_i-\nu_i}, \\ & f_{i,m_i+\nu_i} \ge \cdots \ge f_{iS} \ge 0, \\ & f_{ik} - \alpha_{1i} f_{i,k-2} - \alpha_{2i} f_{i,k-1} - \alpha_{3i} = 0, \\ & \text{for} \quad i = 1,\dots,N, \quad k = 3,\dots,S. \end{array}$$

which assumes a peak activity in dixel i among the positions $k \in [m_i - \nu_i, m_i + \nu_i]$, with m_i and ν_i determined beforehand, say from prior knowledge or by inspecting the projection data. A second possibility, to constrain an increasing-decreasing profile with the peak position assumed among four previously assigned positions, was used in [**22**], and is considered again in the experimental section.

5. Augmented Lagrangian Method

Our first algorithmic approach to (NL) is an augmented Lagrangian method. We consider the partially augmented objective function

$$(13)\quad \ell_c(f,\alpha) := \ell(f,\alpha) + \frac{c}{2}\sum_{i=1}^{N}\sum_{k=3}^{S} e_{ik}(f,\alpha)^2 = \ell(f,\alpha) + \frac{c}{2} e(f,\alpha)^T e(f,\alpha),$$

where the e_{ik} are the equality constraints in (NL),

$$(14)\qquad e_{ik}(f,\alpha) = f_{ik} - \alpha_{1i} f_{i,k-2} - \alpha_{2i} f_{i,k-1} - \alpha_{3i} = 0,$$

and e stands for the vector of e_{ik}. We do *not* include the shape constraints into (13). Due to their linear structure, we prefer to carry them explicitly through the approach. The augmented Lagrangian associated with (13) is then

$$(15)\qquad \mathcal{L}_c(f,\alpha;\lambda,\mu) = \ell_c(f,\alpha) + \lambda^T e(f,\alpha) + \mu^T(Af - b),$$

where $Af \le b$ represents the shape constraints. This suggests the following algorithm:

Algorithm 1

1. Initialize the multiplier estimates λ and $\mu \ge 0$, and the penalty parameter $c > 0$.
2. Given multiplier estimates λ and $\mu \ge 0$, and the penalty parameter c, solve

$$(P)\qquad \begin{array}{ll} \text{minimize} & \mathcal{L}_c(f,\alpha;\lambda,\mu) \\ \text{subject to} & Af \le b \end{array}$$

The solution is (f^+, α^+). Let μ^+ the Lagrange multiplier belonging to the inequality constraint in (P).
3. Update the multiplier estimate through $\lambda^+ = \lambda + c\,e(f^+, \alpha^+)$.
4. Update $c^+ \geq c$.
5. Return to step 2 until convergence.

This scheme is known to have local superlinear convergence if the penalty parameter sequence c_k it generates tends to infinity (cf. [**2**]), which is in practice prohibitive due to ill-conditioning.

Fixing c at a sufficiently large value $\bar{c}$ still ensures local linear convergence (cf. [**2**]). In practice, the correct value $\bar{c}$ is not known, and in order to guarantee numerical stability of the algorithm, after some updates we usually fix c at a moderate size. This means that the solution so obtained, although practically useful, will *not* fully match the difference equation (10). We may therefore consider Algorithm 1 as a relaxation approach, where the regularizer $I[f, \alpha]$ includes, possibly among other terms, the term $(c/2)\, e(f, \alpha)^T e(f, \alpha)$.

We consider a variation of the augmented Lagrangian approach which divides step 2 of the above Algorithm 1 into two steps, a minimization over f, followed by a minimization over α. Given the current estimates f, α, λ and $\mu \geq 0$ as well as the constant $c > 0$, we obtain the update f^+ and the new multiplier $\mu^+ \geq 0$ by minimizing $\mathcal{L}_c(\cdot, \alpha; \lambda, \mu)$ subject to the shape constraints. In a second step, we obtain the new estimate α^+ by solving the problem

$$\begin{array}{ll} \text{minimize} & \displaystyle\sum_{i,k} (f_{ik}^+ - \alpha_{1i} f_{i,k-2}^+ - \alpha_{2i} f_{i,k-1}^+ - \alpha_{3i})^2 \\ \text{subject to} & -1 \leq \alpha_{1i} \leq 0,\ 0 \leq \alpha_{2i} \leq 2,\ \alpha_{2i}^2 + 4\alpha_{1i} \geq 0,\ \alpha_{1i} + \alpha_{2i} \leq 1. \end{array}$$

The multiplier λ^+ is updated as before, as is the constant c^+.

This method is algorithmically easier than optimizing simultaneously in (f, α). Using the constraints from condition 2 in Lemma 1 resp. Lemma 2 is possible here since a posteriori fitting, that is finding α^+ with f^+ given, leads to N problems of size S, and is therefore algorithmically simpler. The drawback of this method is that its convergence properties are not clear. Alternating schemes converge at best with a linear rate, but worse, may even fail to converge. We therefore use this scheme during the initial stages of the optimization only.

6. Lagrangian SQP Method

The canonical way to deal with the equality constraints is the Lagrangian SQP method, which consists in applying the sequential quadratic programming (SQP) method to the augmented objective function (13), cf. [**2**].

Given a current primal estimate (f, α) and Lagrange multiplier estimates λ, and $\mu \geq 0$, the SQP method generates new iterates $(f^+, \alpha^+) = (f, \alpha) + (\Delta f, \Delta\alpha)$ by solving the quadratic program

$$(QP)\quad \begin{array}{ll} \text{minimize} & q_{f,\alpha}(\Delta f, \Delta\alpha) \\ \text{subject to} & A(f + \Delta f) \leq b \\ & M\Delta f + N\Delta\alpha + d = 0 \end{array}$$

choosing as the new multiplier estimates λ^+ and $\mu^+ \geq 0$ the Lagrange multipliers belonging to the linearized constraints in (QP). Here $q_{f,\alpha}$ is the quadratic functional

$$q_{f,\alpha}(\Delta f, \Delta\alpha) =$$
$$\nabla_f \ell_c(f,\alpha)^T \Delta f + \nabla_\alpha \ell_c(f,\alpha)^T \Delta\alpha + \tfrac{1}{2}(\Delta f, \Delta\alpha)\nabla^2 \mathcal{L}_c(f,\alpha;\lambda,\mu)(\Delta f,\Delta\alpha)^T$$

and $\mathcal{L}_c(f,\alpha;\lambda,\mu)$ is the augmented Lagrangian (15). Moreover, the (i,k)th line of the linearized equality constraint $M\Delta f + N\Delta\alpha + d = 0$ is the linearization of (14) about the current (f,α), that is:

$$f_{ik} + \Delta f_{ik} - \alpha_{1i} f_{i,k-2} - \alpha_{1i}\Delta f_{i,k-2} - \Delta\alpha_{1i} f_{i,k-2} - \alpha_{2i} f_{i,k-1}$$
$$-\alpha_{2i}\Delta f_{i,k-1} - \Delta\alpha_{2i} f_{i,k-1} - \alpha_{3i} - \Delta\alpha_{3i} = 0.$$

While having fast local convergence properties (cf. [**2**]), the SQP scheme is certainly demanding in large size applications, as the number of equality constraints is important. We therefore propose a relaxation which bypasses this difficulty.

7. Relaxation

We observe a particular block structure in the equality constraints, which makes it possible to explicitly calculate the least squares solution

$$\Delta\alpha = -(N^T N)^{-1} N^T (M\Delta f + d) =: W\Delta f + v \tag{16}$$

of the linear equality constraints, and to substitute (16) into the objective $q_{f,\alpha}$: $q_f(\Delta f) := q_{f,\alpha}(\Delta f, W\Delta f + v) + \frac{k}{2}\|M\Delta f + N(W\Delta f + v) + d\|^2$.
This suggests the reduced quadratic program (QP')

$$(QP') \quad \begin{array}{ll} \text{minimize} & q_f(\Delta f) \\ \text{subject to} & A(f+\Delta f) \leq b \end{array}$$

where the new penalty term $(k/2)\|M\Delta f + N(W\Delta f + v) + d\|^2$ is intended to keep the feasible variables Δf in (QP') close to producing feasible pairs $(\Delta f, W\Delta f + v)$ for (QP), which we aim at since (QP) presents the correct Newton step ensuring fast local convergence. Since (QP'), nonetheless, allows elements Δf which are *not* admitted in (QP), the problems are not equivalent, and we may at the end expect solutions which do not fully match the equality constraints. Notice that

$$M\Delta f + N(W\Delta f + v) + d = (I - N(N^T N)^{-1} N^T)(M\Delta f + d),$$

with $\Pi := I - N(N^T N)^{-1} N^T$ the projection onto $\ker(N^T)$, so $\ker(\Pi) = R(N)$. Consequently, the penalty term in q_f attributes a high cost to a mismatch in $M\Delta f + N\Delta\alpha + d = 0$, and (QP') may be expected to be close to the original quadratic program (QP).

While the substitution (16) gives a more convenient quadratic subproblem (QP'), no longer involving equality constraints, we recall that ironically we still have to maintain a Lagrange multiplier estimate λ for the linearized equality constraints we have dispensed with. In fact, the terms in $q_{f,\alpha}$ including $L_{f\alpha} = \nabla^2_{f\alpha}\mathcal{L}_c$ require the estimate λ. Due to linearity, this is not the case for the multiplier estimate μ, so maintaining an update of μ is not mandatory here.

The Kuhn-Tucker conditions to (QP') guarantee a multiplier $\nu \geq 0$ such that with the notations $L_{ff} := \nabla^2_{ff}\mathcal{L}(f,\alpha;\lambda,\mu)$, etc.,

$$\begin{array}{ll} (a) & \nabla_f \ell_c + W^T \nabla_\alpha \ell_c + L_{ff}\Delta f + L_{f\alpha} W\Delta f + L_{f\alpha} v + W^T L_{f\alpha}^T \Delta f \\ & \qquad + W^T L_{\alpha\alpha}(W\Delta f + v) + A^T \nu + k\,\Pi(\Delta f + Nv + d) = 0 \\ (b) & A(f+\Delta f) \leq b, \quad \nu^T\big(A(f+\Delta f) - b\big) = 0 \end{array} \tag{17}$$

are satisfied. These have to be matched with the Kuhn-Tucker conditions for the original program (QP), with (16) substituted for $\Delta\alpha$:

$$
(18)\quad
\begin{array}{ll}
(i) & \nabla_f \ell_c + L_{ff}\Delta f + L_{f\alpha}(W\Delta f + v) + A^T\mu^+ + M^T\lambda^+ = 0 \\
(ii) & \nabla_\alpha \ell_c + L_{f\alpha}^T \Delta f + L_{\alpha\alpha}\Delta\alpha + N^T\lambda^+ = 0 \\
(iii) & A(f+\Delta f) \le b, \quad \mu^+(A(f+\Delta f) - b) = 0, \quad \mu^+ \ge 0 \\
(iv) & M\Delta f + N(W\Delta f + v) + d = 0
\end{array}
$$

Matching (a) of (17) with (i) of (18) implies

$$
(19)\quad
\begin{array}{l}
M^T\lambda^+ + A^T\mu^+ = \\
A^T\nu + W^T\big(\nabla_\alpha \ell_c + L_{f\alpha}^T\Delta f + L_{\alpha\alpha}(W\Delta f + v)\big) + k\Pi(\Delta f + Nv + d),
\end{array}
$$

while matching (a) of (17) with (ii) of (18), the latter pre-multiplied by W^T, leads to

$$
(20)\quad
\begin{array}{l}
W^T N^T \lambda^+ = \\
A^T\nu + \nabla_f \ell_c + L_{ff}\Delta f + L_{f\alpha}(W\Delta f + v) + k\, W^T\Pi(\Delta f + Nv + d).
\end{array}
$$

While (19), (20) as they stand would seem difficult to satisfy due to the condition $\mu^+ \ge 0$, $\nu \ge 0$, we may heuristically assume that $\mu^+ = \nu$, as this would be the case if the solution Δf to (QP') matched the constraint $M\Delta f + N\Delta\alpha + d = 0$ exactly, producing a pair $(\Delta f, W\Delta f + v)$ admissible in (QP). With $\mu^+ = \nu$, (19), (20) are conveniently solved for λ^+, possibly in the least squares sense using the same block structure of the constraints. An alternative would be to estimate λ^+ only on the basis of (19), as this is even an explicit update not involving any inversion.

Resuming the above observations, we propose the following algorithm:

Algorithm 2

1. Initialize f, α, λ, and c. Fix $k > 0$.
2. Given the current f, α, λ and c, calculate a step Δf by solving the quadratic program

$$
(QP)\qquad \begin{array}{ll} \text{minimize} & q_f(\Delta f) \\ \text{subject to} & A(f+\Delta f) \le b \end{array}
$$

   Obtain $\Delta\alpha$ through (16).
3. Calculate a step $\Delta\lambda$ by estimating λ^+ through the formulae (19) and (20).
4. Do a line search with an appropriate merit function to find $t > 0$ and $f^+ = f + t\Delta f$, $\alpha^+ = \alpha + t\Delta\alpha$ and $\lambda^+ = \lambda + t\Delta\lambda$.
5. Update $c^+ \ge c$.
6. Return to step 2 until convergence.

We recall that if the correct SQP method using (QP) was used, the second order sufficient optimality conditions at a local minimum would imply fast local convergence with the penalty parameter c fixed at a sufficiently large value (cf. [**2**]). This suggests the same strategy for the relaxed version and fixing c at a moderate value. Since the correct c is not known, we propose to initially use an update rule like in [**2**] until a satisfactory value of c has been found.

We mention that a plausible idea is to apply the same least squares relaxation to the original program (NL) directly, that is, eliminate the unknown variable α in (NL) by calculating for every i the least squares solution $(\alpha_{1i}, \alpha_{2i}, \alpha_{3i})$ to the set of equations $e_{i3} = 0, \dots, e_{iS} = 0$. This is, however, less convenient than the corresponding substitution in the linearized setting, since the functional relationship $\alpha = \alpha(f)$ so obtained is more complicated and renders the gradient and Hessian calculation for the reduced problem difficult.

8. Practical Aspects

Designing a correct compartmental model for a specific situation is usually difficult due to the high complexity of metabolism. As a consequence, reconstructing a dynamic tomographic image with a particular tracer dynamic model imposed may bear the risk to miss or suppress dynamic phenomena apparent in the unknown source, but not included in the chosen compartment model. This risk would be less severe if we were able to reconstruct the dynamic image *before* doing the exponential fit. In Positron Emission Tomography (PET) this strategy *is* valid, since the movie may be reconstructed frame by frame. As we have indicated, this is not possible in SPECT if rotating camera systems are used, because the number of views we could possibly scan during a fixed state of the unknown dynamic source is too small. Nonetheless, with specially equipped triple head cameras one can mimic the situation in PET by doing *fast rotations* which collect sufficiently many views at a single dynamic state of the source, cf. [**15, 25**]. This method suffers from poor data statistics, and we claim that with our present techniques we can improve, doing *slow* rotations.

The approach discussed in [**22**] - similar to the situation in PET - reconstructs a movie prior to exponential fitting by imposing shape constraints only. This seems a crude procedure, but in many cases produces good results, in particular for double and triple head cameras (cf. [**13**]). What we presented here is a more sophisticated approach, which again reconstructs a dynamic image f prior to the definitive exponential fitting, now aiming at reconstructions which are already close to the exponential form (5). Using shape constraints in tandem with the augmented Lagrangian type penalty terms encourages doubly exponential fits.

A second practical aspect is CPU time, which should not be exceedingly large if the method is to be clinically useful. This means that we may perform only a limited number of SQP or augmented Lagrangian steps, probably never reaching any neighborhood of fast local convergence of the method. The approximate solution so obtained may then still be improved by doing an a posteriori fit of the chosen exponential model (5), a procedure which may be repeated several times, as it is fast in comparison to the reconstruction process.

These practical aspects emphasize the idea of regularization versus exact fitting, and one may ask about other possible regularizers for the dynamic emission tomography problem. Reconstructing a movie requires regularization in space and in time. For simplicity we consider regularizers of the form $I[f] = I_s[f] + I_t[f]$, where I_s is a spatial regularizer, I_t a regularizer in time.

The classical choice among the spatial regularizers is of course the Tychonoff functional $I_s[f] = (k/2)\|\nabla f\|^2$, and its offsprings, which we may apply frame by frame. Following [**24**], we prefer a Fourier domain regularizer whose philosophy is based on the Fourier slice theorem for the unattenuated Radon transform: $\hat{f}(\sigma\omega) = (\mathcal{R}f)\hat{}(\sigma,\omega)$. This means that the expected signal f could not possibly have a better spatial resolution than each of the projections. Having decided on a spatial bandwidth b in the s-variable, and after choosing a high-pass filter $\mathcal{H}_b$ with the corresponding cutoff frequency, the spatially regularizing term should be chosen as

$$I_s[f] = (k/2)\|\mathcal{H}_b f\|^2.$$

This filter could be applied either to each frame, or to a subset of frames. The approach should still be qualitatively correct if the transform is attenuated, even though the Fourier slice theorem is no longer true in this case.

Regularizing in the time domain could be done in several ways. A Fourier regularizer could be designed in much the same way as for spatial smoothing. The corresponding dynamic cutoff frequency could be obtained by a prior Fourier analysis of the expected exponential curves (5), as soon as realistic bounds on the eigenvalues λ_i are known.

However, a particular difficulty with Fourier filtering in the time domain, already analyzed in [**23**], is the fact that most practical time profiles are *not* continuous in the sense of the space $C_{\text{per}}[0,T]$, and Fourier filtering would therefore introduce unwanted smoothing at the early and late views. Ways to overcome this have been proposed in [**23**] and could be adapted to build an appropriate temporal high pass filter.

We conclude by presenting a dynamic regularizer based on a different philosophy, motivated by the dynamic heart-in-thorax phantom built at the Vancouver Hospital (cf. [**7**]). The phantom heart consists of several cylindrical containers with a drain and an inflow. Each container is equipped with a mixing propeller to guarantee a homogeneous flow. Tuning the motor which steers the pump could now, at least in principle, produce any possible activity profile in the containers. Consider for instance the situation of washout, where the container is initially filled with activity $f(0)$. Pumping fresh water at flow rate $R(t)$ means that the remaining activity in the container of volume V at time t is

$$f(t) = f(0)\exp\left\{-\int_0^t \frac{R(\tau)}{V}\,d\tau\right\}.$$

The idea is now to avoid highly irregular curves $f(t)$ by penalizing irregular behavior of the motor. And in a sense the most natural profile is a flow at constant rate, that is, $R'(t) = 0$. In terms of the curve f, this leads to $R'/V = -(f'/f)' = -(\log f)'' = 0$, so in the continuous model the dynamic regularizer could be chosen as

$$I_t[f] = \int_0^T \left((\log f(t))''\right)^2 dt.$$

Discretizing the time step and using finite differences to replace time derivatives leads to

$$I_t[f] = \sum_{k=2}^{S-1} (\log f_{k+1} + \log f_{k-1} - 2\log f_k)^2 = \sum_{k=2}^{S-1} \log^2\left(\frac{f_{k+1}f_{k-1}}{f_k^2}\right).$$

The ultimate link with tomography is the hypothesis that each cell, in a very oversimplified way, acts similar to the macroscopic containers of the phantom heart.

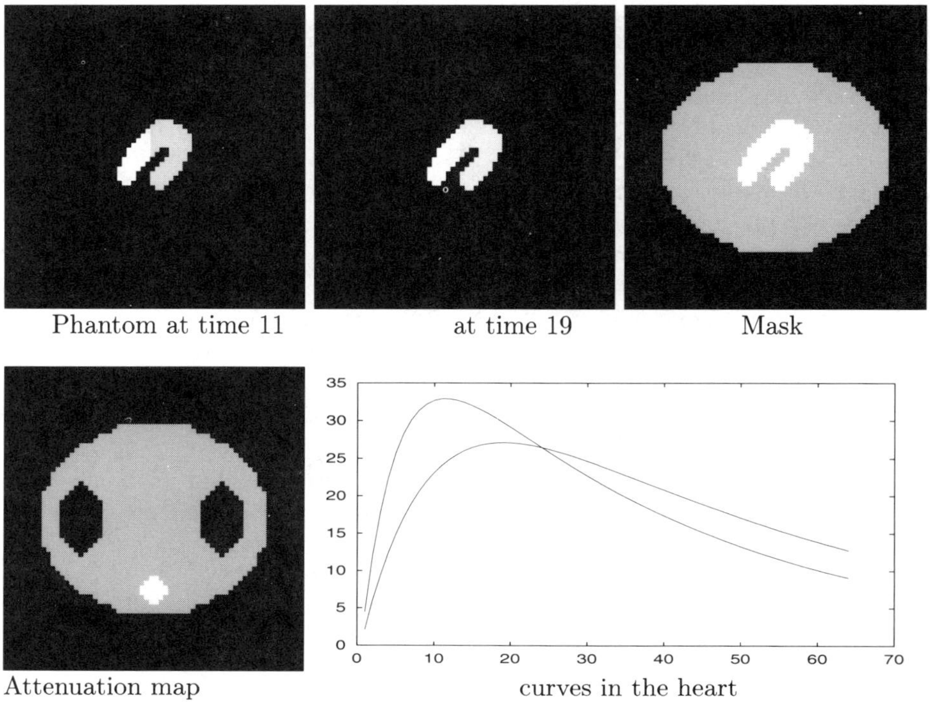

Figure 3a. Simulated phantom and attenuation map.

9. Simulation

The following simulation uses a dynamic heart-in-thorax phantom, displayed in Figure 3a and 3b. The scan assumes 64 time steps, and in line 1 the phantom is displayed at times 11 and 19. The heart has been segmented into two dynamic zones exhibiting different up-take and washout profiles, simulating a stenosis in the left part of the muscle. The other sections of the body contour are assumed static. The ideal time profiles for the two zones are shown in Figure 3a, line 2 (right). From an experimental standpoint, the question is that of knowing whether the different dynamic modes are recognized by our inverse method for ideal and noisy data.

The data acquisition simulates a rotating double head camera with two heads at 90^0, each head rotating over a 90^0 sector, with a total of 180^0 scanned. Each camera head takes 32 angular positions. The ideal sinogram is displayed in Figure 3b (left), while the right hand image shows the data rendered Poisson distributed. Notice that in a static scan, the sinograms of the two heads could usually be assembled to a single sinogram of twice as many steps for one head. Due to changing activity, this is no longer possible in dSPECT, and we emphasis this fact by presenting the two sinograms in reversed order. We assume a realistic tissue attenuation map, displayed in Figure 3a, line 2 (left).

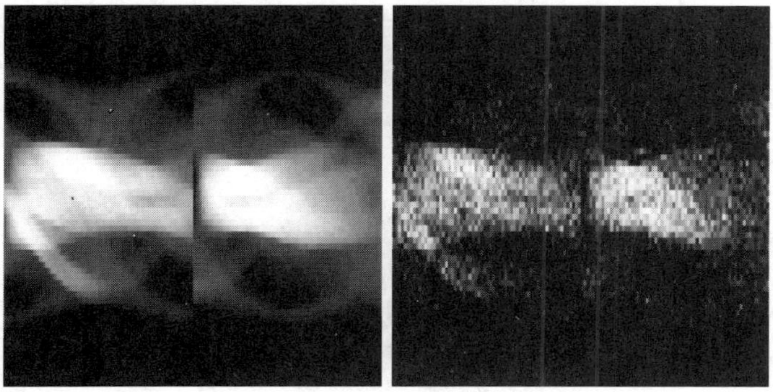

Sinogram Noised sinogram

Figure 3b. Sinograms. The two camera heads are placed at 90^0.

The reconstruction process now starts with a preliminary phase which serves three different purposes. A first reconstruction (a) is obtained by imposing decreasing activities only, a second one (b) imposing increasing activities only. While (a) will show a flat part at the early views, with decay at the later views, (b) will have the opposite form. Matching these reconstructions will provide a first guess for the peak activity position in each pixel.

A third reconstruction (c) is now produced with the assumed peak positions in each pixel, and the reconstruction is used to identify the body contour. Reconstruction (c) also serve the purpose of identifying pixels which are apparently static, and rendering these dixels static reduces the total number of unknown parameters to fit. The resulting mask is shown in Figure 3a, line 1 (right), assuming pixels outside the phantom heart as static. Obtaining the mask may be more difficult in a clinical scan, and some practical issues of this phase, in particular, its automatization, will be discussed elsewhere.

A fourth reconstruction (d) is now obtained on the basis of the previously obtained mask, and is used to initialize the Prony type algorithm. This phase of the reconstruction process uses version (NL_2), either without regularizing term, or with some of the regularizers discussed in section 8. A definitive conclusion on what regularizer behaves best at step (d) has not been drawn as yet, and its choice may depend on the situation.

Reconstruction (d), displayed at times 11 and 19 in Figure 4, line 1, is already of satisfactory visual quality, but the dynamic profiles shown in Figure 5, line 1, are not sufficiently closed to the sought for biexponential form. Step (e) of the procedure now does an a posteriori curve fitting of a biexponential on a pixel-by-pixel basis, translating the model (9) to the form (10), or directly uses the method from section 5. This step is again critical, since some of the reconstructed profiles in (d) may be far from biexponentials and may lead to aberrant parameter estimates.

We address this problem by switching to a local fit which uses the weighted sum of 9 neighboring pixels to a given pixel to stabilize.

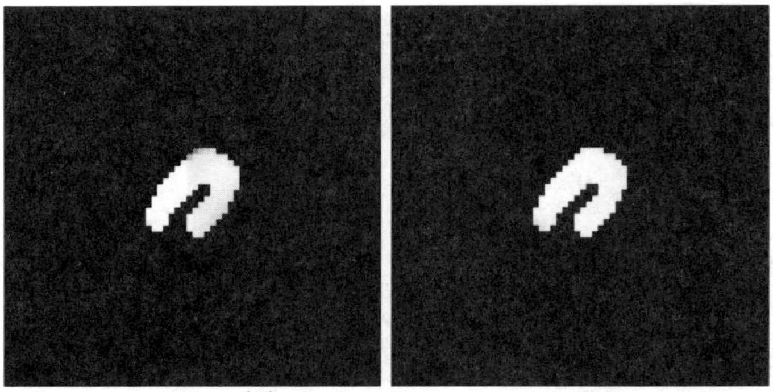

Reconstruction (d): before Prony at times 11 and 19

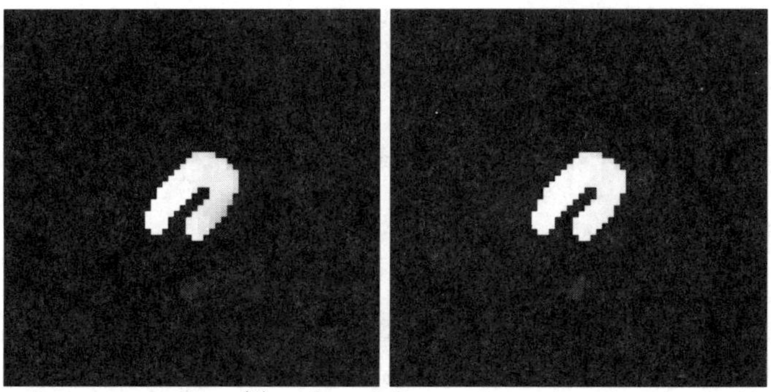

Reconstruction (f): after 4 Prony iterations at times 11 and 19

Figure 4. Noise-free reconstruction. The total CPU was 3 minutes.

Estimate (e), translated into the corresponding $(\alpha_{i1}, \alpha_{i2}, \alpha_{i3})$ information is then used as the initial guess to the Prony type procedure, which in our study is the augmented Lagrangian approach in Section 5 with initial $\lambda = 0$. The result is the reconstruction (f), shown in Figure 4, line 2. As we can see, in the noise-free case a few iterations suffice to produce satisfactory curves which even allow separating the ideal time profiles visually. However, the initial stages of the procedure, shown in Figure 5, seem to indicate oscillatory behavior of the method.

The reconstruction may be terminated by a last step (g) doing a second a posteriori fit, eventually after drawing regions of interest for specific purposes. Notice here that in contrast with other approaches, segmenting is *not* used in the reconstruction process itself, but serves for the interpretation of the results. This is an advantage of our method, since segmentation for reconstruction, usually employed in order to reduce the number of unknown variables, is critical as it may oversimplify or introduce unnecessary bias in the reconstructions.

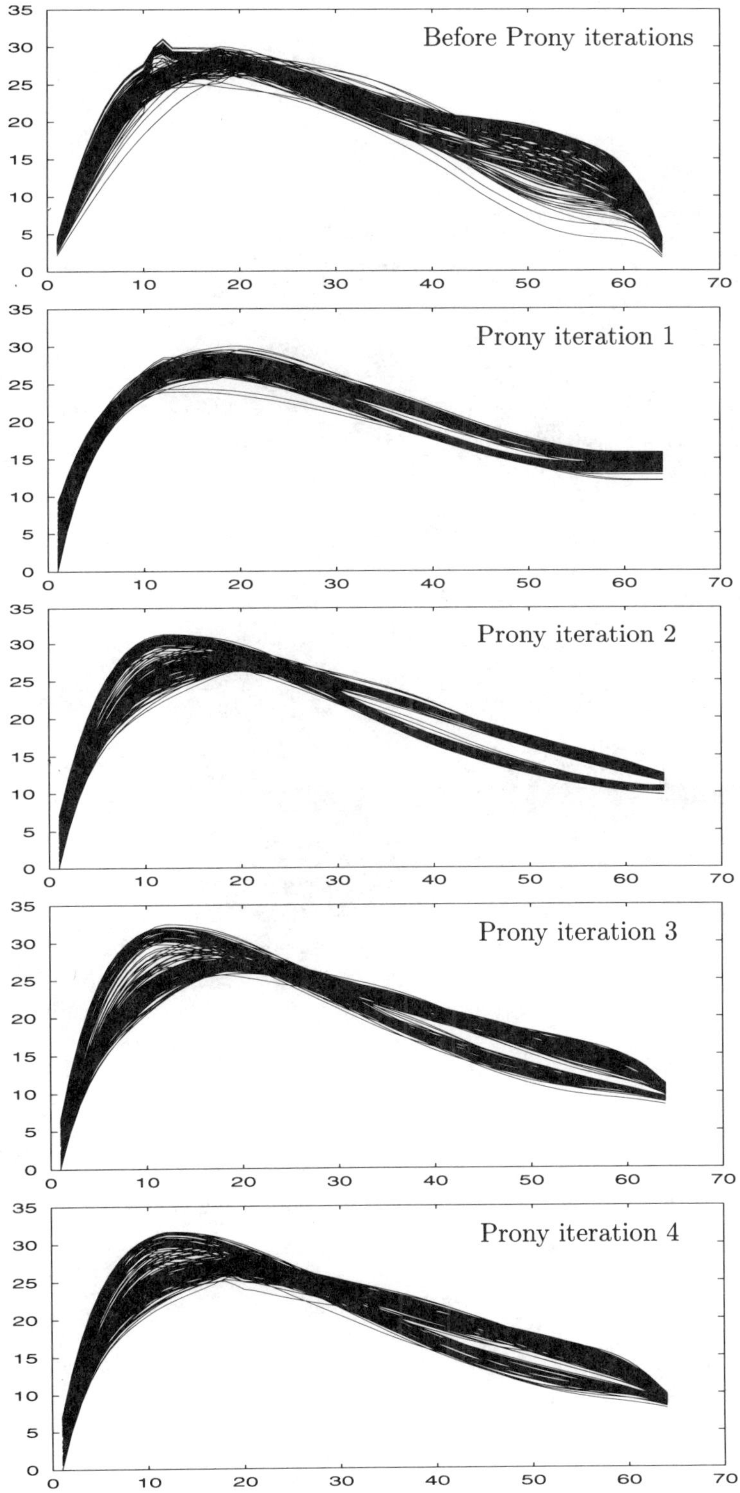

Figure 5. Prony iterates in the heart for the noise free case.

The results in Figure 6 show the same phantom experiment with the noisy sinogram obtained by rendering the ideal data Poisson distributed (Figure 3b, right). Figure 6, lines 1 and 2 show the reconstructed slice, with little to no visual difference between the raw reconstruction and the result obtained after 4 Prony type iterations.

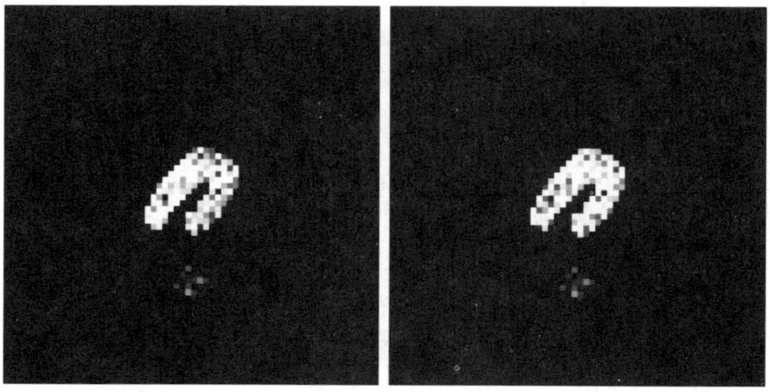

Reconstruction (d): before Prony iterations: images at times 11 and 19

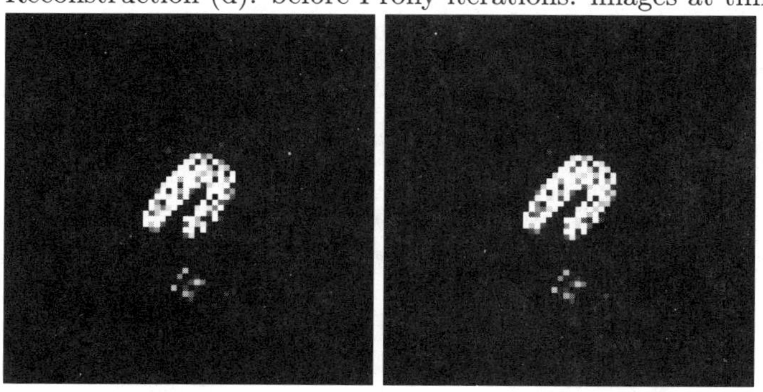

Reconstruction (f): after 4 Prony iterations: images at times 11 and 19

Figure 6. Reconstruction for the noisy case. The total computation time was 3 minutes 30 seconds. Neither the data nor the reconstruction have been filtered.

A visual check of the reconstructed curves as in the noise free case (Figure 5) is no longer possible here. We therefore draw masks for the two parts of the heart we wish to discriminate, shown in Figure 7, line 1. We represent the result for each region by taking the averaged curves. In the left region, this means averaging 46 curves, in the right hand regions 91 curves. The result is displayed in line 3. The correct input is re-displayed in line 2.

As we can see, the reconstruction is capable of representing the cross-over of the two curves apparent in the true image. Equally, the peak positions in each profile are rather accurately represented. The reconstruction is off by about 10% in absolute values. This may not be blamed on attenuation artifacts, as the correct attenuation map has been used in the reconstructions. It is more likely that this is caused by the imperfections of the reconstruction process itself. A refined estimation disregarding outliers in the reconstructed profiles from each region would probably help to improve the accuracy in absolute values.

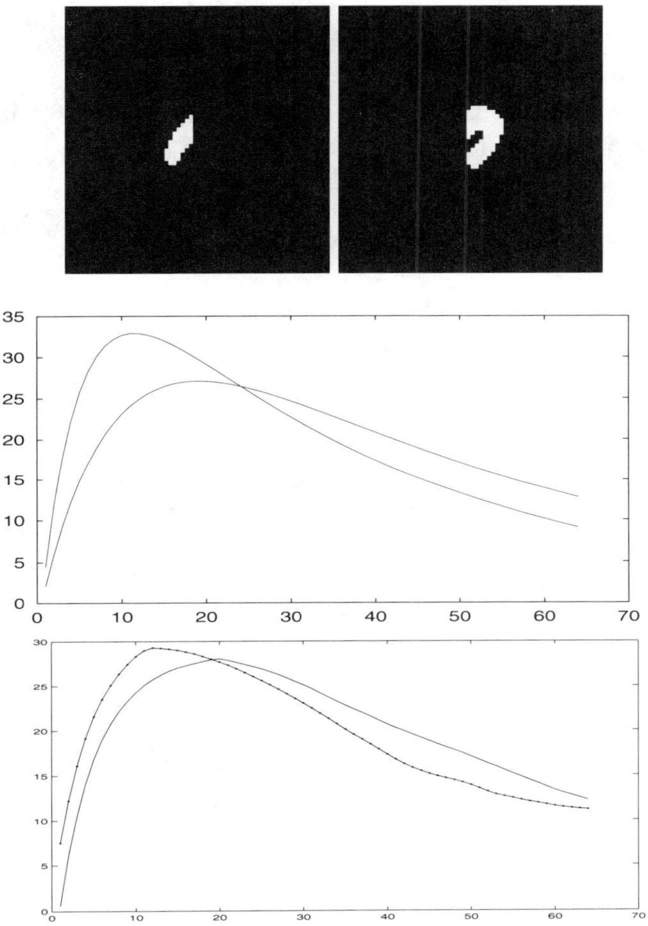

Figure 7. Evaluation of the noisy case. Line 1 shows the masks drawn to distinguish the different dynamic profiles. Line 3 shows the result of the reconstruction after 4 Prony type iterations. We plot the averages of the curves in each region of the heart against each others. Line 2 shows the true curves. The reconstruction is able to represent the cross-over of the curves and the peak activities, but absolute heights around the peaks are off by approximately 10%.

10. Conclusion

We have presented a way of estimating kinetic parameters for compartmental models directly from a dSPECT reconstruction. The approach is an extension of Prony's method and the optimization uses an augmented Lagrangian type procedure. Due to breaking the iteration into two separate steps (section 5), a procedure which accelerates, the method tends to oscillate after some rapid improvements during the early steps. The approach has been validated through a simulated case study. A more refined scheme (described in section 6 and 7) avoiding the oscillatory behavior has been presented but needs to be validated by means of numerical simulations. Several dynamic regularizers have been successfully used, but a qualitative comparison offering a preference for a specific regularizers has not been obtained as yet.

11. Appendix - Transport Model

Dynamic emission tomography is described by the dynamic photon transport equation (compare [**11**, Vol. I, Section 5]):

$$\begin{aligned}\frac{1}{c}\frac{\partial}{\partial t}u(x,\omega,E,t)+\omega\cdot\nabla u(x,\omega,E,t)+\mu(x,E)u(x,\omega,E,t)\\ =\int_{\mathbb{S}^2} b(x,E',E,\omega',\omega)u(x,\omega',E',t)\,d\omega' + f(x,E,t)\end{aligned} \tag{21}$$

Here $u(x,\omega,E,t)$ is the specific intensity or density of the angular transport of photons at position x, time t, in direction $\omega\in\mathbb{S}^2$, and with energy E; c is the speed of light. $f(x,E,t)$ is the unknown radio active source we wish to estimate, and $\mu(x,E)$ is the linear attenuation coefficient which decomposes as $\mu(x,E)=\rho(x)\,\mu_m(E)$ with $\rho(x)$ the unknown density at position x, and $\mu_m(E)$ a known term representing all physical processes which end flights of photons at energy E. For the energies relevant to nuclear medecine, these comprise photo effect, Rayleigh scattering, and Compton scattering, while pair production becomes relevant at higher photon energies encountered in radio therapy. For energies E between 100keV and 1000keV, photo effect and Rayleigh scatter are almost negligeable, and we may model the scattering kernel as $b(x,E,E',\omega,\omega')=\rho(x)\,\beta(E,\theta)$, where $\theta=\widehat{(\omega,\omega')}$ is the angle between ω and ω', and β is the known Klein-Nishina scattering cross-section (cf. [**12**]). Notice here that due to conservation of energy and momentum in the Compton scatter process, the energy E' of the incident photon scattered from direction ω' into the direction ω with energy E is fixed through the formula

$$E=\frac{E'}{1+(E'/mc^2)(1-\cos\theta)} \tag{22}$$

with $\theta=\widehat{(\omega,\omega')}$ and m the electron mass (cf. [**39**]).

Before discussing (21), it is instructive to see the general form of the dynamic emission tomography problem:

Find an input function $c_B(x,t)$, exchange constants $k_{ij}(x)$, and a mass distribution $\rho(x)$ such that with a dynamic source $f(x,E,t)$ satisfying (ii) and (iii) of $(\mathcal{P})$, the photon transport $u(x,\omega,E,t)$ solving (21) matches the projection data $d(t,x,\omega,E)$ acquired at time t, sensor position x, angular direction ω, and energy E.

This cast of the problem, although not very helpful from the algorithmic point of view, reveals it as a control problem with control vector (c_B,K,ρ). In this context, our regularization approach may be understood as an optimal control technique.

Let us continue our investigation of the transport model. Notice that the energy of photons used in nuclear medicine is often monochromatic, so ideally $f(x,E,t)=f(x,t)\,\delta(E-E_0)$ for a fixed energy level E_0. (For example, $E_0=140$keV for ^{99}Tc used in SPECT, and $E_0=511$keV in PET). Assuming that our camera has a perfect energy resolution, we are capable of recognizing photons which have Compton scattered with angle $\theta\neq 0$ due to the loss of energy (22). Therefore, for monochromatic sources, the scattering kernel in (21) becomes irrelevant, and on suppressing the dependence of the variables on E_0, we obtain the simplified transport equation

$$\frac{1}{c}\frac{\partial}{\partial t} u(x,\omega,t) + \omega \cdot \nabla u(x,\omega,t) + \mu(x)\, u(x,\omega,t) = f(x,t), \tag{23}$$

with $\omega \in \mathbb{S}^2$, $x \in \mathbb{R}^3$, $t \geq 0$. Clearly this equation decouples into a series of equations on 2D slices.

In cases where the emission source is not monochromatic, the scattering kernel is required. Even for monochromatic sources, scattering is of practical important due to the finite camera resolution which implies that we collect data over an energy window $E_0 \pm \Delta E$.

Despite these observations it is customary to ignore the scatter contribution or to interpret it as a source of noise in the data. Introducing the cumulated angular transport

$$u(x,\omega,t) = \int_{E_0-\Delta E}^{E_0+\Delta E} u(x,\omega,E,t)\, dE$$

we obtain, on assuming $\mu(x,E) = \mu(x,E_0) = \mu(x)$ energy independent over $E_0 \pm \Delta E$, the same simplified equation (23) in $u(x,\omega,t)$.

Discussing (23) in a slice we parameterize lines by $s \in \mathbb{R}$, and $\omega = (\cos\theta, \sin\theta) \in \mathbb{S}^1$, letting $L(s,\omega) = \{s\omega^\perp + t\omega : t \in \mathbb{R}\}$. Suppose the unknown source and attenuation coefficient are zero outside the unit disk Ω. Then the measured data may be understood as boundary information on Ω. Indeed,

$$u(x_0,\omega,t) = 0 \qquad \text{for } \nu_{x_0} \cdot \omega \leq 0, \quad x_0 \in \partial\Omega \tag{24}$$

($\nu_x =$ outer unit normal to Ω), i.e., there is no incoming radiation. And further, the outgoing radioactive transport is known:

$$u(x_1,\omega,t) =: d(t,s,\omega) \quad \text{for } \nu_{x_1} \cdot \omega \geq 0, \quad x_1 \in \partial\Omega, \tag{25}$$

where

$$d(t,s,\omega) = \int_{E_0-\Delta E}^{E_0+\Delta E} d(t,s,\omega,E)\, dE$$

represents the counts collected over the camera energy window $E_0 \pm \Delta E$ about the expected energy E_0 of the isotope. Here $L(s,\omega) \cap \Omega = [x_0, x_1]$, and we call $x_0 = x_0(s,\omega), x_1 = x_1(s,\omega)$ the entry and exit points.

Assuming that the source $f(x,t)$ and data $d(t,s,\omega)$ decay sufficiently fast as $t \to \infty$ we may Fourier transform equation (23). This leads to the family of equations

$$\omega \cdot \nabla U(x,\omega,\sigma) + (\mu(x) + i\sigma/c) U(x,\omega,\sigma) = F(x,\sigma) \tag{26}$$

with boundary information

$$U(x_0,\omega,\sigma) = 0, \qquad U(x_1,\omega,\sigma) = D(\sigma,s,\omega). \tag{27}$$

Here capital letters refer to the Fourier transforms of the corresponding lower case functions, and σ stands for the Fourier domain variable. Integrating equation (26) using (27) on each segment $L(s,\omega)\cap\Omega=[x_0,x_1]$ separately leads to the formula

$$\int_{x_0}^{x_1} F(x,\sigma)\cdot e^{-i\sigma|x_1-x|/c}\cdot e^{-\int_x^{x_1}\mu(y)\,dy}\,dx \;=\; D(s,\omega,\sigma), \tag{28}$$

which is a Fourier domain counterpart of the known integral equation obtained in the case of stationary SPECT based on the attenuated Radon transform (cf. [**27, 26, 28**]). Multiplying (28) by $e^{i\sigma t}$, Fourier inversion then gives the family of integral equations

$$\int_{x_0}^{x_1} f(x,t-|x_1-x|/c)\cdot e^{-\int_x^{x_1}\mu(y)\,dy}\,dx \;=\; d(t,s,\omega), \tag{29}$$

exhibiting a delay between the data acquired at time t and the source terms emitting at times $t-|x_1-x|/c$. The method of characteristics (cf. [**11**]) would of course have produced the same result without using the Fourier transform. Since photons travel with the speed of light, we may in practice assume

$$t-|x_1-x|/c\approx t,$$

which leads to the approximate equation involving the *dynamic attenuated Radon transform* $\mathcal{R}[\mu,f]$:

$$\mathcal{R}[\mu,f(\cdot,t)](s,\omega):=\int_{x_0}^{x_1} f(x,t)\cdot e^{-\int_x^{x_1}\mu(y)\,dy}\,dx \;=\; d(t,s,\omega). \tag{30}$$

Notice that for monochromatic sources, (30) represents the *correct* transport model, while in more general situations it stands for the simplification obtained by setting $b=0$.

For completeness, let us consider the situation in PET. Assuming again that the relevant information is captured by the disk Ω, we are led to use the PET-geometry which parameterizes the lines in the plane via the entry and exit points x_0,x_1 on $\partial\Omega$. The boundary information being symmetric

$$d(x_0,x_1,t)=d(x_1,x_0,t),$$

and with the incoming radiation assumed zero as before, the formula replacing (28) is

$$e^{-\int_{x_0}^{x_1}\mu(y)\,dy}\cdot e^{-is|x_1-x_0|/c}\cdot\int_{x_0}^{x_1} F(x,s)\,dx \;=\; D(x_0,x_1,s),$$

and Fourier inversion provides the following family of integral equations

$$e^{-\int_{x_0}^{x_1}\mu(y)\,dy}\cdot\int_{x_0}^{x_1} f(x,t-|x_1-x_0|/c)\,dx \;=\; d(x_0,x_1,t),$$

showing a similar type of delay between the source and data terms. In the case of PET a simple change of variables leads to the compact formula

$$\exp\{-\mathcal{R}[\mu](x_0,x_1)\}\cdot\mathcal{R}[f(\cdot,t)](x_0,x_1) \;=\; d(x_0,x_1,t+|x_1-x_0|/c) \tag{31}$$

involving the classical Radon transform parametrized through the lines $L(x_0, x_1)$ spanned by points $x_0, x_1 \in \partial\Omega$. Again we may in practice assume $t \pm |x_1 - x_0|/c \approx t$ in (31), leading to

$$\exp\{-\mathcal{R}[\mu](x_0, x_1)\} \cdot \mathcal{R}[f(\cdot, t)](x_0, x_1) = d(x_0, x_1, t) \tag{32}$$

The interpretation of (32) is that it is justified to do dynamics in PET simply by reconstructing a series of static images, as proposed already in the basic reference for probabilistic approaches to PET [**36**].

References

[1] Bauschke HH, Noll D, Celler A, Borwein JM, An EM-algorithm for dynamic SPEC tomography, IEEE Trans. Med. Imag. 18, no. 3, 1999, 252 – 261.

[2] Bertsekas DP, Constrained Optimization and Lagrange multiplier methods, Academic Press, New York, 1982.

[3] Borwein JM, Sun W, The stability analysis of dynamic SPECT systems, Numerische Mathematik 77, 1997, 283 – 298.

[4] Buck A, Westera G, vonSchulthess GK, Burger C, Modeling alternatives for cerebral carbon-11-iomazenil kinetics, J. Nuclear Med. 37, 1996, 699 – 705.

[5] Byrd RH, Lu P, Nocedal J, Zhu C, A limited memory algorithm for bound constrained optimization, SIAM J. on Scientific Computing 16, 1996, 1190 – 1208.

[6] Carson RE, Lange K, The EM parametric image reconstruction algorithm, J. Amer. Stat. Assoc. 80, 1985, 20 – 22.

[7] Celler A, Farncombe T, Harrop R, Lyster D, Dynamic heart-in-thorax phantom for functional SPECT, IEEE Transactions on Nuclear Science 44, 1997, 1600 – 1605.

[8] Celler A, Sitek A, Harrop R, Reconstruction of multiple line source attenuation maps. In: 1996 IEEE Nuclear Science Symposium Conference Record, pp. 1420 – 1424, November 1996.

[9] Celler A, Sitek A, Stoub E, Lyster D, Dykstra C, Worsley D, Fung A, Development of a multiple line source attenuation array for SPECT transmission scans. J. Nuclear Medicine 38, 1997, 215ff.

[10] Chen K, Huang SC, Feng D, New estimation methods that directly use the time accumulated counts in the input function in quantitative dynamic PET studies, Phys. Med. Biol. 39, 1994, 2073 – 2090.

[11] Dautrey R, Lions JL, Analyse mathématique et calcul numérique, tome 9, Masson, Paris 1984.

[12] Dicken V, Simultaneous activity and attenuation reconstruction in emission tomography, to appear in Inverse Problems.

[13] T. Farncombe, A. Celler, D. Noll, J. Maeght, Dynamic SPECT Imaging Using a Single Camera Rotation (dSPECT), IEEE Transactions on Nuclear Science,vol. 46, no. 4, pp.1055-1061, 1999.

[14] Gould NIM, Hribar ME, Nocedal J, On the solution of equality constrained quadratic programming problems arising in optimization. Preprint.

[15] Gullberg GT, Huesman RH, Ross SG, Di Bella EVR, Zeng GL, Reutter BW, Christian PE, Foresti SA, Dynamic cardiac single photon emission computed tomography, In: Nuclear Cardiology: State of the Art and Future Directions. Mosby-Year Book Inc., Philadelphia, PA. In press.

[16] Gjedde A, Compartmental Analysis. In: Principles of Nuclear Medicine, 2nd edition, eds Wagner HNJr, Szabo Z, Buchanan JW. Saunders, Philadelphia, pp. 451 – 461.

[17] Kammler DW, Least squares approximation of completely monotonic functions by sums of exponentials, SIAM J. Numerical Anal. 16, 1979, 801 – 818.

[18] Kao CM, Yap JT, Mukherjee J, Cooper M, Chen CT, An image reconstruction method for dynamic PET, IEEE NSS/MIC 1995 Conference Record.

[19] Limber MA, Limber MN, Celler A, Barney JS, Borwein JM, Direct reconstruction of functional parameters for dynamic SPECT, IEEE Trans. Nuclear Sciences 42, 1995, 1249 – 1256.

[20] Links JM, Frank TL, Becker LC, Effect of differential tracer washout during SPECT acquisition, Journal of Nuclear Medicine vol 32(12) pp 2253-2257, 1991.

[21] Liu DC, Nocedal J, On the limited memory BFGS method for large scale optimization problems, Mathematical Programming 45, 1989, 503 – 528.

[22] Maeght J, Noll D, Methods in dynamic emission tomography, Technical Report, http://mip.ups-tlse.fr/publi/rapp99/99.26.html
[23] Maeght J, Noll D, Resolution in dynamic emission tomography. To appear in SIAM Journal on Mathematical Analysis.
[24] Maréchal P, Togane D and Celler A, A new reconstruction methodology for computerized tomography: FRECT (Fourier Regularized Computed Tomography), Conference Records of the IEEE Medical Imaging Conference (Seattle, USA), 1999.
[25] Nakajima K, Taki J, Bunko H, Matsudaira M, Muramori A, Matsunari I, Hisada K, Ichihara T, Dynamic acquisition with a three-headed SPECT system: application of Technetium 99m-SQ30217 myocardial imaging, J. Nuclear Medicine 32, 1991, 1273 – 1277.
[26] Natterer F, Computerized tomography with unknown sources, SIAM J. Appl. Math. 43, 1983, 1201 – 1212.
[27] Natterer F, The Mathematics of Computerized Tomography, Teubner Verlag, Stuttgart, 1986.
[28] Natterer F, Determination of tissue attenuation in emission tomography of optically dense media, Inverse Problems 9, 1993, 731 – 736.
[29] O'Connor MK, Cho DS, Rapid radiotracer washout from the heart: Effect on image quality in SPECT performed with a single-headed gamma camera system, J. Nuc. Med., 33(6), 1992, 1146 – 1151.
[30] Oppenheim BE, Krepshaw JD, Dynamic hepatobiliary SPECT: a method for tomography of a changing radioactivity distribution, J. Nuclear Medicine 29, 1988, 98 – 102.
[31] Osborne MR, Smythe GK, A modified prony algorithm for exponential function fitting, SIAM J. Sci. Comput. 16, 1995, 119 – 138.
[32] Ruhe A, Fitting empirical data by positive sums of exponentials, SIAM J. Sci. Stat. Comp. 1, 1980, 481 – 498.
[33] Seber GAF, Wild, CJ, Nonlinear Regression, Wiley and Sons, 1988.
[34] van den Hoff J, Burchert W, Wolpers HG, Meyer GJ, Hundshagen H, A kinetic model for cardiac PET with [1-Carbon-11]-acetate, J. Nuclear Medicine 37, 1996, 521 – 529.
[35] Varah, JM, On fitting exponentials by nonlinear least squares, SIAM J. Sci. Stat. Comput. 6, 1985, 30 – 44.
[36] Vardi Y, Shepp LA, Kaufman L, A statistical model for Positron Emission Tomography, J. Amer. Stat. Assoc. 80, 1985, 8 – 20.
[37] Welch A, Clark R, Natterer F, Gullberg GT, Toward accurate attenuation correction in SPECT without transmission measurements, IEEE Trans. Med. Imag., 16(5), 1997, 532 – 541.
[38] Wells RG, Celler A, Harrop R, Analytic calculation of photon distribution in SPECT projections, IEEE Trans. Nuclear Sci.
[39] Young T, Attenuation correction in SPECT by simultanious reconstruction of emission and intrinsic attenuation information using an ART algorithm. PhD thesis, Graduate School of Syracuse University, 1995.
[40] Zhao S, Welland G, Wang G, Wavelet sampling and localization schemes for the Radon transform in two dimensions, SIAM J. Appl. Math. 57, no. 6, 1997, 1749 – 1762.

Université Paul Sabatier, Mathématiques pour l'Industrie et la Physique, 118 route de Narbonne, 31062 Toulouse, France

E-mail address: `maeght@mip.ups-tlse.fr (Jean Maeght)`

Université Paul Sabatier, Mathématiques pour l'Industrie et la Physique, 118 route de Narbonne, 31062 Toulouse, France

E-mail address: `noll@dumbo.ups-tlse.fr (Dominikus Noll)`

Stanford University, Department of Electrical Engineering, Information Systems Laboratory, Stanford, CA, USA

E-mail address: `boyd@ISL.Stanford.EDU (Stephen Boyd)`

Canadian Mathematical Society
Conference Proceedings
Volume **27**, 2000

Semi-spaces and Bases of Convexity Systems in Complete Lattices

Juan-Enrique Martinez-Legaz and Ivan Singer

Dédié à Jonathan Borwein
à l'occasion de la remise de son titre de Docteur Honoris Causa

ABSTRACT. For a convexity system $\mathcal{B}$ in a complete lattice E we introduce the notion of $\mathcal{B}$-semi-space and find a condition, which we call *property* (P), under which any base of $\mathcal{B}$ contains all $\mathcal{B}$-semi-spaces. We show that if E is Brauerian, then all convexity systems in E have property (P). We also give a sufficient condition for the existence of the smallest base and a necessary and sufficient condition for the smallest base to coincide with the set of all $\mathcal{B}$-semi-spaces. In the particular case when $\mathcal{B}$ is a convexity system in the complete lattice $E = 2^X$ of all subsets of a set X, ordered by containment, we recover some known concepts and results.

1. Introduction

By the classical strict separation theorem, the family of all closed half-spaces in a locally convex linear space X is a base (an intersectional base) of the family of all closed convex subsets of X, i.e., every closed convex set in X is an intersection of closed half-spaces. By another well-known theorem, discovered later, the family of all "semi-spaces" in a linear space X is the smallest base of the family of all convex subsets of X (i.e., every convex set in X is an intersection of semi-spaces), and it is contained in every other base of this set ([**10**], [**3**], [**4**], [**8**]); we recall that, by definition, a *semi-space* is a maximal (with respect to inclusion) convex subset of X which does not contain a given point, or, equivalently, a maximal convex cone in X which does not contain its vertex. These results have been extended to an intersectionally closed family $\mathcal{B}$ of subsets of a set X; such families have been considered e.g. by Moore [**9**] and have been called later "extended topologies" (see e.g. [**3**], [**4**]) or *convexity systems* (see e.g. the references in [**11**]); here we shall use this latter term, and the sets in $\mathcal{B}$ will be called *$\mathcal{B}$-convex sets.* Namely, it has been shown (see e.g. [**4**], [**12**]. [**11**] and the references therein) that under certain assumptions, the family of all "$\mathcal{B}$-semi-spaces" (i.e., of all maximal $\mathcal{B}$-convex sets which do not contain a given point) is the smallest intersectional base of $\mathcal{B}$.

1991 *Mathematics Subject Classification.* Primary: 06A23; secondary: 52A01, 52A05.
Key words and phrases. Complete lattice, convexity system, semi-space.

The above concepts of a convexity system $\mathcal{B}$ and $\mathcal{B}$-convex subsets of a set X have been extended to an arbitrary complete lattice $E = (E, \leq)$ (see e.g. Dolecki and Greco [**2**], who have used a different terminology). Namely, following the terminology of [**11**], we shall say that a subset $\mathcal{B}$ of E is a *convexity system* in E, if it is closed for supremum, i.e., if for any index set I,

$$x_i \in \mathcal{B}\ (i \in I) \Rightarrow \sup_{i \in I} x_i \in \mathcal{B}. \tag{1}$$

The elements $b \in \mathcal{B}$ are called "convex elements" of E. Adopting the usual conventions

$$\sup \emptyset = -\infty, \inf \emptyset = +\infty, \tag{2}$$

where $\emptyset$, $-\infty$ and $+\infty$ denote the empty set, and the least and greatest element of E, respectively, by (1) for $I = \emptyset$ we have

$$-\infty \in \mathcal{B}. \tag{3}$$

In particular, if X is an arbitrary set and $E = (2^X, \supseteq)$, the family of all subsets of X ordered by containment (i.e., $G_1 \leq G_2$ in E if and only if $G_1 \supseteq G_2$), then E is a complete lattice, with

$$-\infty = X, +\infty = \emptyset, \sup = \cap, \inf = \cup, \tag{4}$$

so the concepts of a convexity system $\mathcal{B}$ and $\mathcal{B}$-convex elements in E reduce to those for families of subsets of a set X, mentioned above. Furthermore, the concept of an intersectional base has been extended in [**11**], p. 46, definition 1.8, to the case of a convexity system $\mathcal{B}$ in a complete lattice E, namely, a subset $\mathcal{M}$ of $\mathcal{B}$ is said to be a *base* of $\mathcal{B}$, if each $b \in \mathcal{B}$ is the supremum of some subset $\mathcal{M}'$ of $\mathcal{M}$. In other words, bases yield a way to generate all elements of a convexity system from simpler ones.

In the present paper, our aim is to introduce and study semi-spaces with respect to a convexity system $\mathcal{B}$ in a complete lattice E and to study their relationship with the bases of $\mathcal{B}$.

In Section 2 we introduce a certain "property (P)" of convexity systems, which will be useful in the sequel. We show that in a Brauerian complete lattice every convexity system $\mathcal{B}$ has property (P), and we give an example of a convexity system with property (P) in a non-Brauerian atomic complete lattice.

In Section 3 we introduce $\mathcal{B}$-semi-spaces and show that every base of a convexity system $\mathcal{B}$ with property (P) contains all $\mathcal{B}$-semi-spaces. Hence, if the set $\mathcal{S}$ of all $\mathcal{B}$-semi-spaces is a base of $\mathcal{B}$, then it is the smallest base of $\mathcal{B}$. Also, we show that, for any convexity system $\mathcal{B}$ in an arbitrary complete lattice, a set $\mathcal{M} \subseteq \mathcal{B}$ is a minimal base of $\mathcal{B}$ if and only if it is the smallest base of $\mathcal{B}$.

In Section 4 we show that for a convexity system $\mathcal{B}$ in an atomic complete lattice for which the smallest base exists, every element of the smallest base of $\mathcal{B}$ is a $\mathcal{B}$-semi-space.

In Section 5 we give some sufficient conditions for the existence of the smallest base.

Finally, in Section 6 we show that in the particular case of lattices of subsets, our concepts and results yield some known concepts and results.

We conclude this Introduction by recalling some notations and concepts in a complete lattice $E = (E, \leq)$, which we shall use in the sequel. For $x, y \in E$, the notation $x < y$ means that $x \leq y$ and $x \neq y$. An *atom* is an element $a \in E \backslash \{-\infty\}$

such that

$$x \in E, x \leq a \Rightarrow \text{either } x = -\infty \text{ or } x = a. \tag{5}$$

We shall denote by $\mathcal{A}$ the set of all atoms of E.

A subset $\mathcal{Y}$ of a complete lattice E is called a *supremal generator* of E, if for each $x \in E$ we have

$$x = \sup_{\substack{z \in \mathcal{Y} \\ z \leq x}} z \qquad (x \in E), \tag{6}$$

or, equivalently (see e.g. [**11**], p. 40), if for each $x \in E$ there exists a subset $\mathcal{Y}_x$ of $\mathcal{Y}$ such that

$$x = \sup \mathcal{Y}_x. \tag{7}$$

A complete lattice E is said to be *atomic* if the set of $\mathcal{A}$ of all atoms of E is a supremal generator of E.

A complete lattice E is said to be *Brauerian* (see e.g. [**1**], Ch.2, Section 11 and Ch. 5, theorem 24), if "the inf is completely distributive with respect to sup", i.e., if for any index set I,

$$\inf(\sup_{i \in I} x_i, y) = \sup_{i \in I} \inf(x_i, y) \qquad (\{x_i\}_{i \in I} \subseteq E, y \in E). \tag{8}$$

Dualizing a well-known concept, we say that a subset $\mathcal{B}$ of a complete lattice $E = (E, \leq)$ is $\geq$*-inductive*, if for each chain (i. e., each totally ordered subset) $\mathcal{C}$ of $\mathcal{B}$, we have $\inf \mathcal{C} \in \mathcal{B}$. The axiom of choice is equivalent to the statement that every $\geq$-inductive set has minimal elements (see [**1**], Ch. 8, theorem 27).

For a subset $\mathcal{M}$ of a complete lattice E, an element $x \in E$ is said to be $\mathcal{M}$-*convex* if

$$x = \sup \{m \in \mathcal{M} | \, m \leq x\}. \tag{9}$$

The set of all $\mathcal{M}$-convex elements of E is denoted by $\mathcal{C}(\mathcal{M})$.

For a subset G of R^n we shall denote by $\overline{G}$ and int G the closure and, respectively, the interior of G.

2. Property (P)

DEFINITION 1. We shall say that in a complete lattice E, a convexity system $\mathcal{B}$ has *property* (P), if for each $z \in \mathcal{A}$ and each set $\{b_i\}_{i \in I} \subseteq \mathcal{B}$ such that $\sup_{i \in I} b_i \geq z$ there exists an index $i_0 \in I$ such that $b_{i_0} \geq z$.

The following lemma gives a large class of complete lattices in which every convexity system has property (P).

LEMMA 1. *Let E be a Brauerian complete lattice, $z \in \mathcal{A}$, and $\{m_i\}_{i \in I} \subseteq E$ be such that*

$$\sup_{i \in I} m_i \geq z. \tag{10}$$

Then there exists an index $i_0 \in I$ such that

$$m_{i_0} \geq z. \tag{11}$$

Hence, every convexity system $\mathcal{B}$ in E has properrty (P).

Proof. By (8) and (10) we have

$$\sup_{i\in I}\inf(m_i, z) = \inf(\sup_{i\in I} m_i, z) = z, \tag{12}$$

whence

$$\inf(m_i, z) \leq z \qquad (i \in I). \tag{13}$$

Hence, since $z \in \mathcal{A}$, for each $i \in I$ we must have either $\inf(m_i, z) = -\infty$ or $\inf(m_i, z) = z$. Now, if $\inf(m_i, z) = -\infty$ for all $i \in I$, then $\sup_{i\in I}\inf(m_i, z) = -\infty$, whence, by (12), we obtain $z = -\infty$, in contradiction with $z \in \mathcal{A}$. Therefore there exists an index $i_0 \in I$ such that $\inf(m_{i_0}, z) = z$, i.e., such that (11) holds. □

Let us give now an example of a convexity system with property (P) in a non-Brauerian atomic complete lattice.

EXAMPLE 1. Let E be the set of all open subsets of R, with $\leq$ being $\supseteq$, and with the complete lattice operations

$$\sup G_i = \ int\ (\cap_i G_i),\ \inf G_i = \cup_i G_i. \tag{14}$$

Then E is atomic, namely the atoms are the sets $R\backslash\{x\}$, where $x \in R$. We claim that E is non-Brauerian. Indeed, let

$$C := \{x|\, x < 1\}, D_i := \{x|\, x > \frac{i}{i+1}\}\ (i = 1, 2, ...). \tag{15}$$

Then $\cap_i D_i = \cap_i\{x|\, x > \frac{i}{i+1}\} = \{x|\, x \geq 1\}$, and hence

$$\sup_i D_i = \text{int}\ (\cap_i D_i) = \text{int}\ \{x|\, x \geq 1\} = \{x|\, x > 1\}, \tag{16}$$

so

$$\inf(\sup_{i\in I} D_i, C) = C \cup \text{int}\ (\cap_i D_i) = \{x|\, x < 1\} \cup \{x|\, x > 1\} = R\backslash\{1\}. \tag{17}$$

On the other hand,

$$\inf(D_i, C) = C \cup D_i = \{x|\, x < 1\} \cup \{x|\, x > \frac{i}{i+1}\} = R \qquad (i = 1, 2, ...), \tag{18}$$

and hence

$$\sup_{i\in I}\inf(D_i, C) = \text{int}\ (\cap_i R) = R, \tag{19}$$

which, together with (17), proves the claim that E is non-Brauerian. As an example of a convexity system which has property (P) one can take $\mathcal{B}$ to be the set of all open intervals in R which have integer endpoints, including R itself and the empty set $\emptyset$; indeed, if $B_i \in \mathcal{B}$ $(i \in I)$, $x \in R$ and $\sup_{i\in I} B_i = \text{int}\ \cap_{i\in I} B_i = \cap_{i\in I} B_i \subseteq R\backslash\{x\}$, that is, $x \notin \cap_{i\in I} B_i$, then there exists $i_0 \in I$ such that $x \notin B_{i_0}$, i.e., such that $B_{i_0} \subseteq R\backslash\{x\}$.

3. Semi-spaces and bases of convexity systems

Throughout the sequel, E denotes a complete lattice, and $\mathcal{A}$ the set of all atoms of E.

Let us introduce now

DEFINITION 2. a) Let E be a complete lattice, $\mathcal{B} \subseteq E$ a convexity system and $z \in \mathcal{A}$. We shall say that an element $b \in \mathcal{B}$ is a *$\mathcal{B}$-semi-space at z*, if b is a minimal element of the set $\{y \in B | \, z \leq y\}$, i.e., if $b \in B, z \leq b$ and there exists no element $b' \in B$ such that $z \leq b' < b$).

b) We shall say that an element $b \in \mathcal{B}$ is a *$\mathcal{B}$-semi-space*, if there exists an atom $z \in \mathcal{A}$ such that b is a $\mathcal{B}$-semi-space at z.

REMARK 1. For a convexity system $\mathcal{B}$ in a complete lattice which has atoms, $\mathcal{B}$-semi-spaces need not exist. Indeed, if E and $\mathcal{B}$ are as in example 1 above, then there exists no $\mathcal{B}$-semi-space, since for any given point $x \in R$, its complement $R \backslash \{x\}$ contains no maximal open interval with integer endpoints.

LEMMA 2. *If E is a complete lattice, $\mathcal{B} \subseteq E$ is a convexity system with property* (P), *and $\{b_i\}_{i \in I} \subseteq \mathcal{B}$ is such that $\sup_{i \in I} b_i$ is a $\mathcal{B}$-semi-space, then there exists an index $i_0 \in I$ such that*

$$\sup_{i \in I} b_i = b_{i_0}. \tag{20}$$

Proof. Let $b := \sup_{i \in I} b_i$ be a $\mathcal{B}$-semi-space at some atom $z \in \mathcal{A}$. Since $\mathcal{B}$ has property (P), there exists an index $i_0 \in I$ such that $b_{i_0} \geq z$. If $b \neq b_{i_0}$, then $b = \sup_{i \in I} b_i > b_{i_0} \geq z$, which, since $b_{i_0} \in \mathcal{B}$, contradicts the assumption that b is a minimal element of the set $\{y \in \mathcal{B} | \, z \leq y\}$. Thus, $b = b_{i_0}$. □

THEOREM 3. *Let E be a complete lattice, $\mathcal{B} \subseteq E$ a convexity system with property* (P), *and $\mathcal{M}$ a base of $\mathcal{B}$. Then $\mathcal{M}$ contains all $\mathcal{B}$-semispaces.*

Proof. If there exists no $\mathcal{B}$-semi-space, we are done. Assume now that $b \in \mathcal{B}$ is a $\mathcal{B}$-semi-space at some atom $z \in \mathcal{A}$. Since $\mathcal{M}$ is a base of $\mathcal{B}$, we have $\mathcal{M} \subseteq \mathcal{B}$ and there exists a family $\{m_i\}_{i \in I} \subseteq \mathcal{M}$ such that $b = \sup_{i \in I} m_i$. Then, by lemma 2, there exists an index $i_0 \in I$ such that $b = m_{i_0} \in \mathcal{M}$. □

Theorem 3 says that if E is a complete lattice and $\mathcal{B} \subseteq E$ is a convexity system with property (P), then each base of $\mathcal{B}$ contains all $\mathcal{B}$-semi-spaces. Hence, we obtain

COROLLARY 4. *If E is a complete lattice, $\mathcal{B} \subseteq E$ is a convexity system with property* (P), *and if the set $\mathcal{S}$ of all $\mathcal{B}$-semi-spaces is a base of $\mathcal{B}$, then $\mathcal{S}$ is the smallest base of $\mathcal{B}$ (i.e., every base of $\mathcal{B}$ contains $\mathcal{S}$).*

COROLLARY 5. *Let E be a Brauerian complete lattice, $\mathcal{B} \subseteq E$ a convexity system, and $\mathcal{M}$ a base of $\mathcal{B}$. Then $\mathcal{M}$ contains all $\mathcal{B}$-semispaces.*

Proof. This follows from theorem 3 and lemma 1. □

THEOREM 6. *Let E be a complete lattice and $\mathcal{B} \subseteq E$ a convexity system. For a base $\mathcal{M}$ of $\mathcal{B}$ the following statements are equivalent:*

1°. *$\mathcal{M}$ is a minimal base of $\mathcal{B}$ (i.e., no proper subset of $\mathcal{M}$ is a base of $\mathcal{B}$).*

2°. *$\mathcal{M}$ is the smallest base of $\mathcal{B}$.*

Proof. The implication $2^\circ \Rightarrow 1^\circ$ is obvious.

$1^\circ \Rightarrow 2^\circ$. Assume that a base $\mathcal{M}$ of $\mathcal{B}$ is not the smallest base of $\mathcal{B}$, so there exists a base $\mathcal{N}$ of $\mathcal{B}$ such that $\mathcal{M} \not\subseteq \mathcal{N}$. Let $m \in \mathcal{M} \backslash \mathcal{N}$. Then, since $\mathcal{N}$ is a base of $\mathcal{B}$, there exists a subfamily $\{n_i\}_{i \in I}$ of $\mathcal{N}$ such that $m = \sup\limits_{i \in I} n_i$. But, since $\mathcal{M}$ is a base of $\mathcal{B}$, for each $i \in I$ there exists a subfamily $\{m_{ij}\}_{j \in J_i}$ of $\mathcal{M}$ such that $n_i = \sup\limits_{j \in J_i} m_{ij}$. Hence

$$m = \sup_{i \in I} \sup_{j \in J_i} m_{ij}. \tag{21}$$

We claim that

$$\{m_{ij}\}_{i \in I, j \in J_i} \subseteq \mathcal{M} \backslash \{m\}. \tag{22}$$

Indeed, assume a contrario, that $m = m_{i_0 j_0}$ for some $i_0 \in I, j_0 \in J_{i_0}$. Then, we obtain $n_{i_0} = \sup_{j \in J_{i_0}} m_{i_0 j} \geq m_{i_0 j_0} = m$, whence $m = \sup_{i \in I} n_i \geq n_{i_0} \geq m$, so $m = n_{i_0} \in \mathcal{N}$, in contradiction with $m \in \mathcal{M} \backslash \mathcal{N}$; this proves the claim (22). Hence, since $\mathcal{M}$ is a base of $\mathcal{B}$, it follows that $\mathcal{M} \backslash \{m\}$ is a base of $\mathcal{B}$, so $\mathcal{M}$ is not a minimal base of $\mathcal{B}$. □

REMARK 2. A minimal base (or, equivalently, the smallest base) of $\mathcal{B}$ need not exist. Indeed, if E is an atomic complete lattice and $\mathcal{B}$ is a convexity system in E for which there exists no $\mathcal{B}$-semi-space (see e.g. remark 1), then, by theorem 7 below, $\mathcal{B}$ has no smallest base. For existence of the smallest base of $\mathcal{B}$ in some cases when $\mathcal{B}$-semispaces exist, see theorems 10 and 11 below.

4. Semi-spaces and the smallest base

THEOREM 7. *Let E be an atomic complete lattice, $\mathcal{B} \subseteq E$ a convexity system for which the smallest base $\mathcal{M}$ exists. Then each $m \in \mathcal{M}$ is a $\mathcal{B}$-semispace.*

Proof. Observe first that $-\infty \notin \mathcal{M}$; indeed, if we had $-\infty \in \mathcal{M}$, then $\mathcal{M} \backslash \{-\infty\}$ would be a base of $\mathcal{B}$, with $\mathcal{M} \backslash \{-\infty\} \subset \mathcal{M}$, in contradiction with the minimality of $\mathcal{M}$.

Let $m \in \mathcal{M}$ and assume, a contrario, that m is not a $\mathcal{B}$-semispace (i.e., m is not a minimal element of $\{y \in \mathcal{B} | \ z \leq y\}$, for any atom z). Then for each atom $z \leq m$ there exists $b_z \in \mathcal{B}$ with $z \leq b_z < m$, whence, since E is atomic,

$$m = \sup_{\substack{z \in \mathcal{A} \\ z \leq m}} z = \sup_{\substack{z \in \mathcal{A} \\ z \leq m}} b_z. \tag{23}$$

But, since $\mathcal{M}$ is a base of $\mathcal{B}$ and $b_z \in \mathcal{B}$, there exists a subfamily $\{m_{zi}\}_{i \in I_z}$ of $\mathcal{M}$ such that $b_z = \sup_{i \in I_z} m_{zi}$ (for each atom $z \leq m$). Hence

$$m = \sup_{\substack{z \in \mathcal{A} \\ z \leq m}} b_z = \sup_{\substack{z \in \mathcal{A} \\ z \leq m}} \sup_{i \in I_z} m_{zi}. \tag{24}$$

We claim that

$$\{m_{zi}\}_{z \in \mathcal{A},\ z \leq m,\ i \in I_z} \subseteq \mathcal{M} \backslash \{m\}. \tag{25}$$

Indeed, assume, a contrario, that $m = m_{z_0 i_0}$ for some atom $z_0 \leq m$, $i_0 \in I_{z_0}$. Then, we obtain $m = \sup_{z \in \mathcal{A}, z \leq m} b_z \geq b_{z_0} = \sup_{i \in I_{z_0}} m_{z_0 i} \geq m_{z_0 i_0} = m$, so $m = b_{z_0}$, in contradiction with our choice of b_{z_0}; this proves the claim (25). Hence, since $\mathcal{M}$ is a base of $\mathcal{B}$, from (24) and (25) it follows that $\mathcal{M} \backslash \{m\}$ is a base of $\mathcal{B}$, which contradicts the assumption that $\mathcal{M}$ is the smallest base of $\mathcal{B}$. □

REMARK 3. Let E be a complete lattice. For any supremal generator $\mathcal{Y}$ of E, which does not contain $-\infty$, one can define the concept of $\mathcal{Y}$-$\mathcal{B}$-hemi-space, by saying that an element $b \in \mathcal{B}$ is a $\mathcal{Y}$-$\mathcal{B}$-hemi-space if there exists $z \in \mathcal{Y}$ such that

b is a minimal element of the set $\{y \in \mathcal{B} | \, z \leq y\}$. Then, since the above proof has used only the fact that $\mathcal{A}$ is a supremal generator of E, the same argument shows that *if $\mathcal{Y}$ is a supremal generator of E and $\mathcal{M}$ is the smallest base of $\mathcal{B}$, then each $m \in \mathcal{M}$ is a $\mathcal{Y}$-$\mathcal{B}$-hemi-space.*

$\mathcal{Y}$-$\mathcal{B}$-hemi-spaces (in the sense of remark 3 above) may exist even when E has no atoms, as shown by

EXAMPLE 2. Let E be the set of all closed subsets of R, with $\leq$ being $\supseteq$ and with the complete lattice operations

$$\sup G_i = \cap_i G_i, \inf G_i = \overline{\cup_i G_i}. \tag{26}$$

Then E has no atoms (note that the sets $R \backslash \{x\}$ do not belong to E). Let $\mathcal{Y}$ be the set of all complements of open intervals. Then $\mathcal{Y}$ is a supremal generator of E, i.e., every closed subset of R is an intersection of complements of open intervals (since every open subset of R is a union of open intervals). Let $\mathcal{B} = E$. Then every element $Z \in \mathcal{Y}$, i.e., every subset of R of the form $Z = R \backslash D$, where D is an open interval in R, is a $\mathcal{Y}$-$\mathcal{B}$-hemi-space at Z, i.e., a maximal closed subset of R, not intersecting D.

In the above we have used the term "$\mathcal{Y}$-$\mathcal{B}$-hemi-space" (instead of "$\mathcal{Y}$-$\mathcal{B}$-semi-space") since in standard convexity theory in linear spaces a hemi-space is defined as a maximal convex set not intersecting an affine manifold.

¿From theorems 3 and 7 we obtain

COROLLARY 8. *If E is an atomic complete lattice and $\mathcal{B} \subseteq E$ a convexity system with property* (P), *then a base $\mathcal{M}$ of $\mathcal{B}$ is the smallest base of $\mathcal{B}$ if and only if $\mathcal{M}$ consists of all $\mathcal{B}$-semispaces.*

REMARK 4. From corollary 8 and theorem 6 it follows that if E is an atomic complete lattice and $\mathcal{B} \subseteq E$ a convexity system with property (P), for which there exists no $\mathcal{B}$-semi-space, then $\mathcal{B}$ has no minimal base.

5. Existence of the smallest base

LEMMA 9. *Let E be a complete lattice, and let $\mathcal{B} \subseteq E$ be a $\geq$-inductive convexity sytem. Then for each $b \in \mathcal{B}$ and each $z \in E$ such that $z \leq b$, there exists a minimal element of the set $\{y \in \mathcal{B} | \, z \leq y \leq b\}$.*

Proof. Let $\mathcal{Y} := \{y \in \mathcal{B} | \, z \leq y \leq b\}$. Since $b \in \mathcal{Y}$, we have $\mathcal{Y} \neq \emptyset$. Since $\mathcal{B}$ is $\geq$-inductive, so is $\mathcal{Y}$. Indeed, if $\mathcal{C}$ is any chain in $\mathcal{Y}$, then $\mathcal{C}$ is also a chain in $\mathcal{B}$, so $\inf \mathcal{C} \in \mathcal{B}$; but, since $\mathcal{C} \subseteq \mathcal{Y}$, we have $z \leq c \leq b$ $(c \in C)$, whence $z \leq \inf C \leq b$, so $\inf \mathcal{C} \in \mathcal{Y}$. Hence, $\mathcal{Y}$ has a minimal element. □

THEOREM 10. *Let E be an atomic complete lattice, and let $\mathcal{B} \subseteq E, \mathcal{B} \neq \{-\infty\}$, be a $\geq$-inductive convexity sytem with property* (P). *Then the set $\mathcal{S}$ of all $\mathcal{B}$-semi-spaces is* (*non-empty and*) *the smallest base of $\mathcal{B}$.*

Proof. By corollary 4, it will be enough to show that $\mathcal{S}$ is non-empty and a base of $\mathcal{B}$.

Let $b \in \mathcal{B}$. If $b > -\infty$, then by lemma 9, for any atom $z \in E$ such that $z \leq b$ there exists a minimal element, say s_z, of the set $\{y \in \mathcal{B} | \, z \leq y \leq b\}$. Clearly, s_z is a $\mathcal{B}$-semi-space at z (i.e., a minimal element of $\{y \in \mathcal{B} | \, z \leq y\}$). Consequently, $b = \sup\{s_z \in S | \, s_z \leq b\}$. On the other hand, if $b = -\infty$, then $b = \sup \emptyset$, where $\emptyset$ is the empty subset of $\mathcal{S}$. □

THEOREM 11. *Let E be an atomic complete lattice, and let $\mathcal{B} \subseteq E$ be a $\geq$-inductive convexity sytem with property* (P). *The following statements are equivalent:*

1°. *$\mathcal{B}\backslash\{-\infty\}$ coincides with the smallest base of the convexity system $\mathcal{B}$.*

2°. *$\mathcal{B}\backslash\{-\infty\}$ is a chain and every non-empty subset of $\mathcal{B}\backslash\{-\infty\}$ has a greatest element.*

Proof. $1^\circ \Rightarrow 2^\circ$. Assume 1° and let $b_1, b_2 \in \mathcal{B}\backslash\{-\infty\}$. Then, since by 1° and theorem 7 each $b \in \mathcal{B}\backslash\{-\infty\}$ is a $\mathcal{B}$-semi-space, $\sup(b_1, b_2) \in \mathcal{B}\backslash\{-\infty\}$ is a $\mathcal{B}$-semi-space. Hence, by corollary 2, we have either $b_1 \leq b_2$, or $b_1 \geq b_2$, which proves that $\mathcal{B}\backslash\{-\infty\}$ is a chain.

Now let $\mathcal{Y}$ be a non-empty subset of $\mathcal{B}\backslash\{-\infty\}$. Then, since each $b \in \mathcal{B}\backslash\{-\infty\}$ is a $\mathcal{B}$-semi-space, $\sup \mathcal{Y} \in \mathcal{B}\backslash\{-\infty\}$ is a $\mathcal{B}$-semi-space. Hence, by corollary 2, we have $\sup \mathcal{Y} \in \mathcal{Y}$.

$2^\circ \Rightarrow 1^\circ$. Assume 2°, and let $\mathcal{M} \subseteq \mathcal{B}\backslash\{-\infty\}$ be a base of $\mathcal{B}$, so $\mathcal{B} \subseteq \mathcal{C}(\mathcal{M}) = \mathcal{C}(\mathcal{M} \cup \{-\infty\})$. Then, by 2°, for any $\mathcal{Y} \subseteq \mathcal{M} \cup \{-\infty\}$ we have $\sup \mathcal{Y} \in \mathcal{Y}$, so $\mathcal{M}\cup\{-\infty\}$ is a convexity system. Hence, by [**11**], theorem 1.4, $\mathcal{C}(\mathcal{M}\cup\{-\infty\}) = \mathcal{M}\cup\{-\infty\}$, and therefore $\mathcal{B} \subseteq \mathcal{M} \cup \{-\infty\}$, that is, $\mathcal{B}\backslash\{-\infty\} \subseteq \mathcal{M}$. Consequently, $\mathcal{M} = \mathcal{B}\backslash\{-\infty\}$. □

6. Semi-spaces and bases of convexity systems in lattices of subsets

Now we shall consider the particular case where X is a non-empty set and $E = (2^X, \supseteq)$, the family of all subsets of X ordered by containment (i.e., $G_1 \leq G_2$ in E if and only if $G_1 \supseteq G_2$), with the complete lattice operations

$$\sup G_i = \cap_i G_i, \inf G_i = \cup_i G_i. \tag{27}$$

Then E has atoms, namely the complements of the singletons, i.e., the sets of the form

$$X\backslash\{x\}, \tag{28}$$

where $x \in X$.

Let $\mathcal{B}$ be any convexity system in E, i.e., any family of subsets of X closed for intersections; the sets in $\mathcal{B}$ are called "convex sets". By definition 1, a convexity system $\mathcal{B}$ has property (P) if for each $x \in X$ and each family $\{B_i\}_{i\in I} \subseteq \mathcal{B}$ such that $\cap_{i\in I} B_i \subseteq X\backslash\{x\}$, i.e., such that $x \notin B_i$ for some $i \in I$, there exists an index i_0 such that $B_{i_0} \subseteq X\backslash\{x\}$. Thus, every convexity system $\mathcal{B}$ in E has property (P); this follows also from lemma 1, since $E = (2^X, \supseteq)$ is Brauerian (see e.g. [**1**], Ch.5, Section 10).

By definition 2 above, a set $B \in \mathcal{B}$ is a $\mathcal{B}$-semi-space at an atom $X\backslash\{x\}$ if it is a maximal (with respect to inclusion $\subseteq$) convex subset of $X\backslash\{x\}$, i.e., a $\mathcal{B}$-semi-space at x in the usual sense (see [**11**], p. 55, definition 2.6 and the references on p. 393). Thus, theorems 3, 7 and corollary 8 yield [**11**], p. 55, theorem 2.5 and p. 56, corollary 2.1 respectively. Also, theorem 6 yields [**11**], p. 54, theorem 2.4.

A convexity system $\mathcal{B}$ in $E = (2^X, \supseteq)$ is $\geq$-inductive if for each chain $\mathcal{C}$ of $\mathcal{B}$ we have $\cup_{C\in\mathcal{C}} C \in \mathcal{B}$, i.e., if $\mathcal{B}$ is a so-called "algebraic" convexity system (see e.g. [**11**], theorem 2.10(b) and the references in [**11**], p. 391). Therefore, theorems 10 and 11 yield [**11**], p. 393, theorem 2.11 and proposition 2.11 respectively (see also the references in [**11**], p. 393).

Acknowledgement

The work of the second author has been partially supported by the Dirección General de Investigación Científica y Técnica (DGICYT), Project PB95-0679, and by the Comissionat per Universitats i Recerca de la Generalitat de Catalunya, Grant 1998SGR-00062.

References

[1] G. Birkhoff, Lattice theory (third ed.). Amer. Math. Soc. Colloquim Publ. 25, AMS, New York (1967).
[2] S. Dolecki and G. H. Greco, Cyrtologies of convergences. I. Math. Nachr. 126 (1986), 327-348.
[3] P. C. Hammer, Maximal convex sets. Duke Math. J. 22 (1955), 103-106.
[4] P. C. Hammer, Semispaces and the topology of convexity. In: Convexity (V. Klee, ed.). Proc. Symposia Pure Math. AMS, Providence (1963), 305-316.
[5] P. C. Hammer, Extended topology: Domain finiteness. Indag. Math. 25 (1963), 200-212.
[6] R. E. Jamison, Copoints in antimatroids. Congr. Numer. 29 (1980), 535-544.
[7] R. E. Jamison-Waldner, A perspective on abstract convexity: Classifying alignments by varieties. In: Convexity and related combinatorial geometry (D. C. Kay and M. Breen, eds.). Lecture Notes Pure Appl. Math. 76, M. Dekker, New York (1982), 113-150.
[8] V. Klee, The structure of semispaces. Math. Scand. 4 (1956), 54-64.
[9] E. H. Moore, Introduction to a form of general analysis. Amer. Math. Soc. Coll. Publ. 2, Amer. Math. Soc., New Haven, 1910.
[10] T. S. Motzkin, Linear inequalities. Mimeographed lecture notes, Univ. of California, Los Angeles (1951).
[11] I. Singer, Abstract convex analysis. Wiley-Interscience, New York, 1997.
[12] V. P. Soltan, Introduction to the axiomatic theory of convexity (in Russian). Shtiintza, Khishineu (1984).

CODE and Departament d'Economia i d'Història Econòmica, Universitat Autònoma de Barcelona, 08193 Bellaterra (Barcelona), Spain
E-mail address: `JuanEnrique.Martinez@uab.es`

Institute of Mathematics of the Romanian Academy, P. O. Box 1-764, 70700 Bucharest, Romania
E-mail address: `Ivan.Singer@imar.ro`

Canadian Mathematical Society
Conference Proceedings
Volume **27**, 2000

On Variational Characterizations of Asplund Spaces

Boris S. Mordukhovich and Bingwu Wang

Dedicated to J. M. Borwein

1. Introduction

In 1968, Asplund introduces the class of "strong differentiability" Banach spaces [**1**] (later called Asplund spaces) on which every continuous convex function is generically Fréchet differentiable. This class is sufficiently broad (including, in particular, all reflexive spaces) and convenient for many applications. We refer the reader to the books [**4**] and [**15**] for various properties and characterizations of Asplund spaces, and to [**14**] for recent applications in nonsmooth analysis and optimization. A remarkable characterization of Asplund spaces was obtained by Fabian [**7**] on the base of the smooth variational principle by Borwein and Preiss [**2**] and the method of separable reduction. It is proved in [**7**] that if a Banach space X is Asplund, then for any lower semicontinuous (l.s.c.) function $f\colon X \to (-\infty, \infty]$ the set

$$\widehat{\partial} f(\bar{x}) := \left\{ x^* \in X^* \;\middle|\; \liminf_{x \to \bar{x}} \frac{f(x) - f(\bar{x}) - \langle x^*, x - \bar{x} \rangle}{\|x - \bar{x}\|} \geq 0 \right\} \tag{1.1}$$

is nonempty on a dense set of points $\bar{x} \in \operatorname{dom} f := \{x \in X \mid f(x) < \infty\}$. The opposite implication follows from the earlier work of Ekeland-Lebourg [**6**] and Ioffe [**11**]. Construction (1.1) is usually called the Fréchet subdifferential of f at $\bar{x}$, since it can be treated as a direct set-valued extension of the classical Fréchet derivative to the case of nondifferentiable functions. Other characterizations of Asplund spaces, based on variational principles and involving Fréchet-like normals and ε-normals to closed sets, are obtained in [**8**], [**9**], and [**13**]. In particular, the results therein characterize Asplund spaces as a subclass of Banach spaces where Fréchet-like normal constructions exhibit certain nontriviality or density properties. In this paper we study relationships between density and nontriviality properties in terms of Fréchet-like subgradients, normals, and ε-normals in the framework of general

---

1991 *Mathematics Subject Classification.* Primary 49J52; Secondary 59C20.

Key words and phrases. variational analysis, Asplund spaces, generalized differentiation, variational principles.

Research was partly supported by the National Science Foundation under grant DMS-9704751 and by the Distinguished Faculty Fellowship at Wayne State University.

Banach spaces. We show that all these properties follow from the *extremal principle* established by Mordukhovich and Shao [**13**] as a characterization of Asplund spaces. The latter result can be treated as a variational extension of the classical separation theorem to nonconvex sets; see [**13**] for more references and discussions. It is shown in [**3**] that, for spaces admitting Fréchet smooth renorms, the extremal principle is actually equivalent to a version of the Borwein-Preiss smooth variational principle. In this way we give new proofs of some known characterizations of Asplund spaces and also obtain new characterizations of the Asplund property in terms of ε-normals. The rest of the paper is organized as follows. In Section 2 we present the basic definitions, notation, and some properties used in the sequel. Section 3 contains the main variational characterizations of Asplund properties and their corollaries. All the proofs are given in the last Section 4, where we establish relationships between the properties under consideration.

2. Definitions and Preliminaries

Throughout the paper we use X to denote a Banach space and X^* its norm dual. By $\|\cdot\|$, $B(X)$, and $S(X)$ we denote the norm, the closed unit ball, and the unit sphere of X respectively. Given a nonempty closed subset Ω of X and $\varepsilon \geq 0$, the set of *ε-normals* $\widehat{N}_\varepsilon(\bar{x};\Omega)$ to Ω at $\bar{x} \in \Omega$ is defined by

$$\widehat{N}_\varepsilon(\bar{x};\Omega) = \left\{ x^* \in X^* \;\middle|\; \limsup_{x \xrightarrow{\Omega} \bar{x}} \frac{\langle x^*, x - \bar{x}\rangle}{\|x - \bar{x}\|} \leq \varepsilon \right\}, \tag{2.1}$$

where $x \xrightarrow{\Omega} \bar{x}$ means that $x \to \bar{x}$ with $x \in \Omega$. When $\varepsilon = 0$, the set (2.1) is a cone that is called the *Fréchet normal cone* and is denoted by $\widehat{N}(\bar{x};\Omega)$ for simplicity. When the set Ω is convex, the Fréchet normal cone is reduced to the normal cone of convex analysis. Observe that, for every set Ω and any $\bar{x} \in \Omega$, one has

$$\widehat{N}(\bar{x};\Omega) + \varepsilon B(X^*) \subset \widehat{N}_\varepsilon(\bar{x};\Omega) \quad \text{for all} \quad \varepsilon \geq 0.$$

Thus, $\widehat{N}_\varepsilon(\bar{x};\Omega)$ always contains a nonzero element when $\varepsilon > 0$ in contrast to $\widehat{N}(\bar{x};\Omega)$. In this paper we use the words *normal density* for Ω to mean that the set of points

$$\bar{x} \in \Omega \quad \text{with} \quad \widehat{N}(\bar{x};\Omega) \neq \{0\}$$

is dense in the boundary of Ω. If we only have the set of such $\bar{x}$ nonempty, we call this property the *normal nontriviality* for Ω. Below we also consider the *ε-normal density* and *ε-normal nontriviality* properties formulated in terms of ε-normals in Theorem 3.1. Given an extended-real-valued function $f\colon X \to (-\infty, \infty]$, we denote by (1.1) its *Fréchet subdifferential* at the point $\bar{x} \in \operatorname{dom} f$. Let us mention the well-known relationship

$$\widehat{\partial} f(\bar{x}) = \left\{ x^* \in X^* \;\middle|\; (x^*, -1) \in \widehat{N}((\bar{x}, f(\bar{x})); \operatorname{epi} f) \right\} \tag{2.2}$$

between the Fréchet subdifferential of l.s.c. functions and the normal cone (2.1) to their epigraphical sets

$$\operatorname{epi} f = \{(x,\mu) \in X \times \mathbb{R} \,|\, \mu \geq f(x)\}.$$

We say that the *subdifferential nontriviality* (resp. *subdifferential density*) property holds for f if $\widehat{\partial} f(\bar{x}) \neq \emptyset$ for some $\bar{x} \in \operatorname{dom} f$ (resp. on a dense subset of the domain). In what follows we study relationships between subdifferential, normal, and

ε-normal density and nontriviality properties in the general Banach space framework and relate them to characterizations of Asplund spaces. We show that all these properties follow from necessary conditions for locally extremal points of closed set systems in the form of the *extremal principle*; see [**13**] for more details. Given two closed sets Ω_1 and Ω_2 in a Banach space X, we say that $\bar{x} \in \Omega_1 \cap \Omega_2$ is a *locally extremal point* of the system $\{\Omega_1, \Omega_2\}$ if there exist sequences $\{a_{1k}\}$ and $\{a_{2k}\}$ in X and a neighborhood V of $\bar{x}$, such that $a_{1k} \to 0$ and $a_{2k} \to 0$ as $k \to \infty$ and

$$(\Omega_1 + a_{1k}) \cap (\Omega_2 + a_{2k}) \cap V = \emptyset \text{ for all } \ k = 1, 2, \ldots .$$

In this case we call $\{\Omega_1, \Omega_2\}$ an *extremal system* at $\bar{x}$.

3. Main Characterizations

In this section we formulate the main variational characterizations of Asplund spaces. The proofs of these and some additional results will be given in the next section.

THEOREM 3.1. *The statements (a)–(g) are equivalent: (a) X is an Asplund space. (b)* (**extremal principle**) *Let $\{\Omega_1, \Omega_2\}$ be an extremal system of closed sets in X and let $\bar{x} \in \Omega_1 \cap \Omega_2$ be a locally extremal point. Then for any $\varepsilon > 0$ there exist $x_i \in \Omega_i \cap (\bar{x} + \varepsilon B(X))$ and $x_i^* \in \widehat{N}(x_i; \Omega_i) + \varepsilon B(X^*)$, $i = 1, 2$, such that*

$$x_1^* + x_2^* = 0 \quad \textit{and} \quad \|x_1^*\| + \|x_2^*\| = 1.$$

(c) (**subdifferential density**) *For every l.s.c. function $f : X \to (-\infty, \infty]$, the set of points*

$$\bar{x} \in \operatorname{dom} f \ \textit{ with } \ \widehat{\partial} f(\bar{x}) \neq \emptyset$$

is dense in the domain of f. (d) (**normal density**) *For every proper closed set Ω in X, the set of points*

$$\bar{x} \in \Omega \quad \textit{where} \quad \widehat{N}(\bar{x}; \Omega) \neq \{0\}$$

is dense in the boundary of Ω. (e) (**normal nontriviality**) *For every proper closed set Ω in X, the set of points*

$$\bar{x} \in \Omega \quad \textit{where} \quad \widehat{N}(\bar{x}; \Omega) \neq \{0\}$$

is nonempty. (f) (**ε-normal density**) *For every proper closed set Ω in X, there exists $\lambda_0 > 0$ such that for each $\varepsilon > 0$ the set*

$$\left\{ \bar{x} \in \Omega \,\middle|\, \exists x^* \in \widehat{N}_\varepsilon(\bar{x}; \Omega) \ \textit{ with } \|x^*\| = \lambda_0 \right\}$$

is dense in the boundary of Ω. (g) (**ε-normal nontriviality**) *For every proper closed set Ω in X, there exists $\lambda_0 > 0$ such that for each $\varepsilon > 0$ the set*

$$\left\{ \bar{x} \in \Omega \,\middle|\, \exists x^* \in \widehat{N}_\varepsilon(\bar{x}; \Omega) \ \textit{ with } \|x^*\| = \lambda_0 \right\}$$

is nonempty.

The subdifferential density characterization (c) of Asplund spaces is a well-known result; see Section 1. The extremal characterization (b) was established by Mordukhovich and Shao [**13**] together with the implication (a)$\Rightarrow$(d)[$\Rightarrow$(e)]. The opposite implication (e)$\Rightarrow$(a) was recently justified by Fabian and Mordukhovich [**8**]. In Section 4 we show that, for any Banach space X, the extremal principle (b) in $X \times \mathbb{R}$ implies the subdifferential density (c) in X and that the latter implies the normal density property (d) in $X \times \mathbb{R}$. Furthermore, we show that the fulfillment of the ε-normal nontriviality property in a Banach space $X \times \mathbb{R}$ ensures that X is Asplund. The combination of these results gives us all the statements of Theorem 3.1 with the new proof of the subdifferential characterization (a)$\Leftrightarrow$(c). As a corollary of the main results, we give a new proof of the following characterizations of Asplund spaces obtained in another paper by Fabian and Mordukhovich [**9**]. There

$$\widetilde{N}_\varepsilon(\bar{x};\Omega) = \left\{ x^* \in X^* \,\middle|\, \exists\gamma > 0 \text{ with } \frac{\langle x^*, x - \bar{x}\rangle}{\|x - \bar{x}\|} \le \varepsilon \text{ if } x \in \Omega \cap (\bar{x} + \gamma B(X)) \right\}$$

denoted the *local* ε*-support* of Ω at $\bar{x}$.

COROLLARY 3.2. *Let X be a Banach space . The following assertions are equivalent: (a) X is an Asplund space. (b) For every proper closed subset Ω of X, there exists $\varepsilon > 0$ such that for every $M > \varepsilon$ the set*

$$\left\{ \bar{x} \in \Omega \,\middle|\, \widetilde{N}_\varepsilon(\bar{x};\Omega)\backslash MB(X^*) \neq \emptyset \right\}$$

is dense in the boundary of Ω. (c) For every proper closed subset Ω of X, there exists $\varepsilon > 0$ such that for every $M > \varepsilon$ the set

$$\left\{ \bar{x} \in \Omega \,\middle|\, \widehat{N}_\varepsilon(\bar{x};\Omega)\backslash MB(X^*) \neq \emptyset \right\}$$

is dense in the boundary of Ω. (d) For every proper closed subset Ω of X, there exists $\varepsilon > 0$ such that for every $M > \varepsilon$ there exists $\bar{x} \in \Omega$ with $\widetilde{N}_\varepsilon(\bar{x};\Omega)\backslash MB(X^*) \neq \emptyset$. (e) For every proper closed subset Ω of X, there exists $\varepsilon > 0$ such that for every $M > \varepsilon$ there exists $\bar{x} \in \Omega$ with $\widehat{N}_\varepsilon(\bar{x};\Omega)\backslash MB(X^*) \neq \emptyset$.

REMARK 3.3. The above corollary is presented in a stronger form than Theorem 4.1 in [**9**] ("exists $\varepsilon > 0$" instead of "for every $\varepsilon > 0$" as in [**9**]), but it can be also obtained by the methods of [**9**].

4. Proofs and Auxiliary Results

This section consists of several propositions that establish relationship between the properties listed in Theorem 3.1. They are certainly of independent interest and justify the major steps in the proof of Theorem 3.1. We start with the proof that the extremal principle (b) implies the subdifferential density property (c) in any Banach spaces.

PROPOSITION 4.1. *Let X be a Banach space and let $f\colon X \to (-\infty, \infty]$ be an l.s.c. function. Assume that the extremal principle (b) in Theorem 3.1 holds in the space $X \times \mathbb{R}$. Then for any $\bar{x} \in \operatorname{dom} f$ and $\varepsilon > 0$ there is $\bar{x}_\varepsilon \in \operatorname{dom} f$ such that*

$$\|\bar{x}_\varepsilon - \bar{x}\| < \varepsilon \quad \textit{and} \quad \widehat{\partial} f(\bar{x}_\varepsilon) \neq \emptyset.$$

PROOF. By the lower semicontinuity of f we can find $\gamma < \varepsilon$ such that

$$f(x) \ge f(\bar{x}) - \varepsilon \quad \forall x \in \bar{x} + \gamma B(X).$$

Define the function

$$\widetilde{f}(x) := f(x) + \delta_{\bar{x}+\gamma B(X)}(x), \quad x \in X,$$

where $\delta_\Omega(x)$ denotes the indicator function of the set Ω. Then $\widetilde{f}$ is l.s.c. and $\widetilde{f}(\bar{x}) \le \inf_X \widetilde{f} + \varepsilon$. Choose $m > 1$ such that $\varepsilon/m < \gamma/2$. Applying Ekeland's variational principle [**5**], we find $x_0 \in X$ such that

$$\|x_0 - \bar{x}\| < \gamma/2 \quad \text{and} \quad \widetilde{f}(x_0) < \widetilde{f}(x) + m\|x - x_0\| \quad \forall x \in X\backslash\{x_0\}.$$

Let us form the closed sets in the space $X \times \mathbb{R}$ as follows

$$\Omega_1 := \operatorname{epi} \widetilde{f}, \quad \Omega_2 := \left\{ (x,\mu) \in X \times \mathbb{R} \,\middle|\, \mu \le \widetilde{f}(x_0) - m\|x - x_0\| \right\}.$$

It is easy to check that $(x_0, \widetilde{f}(x_0))$ is a locally extremal point of the set system $\{\Omega_1, \Omega_2\}$. We equip the space $X \times \mathbb{R}$ with the norm $\|(x,\mu)\| := \|x\| + |\mu|$ and observe the dual norm on X^* is given by

$$\|(x^*, \mu^*)\| = \max\{\|x^*\|, |\mu^*|\}.$$

Next we choose a positive constant η such that

$$\eta < \min\left\{ \frac{1}{2(m+1)}, \frac{\gamma}{2} \right\} \tag{4.1}$$

For this η, by the extremal principle, we find $(x_i, \mu_i) \in \Omega_i$ and

$$(x_i^*, \mu_i^*) \in \widehat{N}((x_i, \mu_i); \Omega_i)$$

for $i = 1, 2$ such that

$$\begin{aligned} &\|x_i - x_0\| < \eta, \quad i = 1, 2, \\ &\|(x_1^*, \mu_1^*)\| + \|(x_2^*, \mu_2^*)\| > 1 - \eta, \\ &\|x_1^* + x_2^*\| < \eta, \quad |\mu_1^* + \mu_2^*| < \eta. \end{aligned} \tag{4.2}$$

This implies the estimate

$$\max\{\|x_2^*\|, |\mu_2^*|\} > \frac{1}{2} - \eta \tag{4.3}$$

Due to (2.2) and the constructions above, we directly verify that

$$\mu_2 = \widetilde{f}(x_0) - m\|x_2 - x_0\|, \quad \mu_2^* > 0, \quad \text{and} \quad x_2^*/\mu_2^* \in \widehat{\partial}\phi(x_2),$$

where $\phi(x) := m\|x - x_0\|$. Since $\widehat{\partial}\phi(x_2) \subset mB(X^*)$, we have $\|x_2^*\| \le m\mu_2^*$ that implies

$$\mu_2^* > \frac{1}{2m} - \frac{\eta}{m} \tag{4.4}$$

due to (4.3). Now we derive from the estimates (4.1), (4.2), and (4.4) that

$$\mu_1^* \le \eta + \frac{\eta}{m} - \frac{1}{2m} < 0.$$

The latter yields $\mu_1 = \widetilde{f}(x_1)$. Indeed, assuming $\mu_1 > \widetilde{f}(x_1)$, we get from the definition of $\widehat{N}((x_1, \mu_1); \Omega_1)$ that

$$\limsup_{\mu \to \mu_1} \frac{\langle x_1^*, x_1 - x_1 \rangle + \mu_1^*(\mu - \mu_1)}{|\mu - \mu_1|} \le 0,$$

which implies $\mu_1^* = 0$, a contradiction. Thus $-\dfrac{x_1^*}{\mu_1^*} \in \widehat{\partial}\widetilde{f}(x_1)$ due to (2.2). By the choice of x_1, we see that it is an interior point of $\bar{x} + \gamma B(X)$. So $\widehat{\partial}\widetilde{f}(x_1) = \widehat{\partial} f(x_1)$, and we finish the proof by letting $\bar{x}_\varepsilon := x_1$. □

Now we begin to consider relationships between the other properties listed in Theorem 3.1. To proceed, we first need to give some definitions; cf. [**10**]. Let Ω be a closed subset of a Banach space X. A point $x \in \Omega$ is an *algebraic boundary point* of Ω if there exist $x_0 \in S(X)$ and a sequence $\lambda_k > 0$, $k = 1, 2, \ldots$, such that $\lim_{k\to\infty} \lambda_k = 0$ and $x + \lambda_k x_0 \notin \Omega$ for all $k = 1, 2, \ldots$. We call such an x_0 a *good* *direction* for x and form the set $G(x)$ of all the good directions. We say that $x \in \Omega$ is an *algebraic interior point* of Ω if it is not an algebraic boundary point of this set. Clearly, every algebraic boundary point belongs to the usual norm boundary of Ω. The following proposition easily follows from the definitions.

PROPOSITION 4.2. *Let Ω be a closed subset of a Banach space X. Take arbitrary $\bar{x} \in \Omega$ and $\gamma > 0$. If every point of $\Omega \cap (\bar{x} + \gamma B(X))$ is an algebraic interior point of this set, then $(\bar{x} + \gamma B(X)) \subset \Omega$. Therefore, the set of algebraic boundary points of Ω is dense in the norm boundary of Ω.*

PROOF. Suppose that the conclusion is not true. Since Ω is closed, we find $x_1 \in (\bar{x} + \gamma B(X))\backslash\Omega$ and $\widetilde{\gamma} > 0$ such that $x_1 + \widetilde{\gamma} B(X) \subset (\bar{x} + \gamma B(X))\backslash\Omega$. Let $x_2 \in \Omega$ be a point of $[x_1, \bar{x}] \cap \Omega$ nearest to x_1. Then x_2 can not be an algebraic interior point of the set in question. This is a contradiction. $\square$

The next simple statement is useful in what follows.

PROPOSITION 4.3. *of X with $X = Y \oplus \operatorname{span}\{x_0\}$. Define a new norm on X by*

$$|||x|||: = \|y\| + |\lambda|,$$

where $x = y + \lambda x_0$ with $y \in Y$. Then the norm $|||\cdot|||$ is equivalent to the original one.

PROOF. To see this, we choose $\xi^* \in X^*$ with $\langle \xi^*, x_0\rangle \neq 0$ and put $Y: = \operatorname{Ker}\xi^*$. One can check easily that $\|\cdot\| \le |||\cdot||| \le (2\|\xi^*\| + 1)\|\cdot\|$. $\square$

The following important result establishes relationships between the subdifferential and normal density properties of Theorem 3.1 for any Banach space. It shows, in particular, that the normal density in the space X can be derived from the subdifferential density in its hyperplane built through a good direction of an algebraic boundary point.

PROPOSITION 4.4. *Let X be a Banach space, Ω a closed subset of X, and $\bar{x}$ an algebraic boundary point of Ω. Fix a good direction $x_0 \in G(\bar{x})$ for $\bar{x}$ and take a closed subspace $Y \subset X$ such that $X = Y \oplus \operatorname{span}\{x_0\}$. Assume that the subdifferential density property holds in Y. Then for every $\gamma > 0$ there exist $x_\gamma \in (\bar{x} + \gamma B(X)) \cap \Omega$ and $\xi^* \in \widehat{N}(x_\gamma; \Omega)$ such that $\langle \xi^*, x_0\rangle > 0$.*

PROOF. First we choose $0 < \lambda_0 < \gamma/4$ such that $\bar{x} + \lambda_0 x_0 \notin \Omega$. As Ω is closed, there exists $\widetilde{\gamma} > 0$ such that

$$\bar{x} + \lambda_0 x_0 + \widetilde{\gamma} B(X) \subset X\backslash\Omega.$$

Form the sets

$$Q: = \bar{x} + \lambda_0 x_0 + \frac{\widetilde{\gamma}}{2} B(X) \cap (\bar{x} + \lambda_0 x_0 + Y) \quad \text{and} \quad \Lambda: = \cup_{\lambda \ge 0}(Q - \lambda x_0)$$

and observe that Λ is a closed subset of X due to Proposition 4.3. Without loss of generality, we assume that

$$\Omega \subset \bar{x} + \frac{\gamma}{2} B(X) \cap \Lambda.$$

Then let $\widetilde{\Omega}: = \cup_{\lambda \ge 0}(\Omega - \lambda x_0)$ and construct a function $f: Y \to (-\infty, \infty]$ by

$$f(x): = \inf\left\{\lambda \,|\, x - \lambda x_0 \in \widetilde{\Omega}\right\}.$$

Note that $f(x) = \infty$ if there is no λ with $x - \lambda x_0 \in \widetilde{\Omega}$. Using Proposition 4.3 again, we see that $\widetilde{\Omega}$ is a closed subset of X and that f is a proper l.s.c. function with $\operatorname{epi} f = \widetilde{\Omega}$. Furthermore, $y - f(y)x_0$ is a boundary point of Ω for each $y \in \operatorname{dom} f$. Now we apply the subdifferential density property for the function f and observe that $\widehat{N}(x; \widetilde{\Omega}) \subset \widehat{N}(x; \Omega)$ for every $x \in \Omega$, since $\Omega \subset \widetilde{\Omega}$. In this way we get the desired points x_γ and ξ^* satisfying all the requirements. □

The next proposition is crucial when transfer some properties (in particular, the density properties considered in the paper) from one hyperplane to another.

PROPOSITION 4.5. *Let X be Banach and $x^*, y^* \in X^* \backslash \{0\}$. Then there exist a linear isomorphism ψ:* Ker $x^* \to$ Ker y^* *and two constants $c_1, c_2 > 0$ such that*

$$c_1\|x\| \le \|\psi(x)\| \le c_2\|x\| \quad \textit{for all } x \in \operatorname{Ker} x^*.$$

PROOF. First we note that Ker $x^* \cup$ Ker $y^* \ne X$, so we can choose $h_0 \in S(X)$ such that $h_0 \notin$ (Ker $x^* \cup$ Ker y^*). Define a linear operatorψ: Ker $x^* \to Y^*$ as follows

$$\psi(x) := x - \frac{\langle y^*, x\rangle}{\langle y^*, h_0\rangle} h_0 \quad \text{for all } x \in \operatorname{Ker} x^* \tag{4.5}$$

and observe that $\psi(x) \in$ Ker y^*. From (4.5) we easily conclude that

$$\langle x^*, \psi(x)\rangle = -\frac{\langle y^*, x\rangle}{\langle y^*, h_0\rangle}\langle x^*, h_0\rangle,$$

$$\langle y^*, x\rangle = -\frac{\langle x^*, \psi(x)\rangle}{\langle x^*, h_0\rangle}\langle y^*, h_0\rangle$$

and therefore

$$x = \psi(x) - \frac{\langle x^*, \psi(x)\rangle}{\langle x^*, h_0\rangle} h_0. \tag{4.6}$$

The latter ensures that ψ is an isomorphism. Now employing (4.5) and (4.6), we get the estimates

$$\|\psi(x)\| \le \|x\| + \frac{\|y^*\|\,\|x\|}{|\langle y^*, h_0\rangle|},$$

$$\|x\| \le \|\psi(x)\| + \frac{\|x^*\|\,\|\psi(x)\|}{|\langle x^*, h_0\rangle|}.$$

Finally letting $c_1 := (1 + \|x^*\|/|\langle x^*, h_0\rangle|)^{-1}$ and $c_2 := 1 + \|y^*\|/|\langle y^*, h_0\rangle|$, we complete the proof of the proposition. □

The next result derives the Asplund property of a Banach space X from the ε-normal nontriviality property in the space $X \to \mathbb{R}$.

PROPOSITION 4.6. *Given a Banach space X, we assume that the ε-normal nontriviality property holds in $X \times \mathbb{R}$, i.e., for every proper closed set $\Omega \subset X \times \mathbb{R}$ there exists $\lambda_0 > 0$ such that the set*

$$\{\bar{x} \in \Omega \mid \exists x^* \in \widehat{N}_\varepsilon(\bar{x}; \Omega) \text{ with } \|x^*\| = \lambda_0\}$$

is nonempty for each $\varepsilon > 0$. Then X is an Asplund space.

PROOF. According to Kenderov's result [**12**], a Banach space is Asplund if and only if its dual has the *w^*-Radon-Nikodým property.* So it is sufficient to prove that the ε-normal nontriviality property in $X \times \mathbb{R}$ implies the w^*-Radon-Nikodým property holds in X^*. Recall that the w^*-Radon-Nikodým property means that any norm bounded subset of X^* admits a w^*-slice with an arbitrarily small diameter. Recall that a *w^*-slice* is a set of the form

$$S(A^*, x, \lambda) = \left\{ x^* \in A^* \,\middle|\, \langle x^*, x\rangle \geq \sup_{x_1^* \in A^*} \langle x_1^*, x\rangle - \lambda \right\},$$

where $x \in X$ and $\lambda > 0$. Given a bounded set A^* in X^*, we define the function

$$g(x) := \sup_{x^* \in A^*} \langle x^*, x\rangle, \quad x \in X,$$

which is l.s.c. and convex with $\operatorname{dom} g = X$ since A^* is bounded. Due to its convexity, g is continuous over X, and so is $f = -g$. Therefore, the functions f and g are both locally Lipschitzian. We assume that f is Lipschitz continuous on $\gamma_1 B(X)$ with modulus L for some $\gamma_1 > 0$. Let $\lambda_0 > 0$ be the constant in the ε-normal nontriviality property for the closed set $\Omega := \operatorname{epi} f$ in the space $X \times \mathbb{R}$. Denote

$$\widetilde{\varepsilon} := \min\left\{ \lambda_0^2, \lambda_0, 1, \frac{\sqrt{L^2 + 4\lambda_0(L+1)} - L}{4(L+1)} \right\}$$

and take $0 < \varepsilon < \widetilde{\varepsilon}$. Using the nontriviality assumption, we find $\bar{x} \in X$, $\xi^* \in X^*$, and $\alpha_0 \in \mathbb{R}^*$ such that

$$(\bar{x}, f(\bar{x})) \in ((0, f(0)) + \gamma_1 B(X \times \mathbb{R})) \quad \text{and}$$
$$(\xi^*, \alpha_0) \in \widehat{N}_\varepsilon((\bar{x}, f(\bar{x})); \operatorname{epi} f)$$

with $\max\{\|\xi^*\|, |\alpha_0|\} = \lambda_0$. Since $(\bar{x}, f(\bar{x}) + t) \in \operatorname{epi} f$ for all $t \geq 0$, it is easy to observe from the definition of Fréchet ε-normals that $\alpha_0 \leq \varepsilon$. Let us show that $\alpha_0 \leq -\sqrt{\varepsilon}$. First we assume that $|\alpha_0| \geq \|\xi^*\|$, then $|\alpha_0| = \lambda_0 > \varepsilon$. Therefore, in this case α_0 is negative and $\alpha_0 = -\lambda_0 < \sqrt{\varepsilon}$ by the assumption on ε. Next we assume that $|\alpha_0| < \|\xi^*\|$ and hence $\|\xi^*\| = \lambda_0$. Then we find $h_0 \in S(X)$ such that $\langle \xi^*, h_0\rangle \geq \frac{1}{2}\lambda_0$. By the definition of $\widehat{N}_\varepsilon((\bar{x}, f(\bar{x})); \operatorname{epi} f)$, there exists $\gamma_2 > 0$ such that

$$(\bar{x}, f(\bar{x})) + \gamma_2 B(X \times \mathbb{R}) \subset (0, f(0)) + \gamma_1 B(X \times \mathbb{R})$$

and

$$\frac{\langle \xi^*, h\rangle + \alpha_0(f(\bar{x}+h) - f(\bar{x}))}{\|h\| + |f(\bar{x}+h) - f(\bar{x})|} \leq 2\varepsilon \tag{4.7}$$

for all $(\bar{x}+h, f(\bar{x}+h)) \in (\bar{x}, f(\bar{x})) + \gamma_2 B(X \times \mathbb{R})$. By the continuity of f, there exists $\gamma_3 > 0$ such that

$$(\bar{x}+h, f(\bar{x}+h)) \in (\bar{x}, f(\bar{x})) + \gamma_2 B(X \times \mathbb{R}) \quad \text{for all } h \in \gamma_3 B(X).$$

Substituting λh_0 in (4.7) for h with $0 < \lambda < \gamma_3$, we have

$$\frac{\frac{1}{2}\lambda_0\lambda - |\alpha_0| L\lambda}{\lambda + L\lambda} \leq 2\varepsilon.$$

This implies

$$|\alpha_0| \geq \frac{\lambda_0}{2L} - \frac{2(L+1)}{L}\varepsilon > \sqrt{\varepsilon} > \varepsilon$$

by the assumption on ε. Therefore, α_0 is negative and $\alpha_0 < -\sqrt{\varepsilon}$, which justifies the desired estimate of α_0. Using this estimate together with (4.7), we get

$$\begin{aligned} &\left\langle \frac{1}{\alpha_0}\xi^*, h \right\rangle + (f(\bar{x}+h) - f(\bar{x})) \\ \geq\ & \frac{2\varepsilon}{\alpha_0}(\|h\| + L\|h\|) \geq \frac{2\varepsilon}{-\sqrt{\varepsilon}}(L+1)\|h\| \\ =\ & -2\sqrt{\varepsilon}(L+1)\|h\| \end{aligned} \tag{4.8}$$

for all $h \in \gamma_3 B(X)$. On the other hand, for each $x^* \in S(A^*, \bar{x}, \gamma_3\sqrt{\varepsilon})$ one has

$$\langle x^*, \bar{x} \rangle \geq -f(\bar{x}) - \gamma_3\sqrt{\varepsilon}. \tag{4.9}$$

Since $x^* \in A^*$, we also have

$$\langle x^*, \bar{x}+h \rangle \leq -f(\bar{x}+h) \qquad \text{for all } h \in X. \tag{4.10}$$

It follows from (4.9) and (4.10) that

$$\langle x^*, h \rangle \leq f(\bar{x}) - f(\bar{x}+h) + \gamma_3\sqrt{\varepsilon} \qquad \text{for all } h \in X.$$

Taking (4.8) into account, we get

$$\left\langle x^* - \frac{1}{\alpha_0}\xi^*, h \right\rangle \leq \gamma_3\sqrt{\varepsilon} + 2\sqrt{\varepsilon}(L+1)\|h\|$$

for all $h \in \gamma_3 B(X)$ and $x^* \in S(A^*, \bar{x}, \gamma_3\sqrt{\varepsilon})$. This implies that

$$\left\| x^* - \frac{1}{\alpha_0}\xi^* \right\| \leq (2L+3)\sqrt{\varepsilon}$$

and hence

$$\operatorname{diam} S(A^*, \bar{x}, \delta_3\sqrt{\varepsilon}) \leq 2(2L+3)\sqrt{\varepsilon}.$$

The latter ensures that X^* has the w^*-Radon-Nikodým property. This proves that X is an Asplund space. □

Now combining the above propositions with the well-known fact that the Asplund property is invariant with respect to taking the Cartesian product, we complete the proof of Theorem 3.1.

PROOF OF THEOREM 3.1. Implication (a)⇒(b) is proved in [**13**]. To prove (b)⇒(c), we observe that if X is Asplund, then $X \times \mathbb{R}$ is Asplund as well. So (b)⇒(c) follows from Proposition 4.1. Similarly, implication (c)⇒(d) follows from Proposition 4.4 and the fact that the Asplund property is inherited by subspaces. Implications (d)⇒(e), (d)⇒(f), (e)⇒(g), and (f)⇒(g) are trivial. Finally, implication (g)⇒(a) follows from Proposition 4.6. □

It remains to derive the characterizations in Corollary 3.2 from the corresponding results of Theorem 3.1.

PROOF OF COROLLARY 3.2. It is easy to observe that

$$\widetilde{N}_\varepsilon(x;\Omega) \subset \widehat{N}_\varepsilon(x;\Omega) \subset \widetilde{N}_{\varepsilon+\gamma}(x;\Omega) \qquad \text{for all } \varepsilon > 0 \text{ and } \gamma > 0$$

whenever $x \in \Omega$. Thus (b)⇔ (c) and (d)⇔ (e). So we only need to consider relationships between (a), (c) and (e). Actually, implication (c)⇒(e) is obvious, and (a)⇒(c) easily follows from the normal density property (e) in Theorem 3.1. It remains to show that property (e) in Corollary 3.2 implies that X is Asplund.

To furnish this, let us prove that the latter property in the corollary implies the ε-normal nontriviality property (g) in Theorem 3.1. Since

$$\widehat{N}_{\varepsilon_1}(x;\Omega) \subset \widehat{N}_{\varepsilon_2}(x;\Omega) \text{ if } 0 < \varepsilon_1 < \varepsilon_2,$$

it is sufficient to show that for any $\gamma < \varepsilon$ there exist

$$x \in \Omega \quad \text{and} \quad x^* \in \widehat{N}_\gamma(x;\Omega) \quad \text{with} \quad \|x^*\| = \lambda_0$$

for some fixed $\lambda_0 > 0$. Indeed, we may choose any $\lambda_0 > \varepsilon$ and take $x \in \Omega$ in (e) and $x_1^* \in \widehat{N}(x;\Omega)\backslash MB(X^*)$ with $M := \varepsilon\lambda_0/\gamma$. Putting $x^* := \dfrac{\lambda_0}{\|x^*\|}x^*$, we have $x^* \in \widehat{N}_\gamma(x;\Omega)$ and $\|x^*\| = \lambda_0$. □

REMARK 4.7. By a careful modification of the proof of Proposition 4.1, we can show that the extremal principle in $X \times \mathbb{R}$ implies the following *subdifferential variational principle*, which provides yet another characterization of Asplund spaces:

> *Let $f\colon X \to (-\infty, \infty]$ be l.s.c. and bounded from below. Then for every $\varepsilon, \lambda > 0$, and $x_0 \in \operatorname{dom} f$ with $f(x_0) < \inf_X f + \varepsilon$ there exist $\bar{x} \in X$ and $x^* \in \widehat{\partial} f(\bar{x})$ such that (i) $\|\bar{x} - x_0\| < \lambda$; (ii) $f(\bar{x}) < \inf_X f + \varepsilon$; (iii) $\|x^*\| < \varepsilon/\lambda$.*

We are going to consider this result and its various applications in a separate paper.

References

[1] E. Asplund, Fréchet differentiability of convex functions, *Acta Math.* **121** (1968), 31–47.

[2] J. M. Borwein and D. Preiss, A smooth variational principle with applications to subdifferentiability and to differentiability of convex functions, *Trans. Amer. Math. Soc.* **303** (1987), 517–527.

[3] J. M. Borwein, B. S. Mordukhovich and Y. Shao, On the equivalence of some basic principles in variational analysis, *J. Math. Anal. Appl.* **229** (1999), 228–257.

[4] R. Deville, G. Godefroy and V. Zizler, *Smoothness and Renorming in Banach Spaces*, Wiley, New York, 1993.

[5] I. Ekeland, On the variational principle, *J. Math. Anal. Appl.* **47** (1974), 324–353.

[6] I. Ekeland and G. Lebourg, Generic Fréchet differentiability and perturbed optimization problems in Banach spaces, *Trans. Amer. Math. Soc.* **224** (1976), 193–216.

[7] M. Fabian, Subdifferentiability and trustworthiness in the light of a new variational principle of Borwein and Preiss, *Acta Univ. Carolinae* **30** (1989), 51–56.

[8] M. Fabian, B. S. Mordukhovich, Nonsmooth characterizations of Asplund spaces and smooth variational principles, *Set-Valued Analysis* **6** (1998), 381–406.

[9] M. Fabian and B. S. Mordukhovich, Separable reduction and supporting properties of Fréchet-like normals in Banach spaces, *Canad. J. Math.* **51** (1999), 26–48.

[10] R. B. Holmes, *Geometric Functional Analysis and its Applications*, Springer, New York, 1975.

[11] A. D. Ioffe, On subdifferentiability spaces, *Ann. N. Y. Sci.* **410** (1983), 107–119.

[12] P. S. Kenderov, Monotone operators in Asplund space, *C. R. Acad. Bulgare Sci.* **30** (1977), 963–964.

[13] B. S. Mordukhovich and Y. Shao, Extremal characterizations of Asplund spaces, *Proc. Amer. Math. Soc.* **124** (1996), 197–205.

[14] B. S. Mordukhovich and Y. Shao, Nonsmooth sequential analysis in Asplund spaces, *Trans. Amer. Math. Soc.* **348** (1996), 1235–1380.

[15] R. R. Phelps, *Convex Functions, Monotone Operators and Differentiability*, 2nd edn., Springer, New York, 1993.

DEPARTMENT OF MATHEMATICS, WAYNE STATE UNIVERSITY, DETROIT, MI 48202, USA
E-mail address: boris@math.wayne.edu, wangbw@math.wayne.edu

Canadian Mathematical Society
Conference Proceedings
Volume **27**, 2000

Transposition of Relations

Jean-Paul Penot

To Jonathan Borwein, on the occasion of his award of a Dhc

ABSTRACT. A new notion of transpose of a correspondence is introduced. It corresponds to the classical one in case of a linear mapping or of a convex process. Various calculus rules are established. In particular it is shown that a chain rule can be devised under a qualification assumption.

1. Introduction

During the last decades the role of correspondences (or multimappings or set-valued mappings) has become increasingly important (see for instance [**3**], [**4**], [**6**], [**14**], [**20**], [**21**], [**32**]-[**35**]...) in several fields of mathematics. In optimization, several works of J. Borwein ([**7**]-[**12**]) have been devoted to show the advantages of dealing with a multivalued approach to a number of problems (see also [**2**], [**19**], [**22**], [**26**], [**29**], [**36**]...). A combination of duality theory with the methods of study of correspondences has proved to be useful ([**32**], [**33**], [**16**], [**14**], [**3**], [**4**], [**36**]...). In particular, the important new field of nonsmooth analysis represented by the study of derivatives and coderivatives of multimappings (or relations) relies on both methods (see [**18**], [**27**] and their references). It is the purpose of the present paper to extend the notion of transpose mapping to multimappings which are not necessarily processes (i.e. positively homogeneous). Such an aim has been pursued in the book [**3**] by Aubin and Ekeland. Their proposal relies on the concept of barrier cone of a convex set. We propose another approach using the notion of polar convex set with the purpose of getting chain rules and we compare it to the one in [**3**].

We also compare the transposition we define with the one introduced for fans by Ioffe [**17**]. We exhibit a special class of fans for which the norm of the transpose fan is equal to the norm of the fan. We present estimates for the norms of the transpose of a process which refine results in [**10**], [**12**] and use only separation arguments. We introduce a distance in the space of bounded processes. We compare the distance

1991 *Mathematics Subject Classification.* Primary 49K24, 49K15; Secondary 34A60, 34H05.
Key words and phrases. adjoint, convexity, convex process, correspondence, fan, norm, positively homogeneous, relation, transposed.

of two fans with the distance of their transpose and we do the same for a particular class of convex processes which seems to be of some interest.

In section 3 we give various calculus rules which correspond to several usual operations. For the sake of simplicity we consider only qualification conditions in the case of closed convex subsets of Banach spaces; other assumptions involving openness conditions or finite dimensionality could be given along similar lines. We devote section 4, to a chain rule which is well known for the composition of convex processes with continuous linear maps ([**3**], [**4**]) or when the two multimappings are convex processes ([**10**]). Such a rule applies for densely defined linear operators, an important case in linear functional analysis. The condition we propose is symmetric in such a case; but it is obviously satisfied if one of the operators is everywhere defined.

We also present a condition ensuring that the domain of the transpose is the whole space (Theorem 2.3) and we provide estimates of the norm of the transpose which involve arguments as simple as in the linear case.

Thus the tool of transposition can be fully extended from linear algebra to the category of convex multimappings. Among possible applications are certain models in mathematical economics (see [**23**], [**24**], [**32**]-[**35**]) and duality theory (see section 5 for some elements).

2. Transpose and Bi-transpose

The very definition of the transpose of a multimapping is delicate. As observed by R.T. Rockafellar ([**32**], [**33**]) the formula $\left(F^T\right)^T = F$ does not hold without the introduction of an "orientation". Here we suggest other possibilities. Also it is not clear whether the transpose should be a closed convex process (i.e. a multimapping whose graph is a closed convex cone) as in [**3**] and elsewhere, or just a closed convex multimapping. Here we choose this second viewpoint.

Our analysis of transposition relies on the notion of pairing (or coupling) between topological vector spaces X, X', i.e. a function $c : X' \times X \to \mathbb{R}$ also denoted by $\langle .,. \rangle$ which is bilinear and non degenerate. Given two pairs of coupled vector spaces $(X, X'), (Y, Y')$ and a multimapping $F : X \rightrightarrows Y$ (identified with its graph $Gr\,(F) \subset X \times Y$ when no confusion may arise), we introduce a new pairing between $X \times Y$ and $Y' \times X'$ by

$$\langle (y', x'), (x, y) \rangle_- = \langle x', x \rangle - \langle y', y \rangle \tag{2.1}$$

It is obviously a nondegenerate bilinear map. Then we define the *transpose* of F as the multimapping $F^T : Y' \rightrightarrows X'$ whose graph is the polar set $F^\ominus$ of F with respect to this pairing:

$$\begin{aligned} Gr\left(F^T\right) &= (Gr\,(F))^\ominus \\ &= \{(y', x') \in Y' \times X' : \forall\, (x, y) \in Gr\,(F) \ \ \langle (y', x'), (x, y) \rangle_- \le 1\} \end{aligned}$$

so that

$$F^T(y') = \{x' \in X' : \forall x \in X, \forall y \in F(x) \ \ \langle x', x \rangle - \langle y', y \rangle \le 1\}.$$

Thus $Gr\left(F^T\right)$ belongs to the class $C_0(Y' \times X')$ of weakly closed convex subsets of $Y' \times X'$ which contain the origin. This notion is related to the definition of [**3**] p. 139 in the following way : the adjoint (or transpose) $F^\star$ of F in the Aubin-Ekeland's sense is the shady hull of F in this sense that its graph is the shady hull of the graph of F^T :

$$Gr\left(F^{\star}\right) = [1,\infty)Gr\left(F^T\right);$$

recall that a subset S of a vector space Z is *shady* if $tz \in S$ whenever $z \in S$ and $t \in [1,\infty)$. Since $Gr(F^T)$ is a convex set containing 0, $F^{\star}$ is also the process generated by F^T :

$$Gr\left(F^{\star}\right) = \mathbb{P}Gr\left(F^T\right)$$

($\mathbb{P}$ being the set of positive numbers).

When the graph of F is *shady*, in particular when F is a *process* (i.e. is positively homogeneous in the terminology of [**36**], i.e. when its graph is a cone) both definitions agree and coincide with the one considered in [**10**], [**29**], [**32**]-[**35**]:

$$F^T\left(y'\right) = \{x' \in X' : \forall x \in X, \forall y \in F(x) \quad \langle x', x\rangle \leq \langle y', y\rangle\}.$$

Homogenization brings a link between the two cases. If $F : X \rightrightarrows Y$ is a multimapping, its *homogenization* is the multimapping $\widehat{F} : X \times \mathbb{R} \rightrightarrows Y \times \mathbb{R}$ with domain included in $X \times \mathbb{P}$ given by

$$\widehat{F}(x,t) = tF\left(t^{-1}x\right) \times \{t\}, \qquad (x,t) \in X \times \mathbb{P}.$$

Then one sees that

$$\begin{aligned} x' \in F^T\left(y'\right) \quad &\Longleftrightarrow \quad \forall t' \in \mathbb{P} \ \ (x', t'-1) \in \left(\widehat{F}\right)^T (y', t') \\ &\Longleftrightarrow \quad \exists t' \in \mathbb{P} \quad (x', t'-1) \in \left(\widehat{F}\right)^T (y', t'). \end{aligned}$$

Example. Taking for F the epigraph relation E_f associated with a function $f : X \to \overline{\mathbb{R}}$ given by $E_f(x) := \{r \in \mathbb{R} : r \geq f(x)\}$, one sees that the epigraph relation associated with the convex conjugate function f^* of f is given by

$$(x', t') \in E_{f^*} \Longleftrightarrow (x', 0) \in \left(\widehat{E_f}\right)^T (1, t').$$

Note that we also have for $x' \in X'$, $s > 0$

$$s \in E_{f^*}\left(x'\right) \Longleftrightarrow x' \in s\, E_f^T\left(s^{-1}\right),$$

as easily checked. These equivalences show the connections between transposition of processes and conjugacy; see [**15**] for a use of these connections for continuity properties.

Example. Let A be a linear operator from a subset W of X into Y with closed graph and let F be given by $F(x) = \{A(x)\}$ for $x \in W$, $F(x) = \emptyset$ for $x \in X \backslash W$. Then, the domain of F^T is the set

$$D(F^T) := \{y' \in Y' : \exists x' \in X' \ y' \circ A = x' \mid W\}.$$

Then F^T is a multivalued mapping. In fact, for any $y' \in D(F^T)$ and any $x' \in F^T(y')$ one has $F^T(y') = x' + W^0 = F^T(y') + W^0$ where W^0 is the polar set of W given by $W^0 := \{x' \in X^* : \langle x', x\rangle \leq 1 \ \forall x \in W\}$.

Example. Let A be a linear continuous operator from X into Y, and let D be a nonempty convex subset of X. Then the multimapping F given by $F(x) = A(x)$ for $x \in D$, $F(x) = \emptyset$ otherwise, is a convex multimapping with transpose F^T given by $F^T(y') = A^T(y') + D^0$.

Example. Let A be as in the preceding example and let C be a subset of Y. Then the multimapping F given by $F(x) = A(x) - C$ has its transpose given by $F^T(y') = A^T(y')$ for $y' \in C^0$, $F^T(y') = \emptyset$ otherwise.

Example. The preceding two examples can be combined. A simple expression of the transpose F^T of F defined by $F(x) = A(x) - C$ for $x \in D$, $\emptyset$ otherwise, can be given when C and D are shady, or when one of them is a cone:

$$F^T(y') = A^T(y') + D^0 \text{ for } y' \in C^0, \ \emptyset \text{ otherwise.}$$

Note that in this case the negative polar cone C^- of C given by

$$C^- := \{x' \in X^* : \langle x', x \rangle \leq 0 \, \forall x \in 0\}$$

or the negative polar D^- of D can be substituted to the corresponding polar set. When C and D are convex sets containing 0, one has

$$F^T(y') = \bigcup \{A^T(y') + (1-t)D^0 : \ y' \in tC^0, \ \ t \in [0,1]\},$$

where for a closed convex subset S of a n.v.s. Z and $t = 0$ the set tS is interpreted as the asymptotic cone 0^+S given by

$$0^+S := \{z \in Z : s + tz \in S \, \forall s \in S, \ \forall t \in \mathbb{R}_+\}.$$

When C and D are arbitrary convex sets, $F^T(y')$ is the union of the sets

$$\begin{array}{lcl} \bigcup \{A^T(y') + (1-t)D^0 & : & y' \in tC^0, \ \ t \in [0,1]\} \\ \bigcup \{A^T(y') + (1-t)D^0 & : & y' \in tC^\nabla, \ \ t \in (-\infty, 0)\} \\ \bigcup \{A^T(y') + (1-t)D^\nabla & : & y' \in tC^0, \ \ t \in (1, \infty)\}, \end{array}$$

where for a subset S of a n.v.s. Z

$$S^\nabla := \{z' \in Z' : \ \forall z \in S \ \langle z', z \rangle \geq 1\}.$$

These assertions can be checked directly or by using Lemma 20 and the fact that the graph of F is the image of the product $D \times C$ by the isomorphism $b : (x, y) \mapsto (x, Ax - y)$.

Example. Let A be a subset of the set of continuous linear mappings from X into Y and let F be given by $F(x) = \{A(x) : A \in A\}$. Then $y' \in Y'$ is in the domain of F^T iff the kernel of y' contains the range of $A - B$ for any $A, B \in A$; then $F^T(y') = \{B^T(y')\}$ for an arbitrary $B \in A$. Again, this example can be combined with the previous ones.

Since $Gr\,(F)$ is not necessarily closed, a formula like $F^{**} = F$ cannot be expected without extra assumptions. On the contrary, the bipolar theorem applied to the graph $Gr\,(F)$ of F in $X \times Y$ yields the following result.

PROPOSITION 1. *Suppose the graph $Gr\,(F)$ of $F : X \rightrightarrows Y$ belongs to the class $C_0\,(X \times Y)$ of weakly closed convex subsets of $X \times Y$ containing 0. Then the bi-transpose $F^{TT} = \left(F^T\right)^T$ of F satisfies the relation $Gr\left(F^{TT}\right) = -Gr\,(F)$, or, for each $x \in X$*

$$F^{TT}(x) = -F(-x).$$

Here we define the transpose G^T of $G : Y' \rightrightarrows X'$ in the same manner as in the definition of F^T, i.e. by using the coupling

$$\langle (x, y), (y', x') \rangle_+ = \langle y', y \rangle - \langle x', x \rangle, \tag{2.2}$$

so that

$$G^T(x) = \{y \in Y : \forall (y', x') \in Gr(G) \;\; \langle y', y\rangle - \langle x', x\rangle \leq 1\}.$$

PROOF. Let $C = Gr(F)$, $D = Gr(F^T)$. Since the coupling $\langle .,.\rangle_+$ of (2) is the opposite of the coupling $\langle .,.\rangle_-$ of (1) with an interchange of the spaces, denoting by $D^\ominus$ the polar of D with respect to $\langle .,.\rangle_+$, we have

$$\begin{aligned} Gr\left(F^{TT}\right) &= D^\oplus = (-D)^\ominus \\ &= \left(-C^\ominus\right)^\ominus = -C^{\ominus\ominus} \end{aligned}$$

and since $C^{\ominus\ominus} = C$ by the bipolar theorem, the result follows. □

The simpler formula $F^{TT} = F$ could be obtained by keeping the same coupling $\langle .,.\rangle_-$ in defining $\left(F^T\right)^T$. But then coherence would be lost unless one introduces a notion of orientation as in [**32**], [**33**]. In order to avoid troubles with the choice of couplings, in the sequel we consider only the classical pairing

$$\langle (x', y'), (x, y)\rangle = \langle x', x\rangle + \langle y', y\rangle$$

and we take polars C^0, C^- of subsets C of $X \times Y$ only with respect to this pairing, so that

(2.3) $$x' \in F^T(y') \Longleftrightarrow (x', -y') \in (Gr(F))^0.$$

We observe that when F is odd, closed and convex we recover the equality $F^{TT} = F$; this is the case when F is a linear mapping with closed graph.

When X and Y are Hilbert spaces, the Riesz isomorphisms $R_X : X' \to X$, $R_Y : Y' \to Y$ enables one to consider the multimapping $F^* := R_X \circ F^T \circ R_Y^{-1}$ we call the *adjoint* of F. The study of adjoints is similar to the study of transposed multimappings.

In order to study the domain $D\left(F^T\right)$ of the transpose F^T of $F : X \rightrightarrows Y$, let us introduce the *barrier function* $b_F : X \times Y' \to \mathbb{R} \cup \{-\infty, \infty\}$ associated with F, given by $b_F(x, y') := \inf\langle y', F(x)\rangle$. Equivalently, using the support function of the set $C := F(x)$ given by

(2.4) $$s(y', C) := \sup\langle y', C\rangle := y \in C\sup\langle y', y\rangle,$$

one has

(2.5) $$b_F(y', x) = -s(-y', F(x)).$$

LEMMA 2. *The function b_F is convex (resp. sublinear) in x whenever F is a convex relation (resp. a convex process).*

The proof is easy and uses the fact that for $x, x' \in X$, $t, t' \geq 0$ with $t + t' = 1$ one has $tF(x) + t'F(x') \subset F(tx + t'x')$.

Recall that the *norm* $\|F\|$ (or *remoteness norm*) of a process $F : X \rightrightarrows Y$ is defined by (see [**30**]):

(2.6) $$\|F\| = \sup\{d(0, F(x)) : x \in X, \|x\| \leq 1\}.$$

Here, for a subset S of Y, $d(0,S) := \inf\{d(0,y) : y \in S\}$ (with the convention $\inf\emptyset = +\infty$) is the *remoteness* of S. It is shown in [**30**] that $\|F\|$ is finite whenever X and Y are Banach spaces, F is a closed convex process and $D(F) = X$. Since F

is set-valued, another quantity appears as a generalization of the norm of a linear map; it is given by

$$\|F\|^S = \sup \{e(F(x), 0) : x \in D(F), \|x\| \leq 1\}, \tag{2.7}$$

where, for a subset S of Y,

$$e(S, 0) = \sup \{d(y, 0) : y \in S\}$$

is the excess of S over $\{0\}$. We call $\|F\|^S$ the *spread norm* of F (or the *conorm* of F, a terminology partially justified by the following theorem). Clearly, when $D(F) = X$, one has $\|F\| \leq \|F\|^S$. Note that the following relations, in which F, G are processes from X to Y and $\lambda \in \mathbb{R}$, show that these terminologies have some basis:

$$\begin{aligned} \|F+G\| &\leq \|F\| + \|G\|, & \|F+G\|^S &\leq \|F\|^S + \|G\|^S, \\ \|\lambda F\| &= |\lambda| \|F\|, & \|\lambda F\|^S &= |\lambda| \|F\|^S. \end{aligned}$$

Moreover, the following easy estimates about composition of processes show that remoteness norms and spread norms have intertwined fates:

$$\|G \circ F\| \leq \|G\| \|F\|^S, \quad \|G \circ F\|^S \leq \|G\|^S \|F\|^S.$$

We also observe that the homogeneous procedure does not give a very sensible way of measuring the growth properties of a convex relation: for any convex relation F, the norm of its homogenized process $\widehat{F}$ is infinite as the domain of $\widehat{F}$ is never $X \times \mathbb{R}$; however, we see that $\left\|\widehat{F}\right\| = \max(\|F\|^S, 1)$ if we endow a product space with the supremum norm.

The following statement gives a sufficient condition ensuring that the transpose F^T is everywhere defined and provides estimates for the norm and conorm of F^T. The result is reminiscent to [**10**] Theorem 10.1, [**12**] Theorem 2.1; however, we do not suppose the process is closed and that the spaces are complete. Our arguments do not use CS-closed sets nor open mapping theorems; they simply rely on the Hahn-Banach theorem.

THEOREM 3. *(a) For any process $F : X \rightrightarrows Y$ between n.v.s. one has $\left\|F^T\right\|^S \leq \|F\|$.*

(b) For any process $F : X \rightrightarrows Y$ between n.v.s. one has $\left\|F^T\right\| \geq \|F\|^S$. Thus, whenever $D(F) = X$ and $D(F^T) = Y'$, one has $\left\|F^T\right\| = \left\|F^T\right\|^S = \left\|F^T\right\| = \|F\|^S$.

(c) If F is a convex process with $\|F\|^S$ finite, then $D\left(F^T\right) = Y'$ and F^T has a finite norm verifying $\left\|F^T\right\| = \|F\|^S$.

(d) If F is a convex process with finite norm, then $D\left(F^T\right) = Y'$ and F^T has a finite spread norm verifying $\left\|F^T\right\|^S = \|F\|$.

PROOF. We observe that $x' \in F^T(y')$ iff

$$\forall x \in X \qquad \langle x', x\rangle \leq b_F(x, y').$$

(a) Let us suppose F has finite norm (the case of interest). Then, for any $y' \in Y'$, $x' \in F^T(y')$ one has, for each $x \in X$,

$$\forall x \in X \qquad \langle x', x\rangle \leq b_F(x, y') \leq \inf\{\|y'\| \|y\| : y \in F(x)\} \leq \|y'\| \|F\| \|x\|$$

and one gets $\|x'\| \leq \|F\| \|y'\|$, hence $\left\|F^T\right\|^S \leq \|F\|$.

(b) When $\|F\|^S = 0$, we have $F^T(y') = 0$ for any $y' \in Y'$, and the equality $\|F^T\| = \|F\|^S$ holds. Now, given $c < \|F\|^S$, $c > 0$, we can find $x \in X$, $y \in F(x)$ with $\|x\| = 1$, $\|y\| > c$. The Hahn-Banach theorem yields some $y' \in Y'$ such that $\|y'\| = 1$ and $\langle y', y\rangle < -c$. Then, for any $x' \in F^T(y')$, we have $\langle x', x\rangle \leq \langle y', y\rangle < -c$, hence $\|x'\| > c$. Thus $d(0, F^T(y')) \geq c$ (in a trivial way if $F^T(y')$ is empty), $\|F^T\| > c$ and $\|F^T\| \geq \|F\|^S$.

When $D(F) = X$ and $D(F^T) = Y'$ we get

$$\|F\| \geq \|F^T\|^S \geq \|F^T\| \geq \|F\|^S \geq \|F\|$$

and equality holds.

(c) Now suppose F is a convex process and $\|F\|^S$ is finite. Since for any $x \in X$ and for any given $y' \in X'$ one has

$$-\|y'\|\,\|F\|^S\,\|x\| \leq b_F(x, y')$$

(trivially if $x \notin D(F)$) and since $b_F(\cdot, y')$ is sublinear, the sandwich theorem enables one to find a linear functional x' on X such that

$$\forall x \in X \qquad -\|y'\|\,\|F\|^S\,\|x\| \leq \langle x', x\rangle \leq b_F(x, y').$$

The first inequality ensures that x' is continuous and $\|x'\| \leq \|y'\|\,\|F\|^S$. The second one then implies that $x' \in F^T(y')$, and, as y' is arbitrary, one gets $D(F^T) = Y'$ and $\|F^T\| \leq \|F\|^S$.

(d) Let us prove that $\|F^T\|^S = \|F\|$ when F is a convex process with finite norm. The case $\|F\| = 0$ is trivial by assertion (a). Given $r > 0$ such that $r < \|F\|$, we can find some $u \in X$ such that $\|u\| = 1$ and $\|v\| > r$ for each $v \in F(u)$. The geometric form of the Hahn-Banach theorem enables us to find $y' \in Y'$ such that $\|y'\| = 1$ and

$$\sup\{\langle y', v\rangle : v \in F(u)\} \leq \inf\{\langle y', y\rangle : y \in Y, \|y\| < r\} = -r$$

Then, for any $x' \in F^T(y')$ one has $\langle x', u\rangle \leq -r$, hence $\|x'\| \geq r$ and $\|F^T\| \geq r$. Therefore $\|F^T\| = \|F\|^S$. □

COROLLARY. Suppose X is finite dimensional, Y is strictly convex, $F : X \rightrightarrows Y$ is a closed convex process with finite norm and spread norm. If F is lower semicontinuous and if for each $x \neq 0$ the set $F(x)$ is not a singleton, then the domain $D(F^T)$ of F^T is not Y'.

PROOF. Suppose $D(F^T) = Y'$. Under our assumptions we have $\|F^T\|^S = \|F\| \leq \|F\|^S = \|F^T\| \leq \|F^T\|^S$. Therefore $\|F\|^S = \|F\|$.

Now let us note that the function f given by $f(x) = e(F(x), 0)$ is upper semicontinuous whenever F is lower semicontinuous: if $f(x) < r$ there exists $y \in F(x)$ such that $y \in V$ the open ball with center 0 and radius r; then there exist a neighborhood U of x such that $F(u) \cap V \neq \emptyset$ for each $u \in U$, so that $f(u) < r$ for $u \in U$. When X is finite dimensional, f attains its infimum on the unit sphere of X. If $\overline{x}$ is a maximizer of f, we have $f(\overline{x}) = \|F\| = \|F\|^S \geq \|y\|$ for each $y \in F(\overline{x})$, thus $f(\overline{x}) = \|y\|$ for each $y \in F(\overline{x})$. Since Y is strictly convex and $F(\overline{x})$ is convex, we get that $F(\overline{x})$ is a singleton, a contradiction. □

The preceding result incites to make a comparison with what occurs with fans ([**17**]). Recall that a *fan* $F : X \rightrightarrows Y$ is a process with convex images such that

$$F(x_1 + x_2) \subset cl(F(x_1) + F(x_2)) \quad \forall x_1, x_2 \in X.$$

Let us say that a fan is an *S-fan* (or a saturated fan) if its images are closed and if its support function

$$s_F(y', x) := \sup\langle y', F(x)\rangle \qquad x \in X,\ y' \in Y'$$

is bi-sublinear and bi-lower semicontinuous (i.e. separately sublinear and l.s.c.). Note that for any fan F the function s_F is bi-sublinear and lower semicontinuous (l.s.c.) in its first variable. When the images of F are closed convex, F is determined by s_F; however s_F is not always l.s.c. in its second variable. S-fans are attractive for their properties, especially in terms of transposition. However, the transpose which is natural for such maps is different from the transpose considered above (see [**17**]); for that reason we call it the *S-transpose*. The definition of [**17**] can be extended to an arbitrary multimapping $F : X \rightrightarrows Y$ by

$$F^S(y') := \{x' \in X' : x' \leq s_F(y', \cdot)\}.$$

Then F^S is a fan. When F is a S-fan, in particular when F is a bounded fan in the sense of [**17**], one has

$$s_{F^S}(x, y') = s_F(y', x) \qquad \forall x \in X,\ \forall y' \in Y', \tag{2.8}$$

as easily seen (because for any l.s.c. sublinear function p one has $p = \sup_{x' \leq p} x'$), and F^S is a S-fan (provided that X is reflexive or that X and Y are endowed with topologies compatible with the duality); moreover the domain of F^S is Y'. Since

$$\begin{aligned} \|F\|^S &= \sup\{s_F(y', x) : x \in X,\ y' \in Y',\ \|x\| \leq 1,\ \|y'\| \leq 1\}, \\ \|G\|^S &= \sup\{s_G(x, y') : x \in X,\ y' \in Y',\ \|x\| \leq 1,\ \|y'\| \leq 1\}, \end{aligned}$$

for any multimappings $F : X \rightrightarrows Y$, $G : Y' \rightrightarrows X'$, and any n.v.s. X, Y, we get that

$$\|F^S\|^S = \|F\|^S$$

for any S-fan, in particular for any bounded fan ([**17**] Proposition 5.5). S-transposition has other nice properties (see [**17**] Proposition 3.3). However, we observe that for any multimapping F one $F^S(y') \supseteq F^T(y')$ for any $y' \in Y'$ and that the images of F^S are often much larger than the images of F^T.

Let us introduce a metric on the set of bounded processes from X into Y with nonempty closed images (and an écart on the set of processes) by

$$d(F, G) := \sup_{x \in B_X} d(F(x), G(x)),$$

where B_X is the closed unit ball of X and $d(A, B)$ denotes the Pompeiu-Hausdorff distance of two sets A, B:

$$d(A, B) := \max(e(A, B), e(B, A)), \quad e(C, D) = \sup_{c \in C} d(c, D).$$

Clearly, $\|F\|^S = d(F, 0)$, where 0 denotes the process (and fan) with values $\{0\}$.

In connection with the following results, let us recall that a closed convex process F on a reflexive Banach space has bounded values iff $F(0) = \{0\}$.

THEOREM 4. *Let F, $G : X \rightrightarrows Y$ be S-fans with bounded values, X being reflexive. Then*

$$d(F^S, G^S) = d(F, G).$$

PROOF. Let us recall the fundamental relations linking excesses and Pompeiu-Hausdorff distances with support functions: for any two closed convex bounded subsets A, B of Y one has

$$e(A,B) = \sup\{s_A(y') - s_B(y') : y' \in B_{Y'}\}, \tag{2.9}$$
$$d(A,B) = \sup\{|s_A(y') - s_B(y')| : y' \in B_{Y'}\}, \tag{2.10}$$

where $B_{Y'}$ denotes the closed unit ball of the normed space Y'. The first of these two relations gives slightly more than the statement:

$$\begin{aligned} e(F,G) \quad : &= \sup_{x \in B_X} e(F(x), G(x)) \\ &= \sup_{x \in B_X} \sup_{y' \in B_{Y'}} \left(s_{F(x)}(y') - s_{G(x)}(y')\right) && \text{by (9)} \\ &= \sup_{x \in B_X} \sup_{y' \in B_{Y'}} \left(s_{F^T}(x,y') - s_{G^T}(x,y')\right) && \text{by (8)} \\ &= \sup_{y' \in B_{Y'}} \sup_{x \in B_X} \left(s(x, F^T(y')) - s(x, G^T(y'))\right) \\ &= \sup_{-y' \in B_{Y'}} e(F^T(y'), G^T(y')) = e(G^T, F^T) && \text{by (9).} \end{aligned}$$

The relation $d(F,G) = d(F^T, G^T)$ follows analogously or by symmetrization. □

In order to get a similar result for the transposition we study, we introduce an appropriate class of convex processes. Let us say that a process $F : X \rightrightarrows Y$ is a *T-process* if F is a closed convex process such that its barrier function b_F is l.s.c. in its second variable and does not take the value $-\infty$. The following lemmas give a characterization and a useful sufficient condition.

LEMMA 5. *A closed convex process $F : X \rightrightarrows Y$ is a T-process iff for any $(x, y') \in X \times Y'$ one has*

$$b_F(y', x) = s_{F^T}(x, y') > -\infty. \tag{2.11}$$

PROOF. The condition is obviously sufficient since for each $y' \in Y'$ the function $s_{F^T}(\cdot, y')$ is a supremum of continuous linear forms. It is necessary: when for each $y' \in Y'$ the function $b_F(y', \cdot)$ is l.s.c.and does not take the value $-\infty$, for each $x \in X$ one has

$$\begin{aligned} b_F(y', x) &= \sup\{\langle x', x\rangle : x' \le b_F(y', \cdot)\} \\ &= \sup\{\langle x', x\rangle : x' \in F^T(y')\} = s_{F^T}(x, y'), \end{aligned}$$

since $b_F(y', \cdot)$ is also sublinear. □

LEMMA 6. *Let $F : X \rightrightarrows Y$ be a closed convex process such that for each $x \in X$ and each sequence $(x_n) \to x$ in X, a sequence (y_n) in Y has a weak cluster point whenever $y_n \in F(x_n)$ for each $n \in \mathbb{N}$. Then F is a T-process.*

PROOF. Let $y' \in Y'$ and let $(x_n) \to x$ in X. Suppose that $b_F(y', x_n) \le r$ for each $n \in \mathbb{N}$ for some real number r. Let $y_n \in F(x_n)$ be such that $\langle y', y_n\rangle \le b_F(x_n) - 2^{-n}$. Let y be a weak cluster point of (y_n); since F is closed, y belongs to $F(x)$. Then $\langle y', y\rangle = \liminf_n \langle y', y_n\rangle \le r$, so that $b_F(x) \le r$. □

THEOREM 7. *Let $F,\ G : X \rightrightarrows Y$ be T-processes with bounded values, X being reflexive. Then*

$$d(F^T, G^T) = d(F, G).$$

PROOF. We use again relation (9):

$$\begin{aligned}
e(F,G) &= \sup_{x\in B_X}\sup_{y'\in B_{Y'}} \left(s_{F(x)}(y') - s_{G(x)}(y')\right) && \text{by (9)}\\
&= \sup_{x\in B_X}\sup_{y'\in B_{Y'}} \left(b_G(-y',x) - b_F(-y',x)\right) && \text{by (5)}\\
&= \sup_{x\in B_X}\sup_{y'\in B_{Y'}} \left(s_{G^T}(x,-y') - s_{F^T}(x,-y')\right) && \text{by (4)}\\
&= \sup_{-y'\in B_{Y'}}\sup_{x\in B_X} \left(s(x,G^T(y')) - s(x,F^T(y'))\right) &&\\
&= \sup_{y'\in B_{Y'}} e(G^T(y'),F^T(y')) = e(G^T,F^T) && \text{by (9).}
\end{aligned}$$

The relation $d(F,G) = d(F^T,G^T)$ follows analogously or by symmetrization. □

The following combines a rephrasing of [**30**] Theorem 1 with the relation $\left(F^T\right)^{-1} = -\left(F^{-1}\right)^T$ and Theorem 3. See also [**8**], [**10**] for $(a) \Leftrightarrow (b)$ and [**27**] for $(a) \Rightarrow (d)$ and $(d) \Rightarrow (a)$.

PROPOSITION 8. *Let $F : X \rightrightarrows Y$ be a convex process between n.v.s. Among the following assertions one has*

$$(a) \Longleftrightarrow (b) \Longleftrightarrow (c) \Longrightarrow (d)$$

If moreover X and Y are Banach spaces and if F has a closed convex graph all these assertions are equivalent:

(a) F is open at 0 from X onto Y ;

(b) F^{-1} is lower semicontinuous at 0 from Y into X ;

(c) F^{-1} has finite norm (and $D\left(F^{-1}\right) = Y$);

(d) $\left(F^T\right)^{-1}$ has finite norm and $D\left(\left(F^T\right)^{-1}\right) = X'$.

Let us note that some of the implications between the preceding assertions can be extended to the case of a convex multimapping. However, the concept of a norm for such a relation F is not clear. One can introduce the conorm $n(\widehat{F})$ of the homogenized map $\widehat{F}$ of F; one may also modify the definition of the norm in (6) in order to take the supremum over the domain of the multimapping rather than on the whole space. In such a way, one could overcome the fact that the domain of $\widehat{F}$ is never the whole space.

3. Some Properties of Transposition

Our definition enables to obtain formulas which are close to known ones when applied to linear operators or convex processes.

PROPOSITION 9. *Denoting by $N\left(F^T\right)$ the kernel $\left(F^T\right)^{-1}(0)$ of the transpose F^T of the arbitrary multimapping $F : X \rightrightarrows Y$ one has*

$$\begin{aligned}
F(X)^0 &= -N\left(F^T\right),\\
D(F)^0 &= F^T(0).
\end{aligned}$$

If moreover the graph of F is a closed convex set containing 0 *then*

$$N(F) = -F^T (Y')^0 . \tag{3.1}$$

PROOF. The first two assertions follow immediately from the definitions: $y' \in F(X)^0$ iff $\langle 0, x\rangle - \langle -y', y\rangle \leq 1$ for any $(x, y) \in Gr(F)$ and $x' \in D(F)^0$ iff $\langle x', x\rangle - \langle 0, y\rangle \leq 1$ for any $(x, y) \in Gr(F)$. The last one is a consequence of Proposition 2.1. □

One also gets simple new formulas which arise from the fact that the class of multimappings has stability properties the class of mappings does not enjoy. In particular, the inverse of a convex relation is always a convex relation. Here the inverse $F^{-1} : Y \rightrightarrows X$ of $F : X \rightrightarrows Y$ is given by

$$Gr\left(F^{-1}\right) = \{(y, x) \in Y \times X : (x, y) \in Gr(F)\} .$$

As for convex processes, one has the following obvious formula for the transpose of an inverse.

PROPOSITION 10. *For any multimapping $F : X \rightrightarrows Y$, the transpose $\left(F^{-1}\right)^T$ of the inverse F^{-1} of F satisfies*

$$\left(F^{-1}\right)^T (x') = -\left(F^T\right)^{-1} (-x') .$$

Given two multimappings F_1, $F_2 : X \rightrightarrows Y$ we define their intersection $F_1 \cap F_2$, their union $F_1 \cup F_2$, their sum $F_1 + F_2$, their convex hull $co(F_1, F_2)$ and their closed convex hull $co(F_1, F_2)$ by

$$\begin{aligned}
(F_1 \cap F_2)(x) &= F_1(x) \cap F_2(x) , \\
(F_1 \cup F_2)(x) &= F_1(x) \cup F_2(x) , \\
(F_1 + F_2)(x) &= F_1(x) + F_2(x) , \\
co(F_1, F_2)(x) &= co(F_1(x) \cup F_2(x)) \\
\overline{co}(F_1, F_2)(x) &= \overline{co}(F_1(x) \cup F_2(x))
\end{aligned}$$

respectively, where the convex (resp. closed convex) hull of a subset S of a topological vector space is denoted by $co(S)$ (resp. $\overline{co}(S)$). The last two correspondences must be distinguished from the correspondences $F_1 F_2$ and $F_1 \vee F_2$ whose graphs are $Gr(F_1) + Gr(F_2)$ and $co(Gr(F_1) \cup Gr(F_2))$ respectively.

Assuming F_1 and F_2 are convex, one has

$$\begin{aligned}
&(F_1 \vee F_2)(x) \\
: \quad &= \{t_1 F_1(x_1) + t_2 F_2(x_2) : t_1, t_2 \geq 0,\ t_1 + t_2 = 1,\ t_1 x_1 + t_2 x_2 = x\} .
\end{aligned}$$

If moreover F_1 and F_2 are homogeneous and $0 \in Gr(F_i)$, for $i = 1, 2$, this expression can be simplified :

$$(F_1 \vee F_2)(x) = \{F_1(x_1) + F_2(x_2) : x_1 + x_2 = x\} = (F_1 F_2)(x)$$

Let us introduce the multimappings $F_1 F_2$ and $F_1 \overline{\vee} F_2$ whose graphs are the closures of $Gr(F_1) + Gr(F_2)$ and $co(Gr(F_1) \cup Gr(F_2))$ respectively.

LEMMA 11. *Let F_1 be a convex process and let F_2 be a convex relation such that $0 \in cl(Gr(F_2))$. Then $F_1 \overline{\vee} F_2 = F_1 F_2$.*

PROOF. It suffices to prove the following claim of independent interest. If C is a nonempty convex cone of a normed vector space Z and if D is a convex subset of Z such that $0 \in clD$, then

$$\overline{co}\,(C \cup D) = cl(C + D). \tag{3.2}$$

Since $0 \in clC$ one has $D \subset clC + D \subset cl(C + D)$ and similarly $C \subset cl(C + D)$. Since $cl(C + D)$ is closed convex, we get

$$\overline{co}\,(C \cup D) \subset cl(C + D).$$

On the other hand, for $x \in C$, $y \in D$, one has

$$x + y = \lim_{t \searrow 0} \left[t(t^{-1}x) + (1 - t)\, y\right] \in \overline{co}\,(C \cup D)$$

hence

$$cl(C + D) \subset \overline{co}\,(C \cup D) \subset cl(C + D)$$

and equality holds. □

The transpose of $F_1 \vee F_2$ is easy to compute.

LEMMA 12. *For any multimappings $F_1, F_2 : X \rightrightarrows Y$ one has*

$$(F_1 \vee F_2)^T = F_1^T \cap F_2^T = (F_1 \overline{\vee} F_2)^T.$$

PROOF. Let $C_i = Gr\,(F_i)$ for $i = 1, 2$. Then

$$\begin{aligned} x' \in (F_1 \vee F_2)^T (y') &\iff (x', -y') \in (co\,(C_1 \cup C_2))^0 \\ &\iff (x', -y') \in C_1^0 \cap C_2^0 \\ &\iff x' \in F_1^T (y') \cap F_2^T (y'). \quad \square \end{aligned}$$

It is more interesting to compute $(F_1 \cap F_2)^T$; a qualification condition is required to get a simple expression. Otherwise one has to use the operation $\overline{\vee}$ introduced above.

Let us first recall that the (weak*) *asymptotic cone* $T_\infty C$ of a subset C of a dual space X' is the set

$$T_\infty C = t \to \infty \lim \sup t^{-1} C$$

in the sense of Painlevé-Kuratowski : $v' \in T_\infty C$ iff there exist a net $(t_i, v_i')_{i \in I}$ in $\mathbb{P} \times X'$ with limit (∞, v') such that $t_i v_i' \in C$ for each $i \in I$. When C is a weak* closed convex subset of X', the asymptotic cone coincides with the recession cone 0^+C which is the largest cone Q such that $C + Q \subset C$. This fact (or a simple direct proof) establishes the following characterization.

LEMMA 13. *If A is a subset of X, the asymptotic cone $T_\infty A^0$ of the polar A^0 of A coincides with the negative polar cone of A :*

$$T_\infty A^0 = 0^+ A^0 = A^- := \{x' \in X' : \forall a \in A \ \langle x', a \rangle \leq 0\}.$$

In the sequel, given a multimapping $M : Y'X'$, we denote by 0^+M the multimapping whose graph is $0^+Gr(M)$.

PROPOSITION 14. *Given Banach spaces X, Y and multimappings $F_1, F_2 : X \rightrightarrows Y$ whose graphs are closed, convex and contain 0 one has*

$$(F_1 \cap F_2)^T = F_1^T \overline{\vee} F_2^T.$$

If

$$W := \mathbb{R}_+ Gr(F_1) - \mathbb{R}_+ Gr(F_2)$$

is a closed vector subspace of $X \times Y$, *then*

$$(F_1 \cap F_2)^T = \left(F_1^T \vee F_2^T\right) \cup \left(0^+ F_1^T \ \ F_2^T\right) \cup \left(F_1^T \ \ 0^+ F_2^T\right),$$

or, in other terms,

$$Gr(F_1 \cap F_2)^T = co\left(Gr(F_1^T) \cup Gr(F_2^T)\right) \cup (0^+ Gr F_1^T + Gr F_2^T) \cup (Gr F_1^T + 0^+ Gr F_2^T).$$

In particular, when F_1 *and* F_2 *satisfy the preceding assumptions and* F_1 *or* F_2 *is positively homogeneous, one has*

$$(F_1 \cap F_2)^T (y') = y_1' + y_2' = y' \bigcup F_1^T (y_1') + F_2^T (y_2').$$

PROOF. Since $Gr(F_1 \cap F_2) = Gr(F_1) \cap Gr(F_2)$, these assertions are consequences of the following lemma which generalizes [**26**] Th. 2.3. □

LEMMA 15. *Let* C *and* D *be closed convex subsets of a Banach space* X *which contain* 0. *Then* $(C \cap D)^0 = \overline{co}\left(C^0 \cup D^0\right)$. *If* $W := \mathbb{R}_+ (C - D)$ *is a closed vector subspace of* X *one has*

$$\begin{aligned} (C \cap D)^0 &= (C^- + D^0) \cup (C^0 + D^-) \cup co\,(C^0 \cup D^0) \\ &= (C^- + D^0) \cup (C^0 + D^-) \cup \bigcup_{0<t<1} ((1-t)C^0 + tD^0); \end{aligned}$$

if moreover C^0 *(or* C*) is a cone then*

$$(C \cap D)^0 = C^0 + D^0 = C^- + D^0.$$

PROOF. The first assertion is a consequence of the bipolar theorem and is well-known. To prove the second one, we may suppose without loss of generality that $W = X$ and first observe that the right hand sides are contained in the left hand side. Since the weak* closure of $co\left(C^0 \cup D^0\right)$ coincides with its closure for the bounded weak* topology, it suffices to prove, for each closed ball B' with radius ρ and center 0 in X', that each point in the weak* closure of $B' \cap \left(co\left(C^0 \cup D^0\right)\right)$ belongs to the right hand side. Let x' be the weak* limit of a net $x_i' = s_i y_i' + t_i z_i'$ with $s_i, t_i \in [0,1]$, $s_i + t_i = 1$, $y_i' \in C^0$, $z_i' \in D^0$, $x_i' \in B'$. Without loss of generality we may assume $(t_i)_{i \in I}$ converges to some $t \in [0,1]$. Let us prove that $(s_i y_i')$ and $(t_i z_i')$ are bounded nets. For each $x \in X$ we can find $c \in C$, $d \in D$, q, $r \in \mathbb{R}_+$ such that $x = qc - rd$. Then

$$\begin{aligned} \langle s_i y_i', x \rangle &\leq s_i q - \langle s_i y_i', rd \rangle \\ &\leq s_i q - \langle x_i', rd \rangle + \langle t_i z_i', rd \rangle \\ &\leq q + \rho r \|d\| + r \end{aligned}$$

so that $(s_i y_i')$ is bounded by the Banach-Steinhaus theorem. It follows that $(t_i z_i')$ is bounded too. If $0 < t < 1$, taking subnets if necessary, we may assume (y_i') and (z_i') converge to some y' and z' in C^0 and D^0 respectively. Thus $x' = (1-t)\,y' + tz' \in co\left(C^0 \cup D^0\right)$. If $t = 0$, $(s_i) \to 1$ and taking subnets if necessary, we may assume (y_i') and $(t_i z_i')$ converge to some y' and w' in C^0 and $T_\infty D$ respectively, so that $x' = y' + w' \in C^0 + D^-$. Similarly, when $t = 1$ we get $x' \in C^- + D^0$.

When C^0 is a cone, a fortiori when C is a cone, one has $C^- = C^0$ and since C^0 and D^0 are star-shaped the last assertion follows. □

Remark: The preceding result is also a consequence of the Attouch-Brézis theorem ([**1**], [**5**] Theorem V.43). It is easy to show (since $(C\cap D)^0 = s_{C\cap D}^{-1}((-\infty,1]) = (s_C \Box s_D)^{-1}((-\infty,1])$ and the infimum convolution is exact) that the assumption $0 \in C \cap D$ can be replaced with $C \cap D \neq \emptyset$) provided the conclusion is changed into

$$(3.3)\, (C\cap D)^0 = \left(\bigcup_{t\in[0,1]} ((1-t)C^0 + tD^0)\right) \cup \left(\bigcup_{r<0} (rC^{\nabla} + (1-r)D^0)\right) \cup \left(\bigcup_{s<0} ((1-s)C^0 + sD^{\nabla})\right).$$

It is easy to write down the corresponding modification of Theorem 14 which allows to drop the assumption $0 \in Gr(F_1) \cap Gr(F_2)$.

The transposed of a sum can be computed under a qualification condition which is satisfied when F_1 and F_2 are closed linear operators, one of which being everywhere defined, as in [**4**] Cor. 2.5.4. The result generalizes the case of two convex processes treated in [**10**] Theorem 7.4 (in which another condition is also given).

THEOREM 16. *Let $F_1, F_2 : X \rightrightarrows Y$ be multimappings, the graph of one of them being shady. Then*

$$F_1^T + F_2^T \subset (F_1 + F_2)^T.$$

If moreover X and Y are Banach spaces, if the graphs of F_1 and F_2 are closed convex and if

$$V := \mathbb{R}_+(D(F_1) - D(F_2))$$

is a closed vector subspace, then

$$(F_1 + F_2)^T = F_1^T + F_2^T.$$

PROOF. Suppose F_1 is *shady* i.e. that its graph is shady. Given $y' \in Y'$, $x_1' \in F_1^T(y')$, $x_2' \in F_2^T(y')$, $x' = x_1' + x_2'$, for any $x \in X$, $y_1 \in F_1(x)$, $y_2 \in F_2(x)$ we have

$$\begin{aligned} \langle x_1', x\rangle - \langle y', y_1\rangle &\leq 0 \\ \langle x_2', x\rangle - \langle y', y_2\rangle &\leq 1 \end{aligned}$$

hence, by summing, $(x', -y') \in Gr(F_1+F_2)^0$ or $x' \in (F_1+F_2)^T(y')$.

Conversely, given $y' \in Y'$, $x' \in (F_1+F_2)^T(y')$, for each $(x, y_1, x, y_2) \in C := C_1 \times C_2$ where $C_i = Gr(F_i)$ for $i = 1,2$ we have

$$\langle x', x\rangle - \langle y', y_1 + y_2\rangle \leq 1.$$

This means that $(x', -y', 0, -y')$ belongs to the polar set of $C \cap D$, where

$$D = \{(x, y_1, x, y_2) : x \in X,\, y_1, y_2 \in Y\}.$$

In view of our qualification condition, given $(u_i, v_i) \in X \times Y$ for $i = 1,2$ such that $u_1 - u_2 \in V$, we can find $x \in X$, $r \in \mathbb{R}_+$, $z_1, z_2 \in Y$, $(x_i, y_i) \in C_i$ for $i = 1,2$ such that

$$(u_1, v_1, u_2, v_2) = (x, z_1, x, z_2) - r(x_1, y_1, x_2, y_2):$$

it suffices to take $r \in \mathbb{R}_+$, $(x_i, y_i) \in C_i$ such that

$$r(x_1 - x_2) = u_2 - u_1$$

and set $x = u_1 + rx_1 = u_2 + rx_2$, $z_i = v_i + ry_i$, $i = 1, 2$. Therefore

$$\mathbb{R}_+ C - D = W := \{(u_1, v_1, u_2, v_2) \in (X \times Y)^2 : u_1 - u_2 \in V\},$$

a closed vector subspace, and, as D is a vector subspace, Lemma 15 ensures that $(C \cap D)^0 = C^0 + D^0$. It follows that we can find $\left(u_1', v_1', u_2', v_2'\right) \in C^0 = C_1^0 \times C_2^0$ (since C_1 is a cone), $w' \in X'$ such that, by an easy computation of D^0,

$$(x', -y', 0, -y') = (u_1', v_1', u_2', v_2') + (w', 0, -w', 0).$$

Thus $v_2' = -y' = v_1'$, $u_2' = w'$,

$$x' = u_1' + w' = u_1' + u_2' \in F_1^T(y') + F_2^T(y'),$$

and the result is proved. □

Let us end this section with a study of another operation on convex relations. It stems from the harmonic sum of convex sets considered in [**37**] and [**28**]. Recall that if A and B are two subsets of a n.v.s. Z their harmonic sum $A \Diamond B$ is given by

$$A \Diamond B := \left(\bigcup_{\lambda \in]0,1[} \lambda A \cap (1-\lambda) B \right) \cup (0^+ A \cap B) \cup (A \cap 0^+ B).$$

This operation (which is closely related to the inverse sum considered in [**34**]) allows to introduce an operation on multimappings by defining, for two multimappings $F, G : X \rightrightarrows Y$ their harmonic sum $F \Diamond G$ as the multimapping with graph $Gr(F) \Diamond Gr(G)$.

PROPOSITION 17. *Let $F, G : X \rightrightarrows Y$ be two convex multimappings whose graphs are not disjoint. Then $(F \Diamond G)^T = F^T G^T$. If $0 \in F(0) \cap G(0)$ then $(FG)^T = F^T \Diamond G^T$.*

PROOF. Both assertions are immediate consequences of [**28**] Proposition 4.3. □

4. Transpose of a Composition

We devote this section to a chain rule for transposes and to a formula for the polar set of an inverse image by a convex relation. Both questions are intimately linked. We start with the later one.

PROPOSITION 18. *Let X and Y be Banach spaces let $F : X \rightrightarrows Y$ be a closed convex multimapping with $(0,0) \in Gr(F)$, and let D be a closed convex subset of Y containing 0. Suppose*

$$Z := \mathbb{R}_+ (F(X) - D)$$

is a closed vector subspace of Y. Then $x' \in \left(F^{-1}(D)\right)^0$ iff $x' \in F^T(D^-)$ or $x' \in 0^+ F^T(D^0)$ or there exists $y' \in D^0$, $t \in (0,1]$ such that $x' \in tF^T\left(t^{-1} y'\right)$. In particular $\left(F^{-1}(D)\right)^0 \supset F^T\left(D^0\right)$. If furthermore D or $Gr(F)$ is a cone one has

$$\left(F^{-1}(D)\right)^0 = F^T\left(D^0\right).$$

PROOF. Let p be the canonical projection of $X \times Y$ onto X, and let $C := F^{-1}(D)$, so that $C = p(F \cap X \times D)$ and, since p is surjective and open,

$$C^0 = \left(p^T\right)^{-1}\left((F \cap (X \times D))^0\right).$$

The assumption ensures that

$$\mathbb{R}_+ F - \mathbb{R}_+ (X \times D) = X \times Z$$

so that, by Lemma 15,

$$(F \cap (X \times D))^0 = \left(F^0 + \{0\} \times D^-\right) \cup \left(F^- + \{0\} \times D^0\right) \cup co\left(F^0 \cup \{0\} \times D^0\right).$$

where F^- is the negative polar cone of F considered as a subset of $X \times Y$. Therefore $x' \in C^0$ iff $(x', 0)$ belongs to one of the three sets in the right hand side. This circumstance happens iff one of the following three cases occurs : 1) there exists $y' \in D^-$ such that $(x', -y') \in F^0$; 2) there exists $y' \in D^0$ such that $(x', -y') \in F^-$; 3) there exist $t \in (0,1)$, $(u', v') \in F^0$, $w' \in D^0$ such that

$$(x', 0) = t(u', v') + (1-t)(0, w').$$

When $x' \neq 0$, so that $t \neq 0$, we get $v' = -t^{-1}y'$ for $y' = (1-t)w'$ and

$$x' = tu' \in tF^T(-v') = tF^T\left(t^{-1}y'\right),$$

with $y' \in D^0$ as D^0 is convex and contains 0.

When D or GrF is a cone we have $(F \cap X \times D)^0 = F^0 + \{0\} \times D^0$; then

$$(x', 0) = (u', v') + (0, w')$$

and $x' = u' \in F^T(-v') = F^T(w')$ with $w' \in D^0$. □

Remark. The remark following Lemma 15 allows to consider the case $0 \notin Gr(F)$ or $0 \notin D$. We leave the task of writing the corresponding conclusion to the interested reader.

The final assertion of the following chain rule is given in [**10**] for the case of two convex processes (see also [**4**] Theorem 2.5.3 for the case of composition with continuous linear maps); the general case is more involved.

THEOREM 19. *Let X, Y, Z be Banach spaces and let $F : X \rightrightarrows Y$, $G : Y \rightrightarrows Z$, be multimappings with closed convex graphs containing* 0.

Suppose

$$V := \mathbb{R}_+\left(F(X) - G^{-1}(Z)\right)$$

is a closed vector subspace of Y. Then for $z' \in Z'$ one has

$$(G \circ F)^T(z') = \left\{x' \in X' : \exists y' \in Y', (x', -y', y', -z') \in (GrF \times GrG)^0\right\}.$$

If moreover F or G is a process, one has

$$(G \circ F)^T = F^T \circ G^T.$$

PROOF. Let $K : X \times Z \rightrightarrows X \times Y^2 \times Z$ be the convex process given by

$$K(x, z) = \{(x, y, y, z) : y \in V\}$$

so that $H := G \circ F$ (or rather its graph) satisfies

$$H = K^{-1}(F \times G).$$

Since the graph of K is a closed vector subspace and since our assumption is equivalent to $\mathbb{R}_+\left(F(X)-G^{-1}(Z)\right)=V$ or

$$K(X\times Z)-\mathbb{R}_+(F\times G)=X\times V\times Z,$$

by the preceding result we get

$$H^0=K^T\left((F\times G)^0\right).$$

Since $K^T(x',y_1',y_2',z')=(x',z')$ if $y_1'=-y_2'$, $\emptyset$ otherwise, the first part of the result follows. The last assertion is a consequence of the following lemma or of a direct easy computation. □

In order to give a more precise expression for $(G\circ F)^T$ we are led to compute the polar set of a product. Such a computation may have its own interest.

LEMMA 20. *Let X and Y be normed vector spaces and let A (resp. B) be a subset of X (resp. Y).*

(a) If $cl(A)$ and $cl(B)$ contain 0, then, with the convention $0S=0^+S$, one has

$$(A\times B)^0=t\in[0,1]\bigcup(1-t)A^0\times tB^0. \tag{4.1}$$

In particular, if A is a cone, one has $(A\times B)^0=A^0\times B^0$.

(b) If $cl(B)$ contains 0, one has

$$(A\times B)^0=\left(t\in[0,1]\bigcup(1-t)A^0\times tB^0\right)\cup\left(s\in(-\infty,0)\bigcup sA^\nabla\times(1-s)B^0\right).$$

(c) If A is shady one has

$$(A\times B)^0=\left(A^-\times B^0\right)\cup\left(s\in(-\infty,0)\bigcup sA^\nabla\times(1-s)B^0\right)$$

(d) In the general case one has

$$\begin{aligned}(A\times B)^0 &= \left(t\in[0,1]\bigcup(1-t)A^0\times tB^0\right)\\ &\cup\left(s\in(-\infty,0)\bigcup sA^\nabla\times(1-s)B^0\right)\cup\left(t\in(-\infty,0)\bigcup(1-t)A^0\times tB^\nabla\right).\end{aligned}$$

PROOF. (a) Let $C=A\times B$ and let D be the right hand side. Taking into account the relation $0^+A^0=A^-$, the inclusions

$$\begin{aligned}0^+A^0 &\times B^0\subset C^0 \quad (\text{resp. } A^0\times 0^+B^0\subset C^0)\\ (1-t)A^0 &\times tB^0\subset C^0\end{aligned}$$

for $t\in[0,1]$ are obvious, so that $D\subset C^0$.

Conversely, if $(x',y')\in C^0$, then, for

$$\begin{aligned}s &= s_A(x'):=x\in A\sup\langle x',x\rangle\\ t &= s_B(y'):=y\in B\sup\langle y',y\rangle\end{aligned}$$

we have $s+t\le 1$ and, assuming $0\in B$, $t\ge 0$. If $t\in(0,1)$ we have $t^{-1}y'\in B^0$ and $(1-t)^{-1}x'\in A^0$ as $(1-t)^{-1}s\le 1$ and A^0 is starshaped. If $t=0$ we have $y'\in 0^+B^0$ and $x'\in A^0$, while if $t=1$ we have $x'\in 0^+A^0$ and $y'\in B^0$. If $0\in A$ we have $s\ge 0$, hence $t\le 1$ and we get assertion (a). In the case $0\notin A$ we may have $s<0$. Then $\langle s^{-1}x',a\rangle\ge 1$ for each $a\in A$, i.e. $s^{-1}x'\in A^\nabla$ and $\langle(1-s)^{-1}y',b\rangle\le 1$ for each $b\in B$, i.e. $(1-s)^{-1}y'\in B^0$. The case $t<0$ is similar, and since the right hand side of the relations in (c) and (d) are contained in C^0, these assertions are proved.

If A is a cone, one has $0^+A^0 = A^0$ and since $0^+B^0 \subset B^0$, $tB^0 \subset B^0$ for $t \in [0,1]$ the right hand side of relation (15) coincides with $A^0 \times B^0$. □

Since the general case can be reduced to the case X and Y are complete and C and D are closed convex subsets, the preceding lemma can be deduced from Lemma 15 by observing that $A \times B = C \cap D$, where $C := A \times Y$, $D := X \times B$.

COROLLARY. Let X, Y, Z be Banach spaces and let $F : X \rightrightarrows Y$, $G : Y \rightrightarrows Z$ be multimappings with closed convex graphs containing 0. Suppose

$$\mathbb{R}_+ \left(F(X) - G^{-1}(Z)\right) = Y.$$

Then, denoting by 0^+F^T (resp. 0^+G^T) the multimapping whose graph is $0^+ \left(Gr\left(F^T\right)\right)$ (resp. $0^+Gr(G^T)$), for any $z' \in Z'$ one has $x' \in (G \circ F)^T(z')$ iff there exists $t \in (0,1)$ such that $x' \in (1-t)F^T\left(t(1-t)^{-1}G^T\left(t^{-1}z'\right)\right)$ or $x' \in 0^+F^T\left(G^T(z')\right)$ or $x' \in F^T\left(0^+G^T(z')\right)$.

An alternative proof consists in observing that (identifying a multimapping with its graph)

$$H = q(M) \text{ with } M = F \times Z \cap X \times G$$

where $q(x,y,z) = (x,z)$, in using the remark following Lemma 15 for computing the polar set $(F \times Z \cap X \times G)^0$ and in applying the relation $(q(M))^0 = \left(q^T\right)^{-1}(M^0)$. Then, denoting by K^∇ the multimapping whose graph is $(Gr(K))^\nabla$, and without assuming that the graphs of F and G contain the origin, we get that $x' \in (G \circ F)^T(z')$ iff there exists $t \in (0,1)$ such that $x' \in (1-t)F^T\left(t(1-t)^{-1}G^T\left(t^{-1}z'\right)\right)$ or $x' \in 0^+F^T\left(G^T(z')\right)$ or $x' \in F^T\left(0^+G^T(z')\right)$ or there exists $r < 0$ such that $x' \in rF^\nabla\left((1-r)r^{-1}G^T\left(r^{-1}z'\right)\right)$ or $x' \in (1-r)F^T\left(r(1-r)^{-1}G^\nabla\left(r^{-1}z'\right)\right)$.

Remark. Conversely, Theorem 4.1 yields Proposition 4.1. It suffices to observe that if $G : Y \rightrightarrows \{0\}$ is the multimapping with graph $D \times \{0\}$ one has $F^{-1}(D) = D(H)$, where $H := G \circ F$, hence

$$\left(F^{-1}(D)\right)^0 = H^T(0).$$

5. Problems in duality

Let us consider the following problem:

$$(P) \text{ minimize } f(x,y) \quad : y \in F(x),$$

where $F : X \rightrightarrows Y$ is a multimapping, and $f : X \times Y \to R \cup \{\infty\}$ is a function. Let F^∇ be the multimapping given by $F^\nabla(y') := \{x' \in X' : (x', -y') \in (Gr(F))^\nabla\}$ and let f^* be the Fenchel conjugate of f. We associate to (P) the dual problems

$$(D_+) \text{ maximize } -f^*(-x',y') - t \quad : x' \in tF^T(t^{-1}y'),\ t > 0,\ y' \in Y'$$

$$(D_0) \text{ maximize } -f^*(-x',y') \quad : x' \in 0^+F(y'),\ y' \in Y'$$

$$(D_-) \text{ maximize } -f^*(-x',y') - t \quad : x' \in tF^\nabla(t^{-1}y'),\ t < 0,\ y' \in Y'$$

whose values are denoted by d_+, d_0, d_- respectively. Setting $C := Gr(F)$, $z' := (x', -y')$ and observing that for $z := (x,y) \in C$, $z' \in tC^0$ we have $f(z) + f^*(-z') \geq \langle -z', z\rangle \geq -t$, we get that $p := \inf(P) \geq d_+$. Interpreting $0C$ as 0^+C, we obtain similarly $p \geq d_0$. Now, for $t < 0$, $z := (x,y) \in C$, $z' \in tC^\nabla$ we have $f(z) + f^*(-z') \geq -t$ since $\langle -z', z\rangle \geq t$, hence $\langle -z', z\rangle \geq t$ and $p \geq \max(d_+, d_0, d_-)$ (weak duality). Under a qualification condition we can get a strong duality result.

THEOREM 21. *Let $F : X \rightrightarrows Y$ be a closed convex multimapping between Banach spaces, and let $f : X \times Y \to R \cup \{\infty\}$ be a closed convex function. Suppose $\mathbb{R}_+(dom f - Gr(F)) = X \times Y$. Then $p = \max(d_+, d_0, d_-)$.*

We derive this statement from the following duality result.

LEMMA 22. *Let Z be a Banach space, let $f : Z \to R \cup \{\infty\}$ be a closed convex function and let C be a closed convex subset of Z. Suppose $\mathbb{R}_+ (dom(f) - C)$ is a closed vector subspace of Z. Then*

$$\inf_{z \in C} f(z) = \max_{z' \in Z'} -f^*(-z') - s_C(z').$$

PROOF. Let ι_C be the indicator function of C ($\iota_C(z) = 0$ for $z \in C$, $+\infty$ else), so that $\iota_C^* = s_C$. Since $\mathbb{R}_+ (dom(f) - dom(\iota_C))$ is a closed vector subspace of Z, the Attouch-Brézis duality theorem yields

$$\inf_{z \in C} f(z) = -(f + \iota_C)^*(0) = -(f^* \Box \iota_C)(0)$$

which is the announced relation, with the supremum attained when finite. □

The theorem derives from this result by setting $C = Gr(F)$, $z' := (x', -y') \in Z'$ $t = s_C(z')$. For $t > 0$, we have $x' \in tF^T(t^{-1}y')$. For $t = 0$ we have $x' \in 0^+F^T(y')$. For $t < 0$ we have $z' \in tC^\nabla$. In view of the weak duality obtained above, equality holds. □

When $0 \in Gr(F)$ the case $t = s_C(z') < 0$ is excluded and we obtain a simpler duality result.

COROLLARY. Let F and f be as in Theorem 21. Suppose $0 \in Gr(F)$. Then $p = \max(d_+, d_0)$: interpreting the inclusion $x' \in tF^T(t^{-1}y')$ as $x' \in 0^+F^T(y')$ when $t = 0$, one has

$$\inf\{f(x,y) : y \in F(x)\} = \max\{-f^*(-x', y') - t : x' \in tF^T(t^{-1}y'),\ t \in \mathbb{R}_+,\ y' \in Y'.$$

When F is a process and when f is a separable function, i.e. $f(x,y) = g(x) + h(y)$, we recover the Fenchel type duality theorem of [**10**] Theorem 8.1. If moreover h is the indicator function of $\{0\}$, we recover the geometric duality of [**10**] Theorem 5.2.

COROLLARY. Let F and f be as in Theorem 21. Suppose F is shady. Then $p = \max(d_-, d_0)$: interpreting the inclusion $x' \in tF^\nabla(t^{-1}y')$ as $x' \in 0^+F^T(y')$ when $t = 0$, one has

$$\inf\{f(x,y) : y \in F(x)\} = \max\{-f^*(-x', y') - t : x' \in tF^\nabla(t^{-1}y'),\ t \in \mathbb{R}_+,\ y' \in Y'$$

References

[1] H. Attouch and H. Brézis, Duality for the sum of convex functions in general Banach spaces, in "*Aspects of Mathematics and its Applications*", J. Barroso, ed., North-Holland (1986), pp. 125-133.

[2] J.-P. Aubin, Lipschitz behavior of of solutions to convex minimization problems, Math. Oper. Research 8 (1984), 87-111.

[3] J.-P. Aubin and I. Ekeland, *Applied Nonlinear Analysis*, Wiley, New York, 1984.

[4] J.-P. Aubin and H. Frankowska, *Set-Valued Analysis*, Birkhäuser, Basel, 1990.

[5] D. Azé, *Eléments d'analyse convexe et variationnelle*, Ellipse, Paris, 1997.

[6] C. Berge, *Espaces topologiques et fonctions multivoques*, Dunod, Paris, 1959.

[7] J.-M. Borwein, Multivalued convexity and optimization: a unified approach to inequality and equality constraints, Math. Programming 13 (1977), 183-199.

[8] J.-M. Borwein, A Lagrange multiplier theorem and a sandwich theorem for convex relations, Math. Scand. 48 (1981), 189-204.
[9] J.-M. Borwein, Convex relations in Optimization and Analysis, "*Generalized Concavity in Optimization and Economics*", S. Schaible and W. Ziemba (eds.), Academic Press, New York (1981).
[10] J.-M. Borwein, Adjoint process duality, Math. Oper. Research 8 (3), (1983), 403-434.
[11] J.-M. Borwein, Stability and regular points of inequality systems, J. Optim. Th. Appl. 48 (1) (1986), 9-52.
[12] J.-M. Borwein, Norm duality for convex processes and applications, J. Optim. Th. Appl. 48 (1) (1986), 53-64.
[13] J.-M. Borwein, Automatic continuity and openness of convex relations, Proc. Amer. Math. Soc. 99 (1), (1987), 49-55.
[14] C. Castaing and M. Valadier, *Convex analysis and measurable multifunctions*, Lecture Notes in Math. 580, Springer Verlag, New York, 1977.
[15] L. Contesse and J.-P. Penot, Continuity of the Fenchel correspondence and continuity of polarities, J. Math. Anal. Appl. 156 (2) (1991), 305-328.
[16] K. Fan, Asymptotic cones and duality of linear relations, J. Approx. Th. 2 (1969), 152-159.
[17] A.D. Ioffe, Nonsmooth analysis: differential calculus of nondifferentiable mappings, Trans. Amer. Math. Soc. 266 (1) (1981), 1-56.
[18] A.D. Ioffe, Codirectional compactness, metric regularity and subdifferential calculus, preprint, Technion, Haifa, 1996.
[19] G. Isac, Processus convexes et inéquations dans les espaces de dimension infinie, Bull. Polish Acad. Sci., Mathematics 32 (1-2) (1984), 101-112.
[20] Klein and Thompson, Theory of correspondences, Wiley, New York.
[21] Kuratowski, *Topologie*, Panstowowe Wyd Nauk, Warszawa English translation, Academic Press, New York, 1966.
[22] A.S. Lewis, Ill-conditioned convex processes and conic linear systems, preprint, Univ. of Waterloo, Sept. 1999.
[23] J. Los and M.W. Los, *Mathematical Models in Economics*, North Holland, 1974.
[24] V.L. Makarov and A.M. Rubinov, *Mathematical theory of economic dynamics and equilibria*, Springer-Verlag, New York, 1977.
[25] R. Nozawa, Programmings with constraints of convex processes, Hiroshima Math. J. 11 (1981), 369-378.
[26] J.-P. Penot, Regularity conditions in mathematical programming, Math. Programming Study 19 (1982), 167-199.
[27] J.-P. Penot, Compactness properties, openness criteria and coderivatives, preprint Univ of Pau, Sept.1995, Set-Valued Anal. 6 (1998), 363-380.
[28] J.-P. Penot and C. Zalinescu, Harmonic sum and duality, submitted.
[29] B.N. Pshenichnii, Convex multivalued maps and their conjugate, Kibernetika 3 (1972), 94-102.
[30] S.M. Robinson, Normed convex processes, Trans. Amer. Math. Soc. 174 (1972), 127-140.
[31] S.M. Robinson, Regularity and stability for convex multivalued functions, Math. Oper. Res. 1 (1976), 130-143.
[32] R.T. Rockafellar, A monotone convex analogue of linear algebra, in : *Proceedings of the Colloquium on Convexity*, W. Fenchel (ed), Copenhagen, 1965, 261-276.
[33] R.T. Rockafellar, *Monotone processes of convex and concave type*, Memoir Amer. Math. Soc., No 77, Amer. Math. Soc. Providence, 1967.
[34] R.T. Rockafellar, *Convex Analysis*, Princeton University Press, Princeton, N.J. 1970.
[35] R.T. Rockafellar, Convex algebra and duality in dynamic models of production, in "*Mathematical Models in Economics*", J. Los and M.W. Los, eds., North Holland, 1974, 351-378.
[36] R.T. Rockafellar and R. J-B. Wets, *Variational Analysis*, Springer-Verlag, Berlin, 1997.
[37] A. Seeger, Direct and inverse addition in convex analysis and applications, J. Math. Anal. Appl. 148 (1990), 317-349.

DÉPARTEMENT DE MATHMATIQUES, FACULTÉ DES SCIENCES, AV. DE L'UNIVERSIT 64000 PAU, FRANCE

E-mail address: jean-paul.penot@univ-pau.fr

Canadian Mathematical Society
Conference Proceedings
Volume **27**, 2000

On a classical existence theorem for nonlinear elliptic equations

Biagio Ricceri

There is no doubt that the so called *mountain pass theorem* by Ambrosetti and Rabinowitz ([**1**]) is, since 25 years, one of the most often used tools in dealing with nonlinear differential equations, once the variational point of view is assumed. For our discussion, it is useful to recall its statement. A Gâteaux differentiable functional J on a real Banach space X is said to satisfy the Palais-Smale condition if each sequence $\{x_n\}$ in X such that $\sup_{n\in\mathbf{N}} |J(x_n)| < +\infty$ and $\lim_{n\to+\infty} \|J'(x_n)\|_{X^*} = 0$ admits a strongly converging subsequence.

THEOREM 1. ([**1**], Theorem 2.1) - *Let X be a real Banach space, and let $J : X \to \mathbf{R}$ be a continuously Gâteaux differentiable functional satisfying the Palais-Smale condition. Assume that there are a number $r > 0$ and two points $x_0, x_1 \in X$, with $\|x_1 - x_0\| > r$, such that*

$$\inf_{\|x-x_0\|=r} J(x) > \max\{J(x_0), J(x_1)\} .$$

Then, the functional J has a critical point distinct from x_0 and x_1.

Note that the formulation of Theorem 1 has a drawback. Namely, due to the strict inequality required, it cannot cover the case where one knows simply that $x_0 \neq x_1$, $J(x_1) \leq J(x_0)$ and x_0 is a local minimum of J. To cover such a case (which is of interest for us later), one needs a variant of Theorem 1 which was obtained by Pucci and Serrin in [**3**].

THEOREM 2. ([**3**], Theorem 1) - *Let X be a real Banach space, and let $J : X \to \mathbf{R}$ be a continuously Gâteaux differentiable functional satisfying the Palais-Smale condition. Assume that there are a two numbers r, s, with $0 < r < s$, and two points $x_0, x_1 \in X$, with $\|x_1 - x_0\| \geq s$, such that*

$$\inf_{r<\|x-x_0\|<s} J(x) \geq \max\{J(x_0), J(x_1)\} .$$

Then, the functional J has a critical point distinct from x_0 and x_1.

To begin to show the usefulness of Theorem 1, Ambrosetti and Rabinowitz, again in [**1**], proposed Theorem 3 below which is by now one the most classical existence results in the field of nonlinear elliptic equations.

1991 *Mathematics Subject Classification.* 35J20; 58E30.
Key words and phrases. Local minima; variational methods; nonlinear elliptic equations.

In the sequel, $\Omega \subset \mathbf{R}^n$ is a bounded open set, with smooth boundary, and $f : \Omega \times \mathbf{R} \to \mathbf{R}$ is a Carathéodory function.

THEOREM 3. ([**1**], Theorem 3.10; [**4**], Theorem 2.15; [**7**], Theorem 6.2; [**2**], Theorem 6.5.1) - *Assume that:*
(1) *there are two positive constants* a, q, *with* $q < \frac{n+2}{n-2}$ *if* $n \geq 3$, *such that*

$$|f(x,\xi)| \leq a(1 + |\xi|^q)$$

for all $(x, \xi) \in \Omega \times \mathbf{R}$;
(2) *there are constants* $r \geq 0$ *and* $c > 2$ *such that*

$$0 < c \int_0^{\xi} f(x,t)dt \leq \xi f(x,\xi)$$

for all $(x, \xi) \in \Omega \times \mathbf{R}$ *with* $|\xi| \geq r$;
(3) *one has*

$$\lim_{\xi \to 0} \frac{f(x,\xi)}{\xi} = 0$$

uniformly with respect to x.
Then, the problem

$$\begin{cases} -\Delta u = f(x,u) & in\ \Omega \\ \\ u_{|\partial\Omega} = 0. \end{cases}$$

has a non zero weak solution.

Recall that a weak solution of this problem is any $u \in W_0^{1,2}(\Omega)$ such that

$$\int_\Omega \nabla u(x) \nabla v(x) dx - \int_\Omega f(x, u(x)) v(x) dx = 0$$

for all $v \in W_0^{1,2}(\Omega)$.

A very sharp analysis of Theorem 3 can be found in the Introduction of [**6**]. There, Schecther was primarily concerned with the question of what can be said when, in Theorem 3, condition (2) is removed at all, or, at least, replaced by a much weaker one.

In the present paper, we are interested in removing condition (3) only.

Our result is as follows:

THEOREM 4. - *Assume that conditions* (1) *and* (2) *hold.*
Then, for each $\rho > 0$ *and each* μ *satisfying*

$$\mu > \inf_{u \in B_\rho} \frac{\sup_{v \in B_\rho} \int_\Omega \left(\int_0^{v(x)} f(x,\xi)d\xi \right) dx - \int_\Omega \left(\int_0^{u(x)} f(x,\xi)d\xi \right) dx}{\rho - \int_\Omega |\nabla u(x)|^2 dx},$$

where

$$B_\rho = \left\{ u \in W_0^{1,2}(\Omega) : \int_\Omega |\nabla u(x)|^2 dx < \rho \right\},$$

the problem

$$\begin{cases} -\Delta u = \frac{1}{2\mu} f(x,u) & in\ \Omega \\ \\ u_{|\partial\Omega} = 0. \end{cases}$$

has at least two weak solutions one of which lies in B_ρ.

To prove Theorem 4, we use Theorem 6 below, a new two critical points result which comes out from a joint application of Theorem 2 with our recent Theorem 2.5 of [**5**]. Let us recall this latter.

THEOREM 5. ([**5**], Theorem 2.5) - *Let X be a reflexive real Banach space, and let $\Phi, \Psi : X \to \mathbf{R}$ be two sequentially weakly lower semicontinuous functionals. Assume also that Ψ is (strongly) continuous and satisfies $\lim_{\|x\|\to+\infty} \Psi(x) = +\infty$.*
Then, for each $\rho > \inf_X \Psi$ and each μ satisfying

$$\mu > \inf_{x\in\Psi^{-1}(]-\infty,\rho[)} \frac{\Phi(x) - \inf_{\overline{(\Psi^{-1}(]-\infty,\rho[))}_w} \Phi}{\rho - \Psi(x)} ,$$

where $\overline{(\Psi^{-1}(]-\infty,\rho[))}_w$ is the closure of $\Psi^{-1}(]-\infty,\rho[)$ in the weak topology of X, the functional $\Phi + \mu\Psi$ has a local minimum which lies in $\Psi^{-1}(]-\infty,\rho[)$.

THEOREM 6. - *Assume that the hypotheses of Theorem 5 are satisfied. In addition, assume that, for each $\lambda > 0$, the functional $\Phi + \lambda\Psi$ is continuously Gâteaux differentiable and satisfies the Palais-Smale condition.*
Then, for each $\rho > \inf_X \Psi$ and each μ satisfying

$$\mu > \inf_{x\in\Psi^{-1}(]-\infty,\rho[)} \frac{\Phi(x) - \inf_{\overline{(\Psi^{-1}(]-\infty,\rho[))}_w} \Phi}{\rho - \Psi(x)} ,$$

the following alternative holds: either the functional $\Phi + \mu\Psi$ has a strict global minimum which lies in $\Psi^{-1}(]-\infty,\rho[)$, or the same functional has at least two critical points one of which lies in $\Psi^{-1}(]-\infty,\rho[)$.

PROOF. Let ρ and μ be as in the statement. Then, by Theorem 5, the functional $\Phi+\mu\Psi$ has a local minimum, say x_0, which lies in $\Psi^{-1}(]-\infty,\rho[)$. Assume that there is no strict global minimum of $\Phi+\mu\Psi$ lying in $\Psi^{-1}(]-\infty,\rho[)$. Consequently, there is some $x_1 \in X$, with $x_1 \neq x_0$, such that

$$\Phi(x_1) + \mu\Psi(x_1) \leq \Phi(x_0) + \mu\Psi(x_0) .$$

Next, choose $s \in]0, \|x_0 - x_1\|[$ so that

$$\Phi(x_0) + \mu\Psi(x_0) \leq \Phi(x) + \mu\Psi(x)$$

for all $x \in X$ satisfying $\|x - x_0\| \leq s$. Then, if we further choose $r = \frac{s}{2}$, we see that the functional $\Phi + \mu\Psi$ satisfies the assumptions of Theorem 2. Consequently, it has a critical point distinct from x_0 and x_1. Since, of course, x_0 is another critical point of $\Phi + \mu\Psi$, the conclusion follows. □

Proof of Theorem 4. Let $\rho > 0$ and μ be as in the statement. Let us apply Theorem 6, taking $X = W_0^{1,2}(\Omega)$, with the norm $\|u\| = (\int_\Omega |\nabla u(x)|^2 dx)^{\frac{1}{2}}$, and setting

$$\Phi(u) = -\int_\Omega \left(\int_0^{u(x)} f(x,\xi)d\xi \right) dx$$

and

$$\Psi(u) = \int_\Omega |\nabla u(x)|^2 dx$$

for all $u \in X$. By a classical result ([**7**], Theorem C.1), condition (1) implies that the functionals Φ and Ψ are continuously Gâteaux differentiable on $W_0^{1,2}(\Omega)$, the weak solutions of our problem being precisely the critical points of $\Phi+\mu\Psi$. Furthermore, Ψ is clearly weakly lower semicontinuous and coercive, and Φ is even sequentially weakly continuous, thanks to the Rellich-Kondrachov theorem. Moreover, from the standard proof of Theorem 3, we know that conditions (1) and (2) imply that the functional $\Psi + \mu\Phi$ is unbounded below and satisfies the Palais-Smale condition as well. Hence, Theorem 6 applies, and so the functional $\Phi + \mu\Psi$ has at least two critical points one of which lies in B_ρ. □

References

[1] A. AMBROSETTI and P. H. RABINOWITZ, *Dual variational methods in critical point theory and applications*, J. Funct. Anal., **14** (1973), 349-381.
[2] J. CHABROWSKI, *Variational methods for potential operator equations*, Walter de Gruyter, 1997.
[3] P. PUCCI and J. SERRIN, *A mountain pass theorem*, J. Differential Equations, **60** (1985), 142-149.
[4] P. H. RABINOWITZ, *Minimax methods in critical point theory with applications to differential equations*, CBMS Reg. Conf. Ser. in Math., **65**, Amer. Math. Soc., Providence, 1986.
[5] B. RICCERI, *A general variational principle and some of its applications*, J. Comput. Appl. Math., special issue on "Fixed point theory with applications in nonlinear analysis", to appear.
[6] M. SCHECHTER, *The mountain cliff theorem*, in *Differential equations and mathematical physics* (Birmingham, AL, 1990), Math. Sci. Engineer., **186**, Academic Press, 1992.
[7] M. STRUWE, *Variational methods*, Springer, 1996.

DEPARTMENT OF MATHEMATICS, UNIVERSITY OF CATANIA, VIALE A. DORIA 6, 95125 CATANIA, ITALY
E-mail address: ricceri@dipmat.unict.it

Canadian Mathematical Society
Conference Proceedings
Volume **27**, 2000

Limiting convex subdifferential calculus with applications to integration and maximal monotonicity of subdifferential

Lionel Thibault

Dédié à Jonathan Borwein
à l'occasion de la remise de son titre de Docteur Honoris Causa

ABSTRACT. Nonsmooth analysis techniques used to prove fuzzy sum rules in smooth spaces are adapted to provide, in general Banach spaces, several forms of generalized sequential formulas for subdifferential calculus of convex functions. Various important applications to the integration and maximal monotonicity of subdifferential are given.

1. Introduction

One knows that a qualification condition is required for any subdifferential calculus rule, expressing the subdifferential $\partial(f_1 + f_2)(x)$ of the sum of two convex functions f_1 and f_2 at a point x in a Banach space X in terms of the subdifferentials $\partial f_1(x)$ and $\partial f_2(x)$ of the functions f_1 and f_2 at the *same point* x. Such a formula is generally called an exact subdifferential formula. Subdifferential formulas for the sum of two convex functions that do not require any constraint qualification began with the papers by Hiriart-Urruty and Phelps [7] and Attouch, Baillon and Théra [1]. The paper by Hiriart-Urruty and Phelps (see also [8]) proves, for locally convex spaces, a general formula expressing $\partial(f_1 + f_2)(x)$ in terms of the approximate ε-subdifferentials of f_1 and f_2 at the *same point* x. One year later, Attouch, Baillon and Théra obtained another formula without any constraint qualification, when they showed in the context of Hilbert spaces that $\partial(f_1 + f_2)(x)$ can be described with some kinds of limits of sums of subgradients of f_1 and f_2 at neighboring points. The proof of Hiriart-Urruty and Phelps formula given in [7] deals with expressions for Legendre-Fenchel conjugates of some convex functions, and the proofs offered in [8] are based on some tricks using the support functions of the approximate subdifferentials of f_1 and f_2 at the point x. The proof by Attouch, Baillon and Théra is completely different. They obtained their result as a consequence of general results about variational sums of maximal monotone operators on Hilbert spaces,

---

1991 *Mathematics Subject Classification.* Primary 90C25; Secondary 52A41, 26E15.

Key words and phrases. Convex function, subdifferential, ε-subdifferential, subdifferential calculus, integration of subdifferentials, maximal monotonicity.

taking into account that the subdifferential of a proper lower semicontinuous convex function is maximal monotone. These papers motivated us in our study in [17, 18] to establishing some generalized sequential subdifferential formulas characterizing, in the setting of Banach spaces, the subdifferential $\partial(f_1 + f_2)(x)$ as some special limits of sums of points in the subdifferentials of f_1 and f_2 at neighboring points of x. We also showed how exact subdifferential formulas can be derived whenever any qualification condition holds. Our method in [17, 18] is based on Hiriart-Urruty and Phelps formula and a version of Brønsted-Rockafellar theorem.

The aim of this paper, dedicated to J.M. Borwein, is to show how some recent techniques in smooth spaces by Borwein and Ioffe [3] and Borwein and Zhu [4] can be used, in the setting of general Banach spaces, for the study of generalized sequential formula for the subdifferential $\partial(f_1+f_2)(x)$. In so doing, we do not rely on the formula in [7] but instead we derive it as an application. The method here is so strong that it allows us to establish a new generalized sequential characterization (the form i) in Theorem 1) that did not appear in the preceding sequential formulas and that does not seem either to be a direct consequence of these formulas. In section 2, we prove the main result describing in several different generalized sequential forms the subdifferential $\partial(f_1 + f_2)(x)$ as special limits of sums of subgradients of f_1 and f_2 at neighboring points of x. All the forms are used to give several applications. In the process, the formula in [7] as well as the famous Rockafellar integration theorem (of subdifferentials of convex functions) are derived in a very simple way. As another important application, the maximal monotonicity of the subdifferential operator is also obtained in a similar manner.

2. Limiting convex subdifferential calculus

In all the paper $\mathbb{B}_X$ will denote the closed unit ball of a Banach space X and ψ_S will denote the indicator function of a subset $S \subset X$, that is $\psi_S(x) = 0$ if $x \in S$ and $\psi_S(x) = +\infty$ otherwise. If $f : X \to \mathbb{R} \cup \{+\infty\}$ is a convex function, $\partial f(x)$ will denote the subdifferential of f at $x \in X$ in the sense of Convex Analysis, that is

$$\partial f(x) = \{x^* \in X^* : \langle x^*, y - x\rangle \leq f(y) - f(x), \ \forall y \in X\}$$

if $f(x)$ is finite and $\partial f(x) = \emptyset$ otherwise. Here X^* denotes the topological dual of the Banach space X. Instead of $x^* \in \partial f(x)$, we will sometimes write $(x, x^*) \in \partial f$. Recall also that, when f is finite at some point, one says that f is proper and one writes

$$\operatorname{dom} f := \{x \in X : f(x) < \infty\},$$

the effective domain f. Similarly, one puts

$$\operatorname{dom} \partial f := \{x \in X : \partial f(x) \neq \emptyset\}.$$

The following lemma will be used in the sequel. It is due to Borwein and Zhu [4]. In fact, their result (Lemma 2.8 in [4]) is more general. It has been first stated and used in a different form in Borwein and Ioffe [3] and it is an infinite dimensional extension of the smooth penalization result given in Lemma 3.1. in Crandall, Ishii and Lions [6]. Similar variants go back to the early days of fuzzy subdifferential calculus rules for nonconvex functions, in Ioffe [9].

LEMMA 1. *Let X be a Banach space and $f_i : X \to \mathbb{R} \cup \{+\infty\}$ with $i = 1, 2, 3$ be lower semicontinuous functions which are finite at some point in a closed subset $S \subset$*

X. Assume that $f_3(x) = +\infty$ outside of some finite dimensional vector subspace of X and put for $(x_1, x_2, x_3) \in X^3$

$$\Delta(x_1, x_2, x_3) := \frac{1}{2}(\|x_1 - x_2\|^2 + \|x_2 - x_3\|^2 + \|x_3 - x_1\|^2)$$

and

$$w_t(x_1, x_2, x_3) := f_1(x_1) + f_2(x_2) + f_3(x_3) + t\Delta(x_1, x_2, x_3).$$

Then

$$\inf_{(x_1,x_2,x_3)\in S^3} w_t(x_1, x_2, x_3) \underset{t\to\infty}{\longrightarrow} \inf_{x\in S}(f_1(x) + f_2(x) + f_3(x)).$$

Furthermore, for any $(x_{1,t}, x_{2,t}, x_{3,t}) \in S^3$ satisfying

$$w_t(x_{1,t}, x_{2,t}, x_{3,t}) - \inf_{(x_1,x_2,x_3)\in S^3} w_t(x_1, x_2, x_3) \underset{t\to+\infty}{\longrightarrow} 0$$

one has

$$t\Delta(x_{1,t}, x_{2,t}, x_{3,t}) \underset{t\to+\infty}{\longrightarrow} 0.$$

The next lemma follows the proof of Theorem 2.9 in Borwein and Zhu [4]. We use the Ekeland variational principle and we adapt some techniques of the Borwein and Zhu theorem to the case of convex functions that are defined on an arbitrary Banach space which is not assumed to admit equivalent differentiable renorms. The assumption on the differentiability of some equivalent renorm can be removed here because of the convexity assumption of the functions.

LEMMA 2. *Let X be a Banach space and $f_i : X \to \mathbb{R} \cup \{+\infty\}$, with $i = 1, 2, 3$, be lower semicontinuous convex functions that are finite at $\bar{x} \in X$. Assume that $f_3(x) = +\infty$ outside of a finite dimensional vector subspace of X and assume that $\bar{x}$ is a minimum point for $f_1 + f_2 + f_3$ over X with $f_1(\bar{x}) + f_2(\bar{x}) + f_3(\bar{x}) < \infty$. Then for any real number $\varepsilon > 0$, there exist $x_i^* \in \partial f_i(x_i)$ such that*

$$\|x_i - \bar{x}\| \le \varepsilon, |f_i(x_i) - f_i(\bar{x})| \le \varepsilon, \|x_1^* + x_2^* + x_3^*\| \le \varepsilon$$

and

$$\|x_i^*\| \ \max(\|x_1 - x_2\|, \|x_2 - x_3\|, \|x_3 - x_1\|) \le \varepsilon.$$

Proof. Fix $\varepsilon > 0$ and keep the notations in the statement of Lemma 1. Applying this lemma with the closed subset $S := \bar{x} + \varepsilon \mathbb{B}_X$ we obtain

$$\inf_{(x_1,x_2,x_3)\in S^3} w_t(x_1, x_2, x_3) \underset{t\to\infty}{\longrightarrow} f_1(\bar{x}) + f_2(\bar{x}) + f_3(\bar{x})$$

and hence for each integer $n \ge 1$ there exists some real number $t_n > n$ such that

$$f_1(\bar{x}) + f_2(\bar{x}) + f_3(\bar{x}) < \inf_{(x_1,x_2,x_3)\in S^3} w_{t_n}(x_1, x_2, x_3) + \frac{\varepsilon^2}{n}. \tag{1}$$

As $\Delta(\bar{x}, \bar{x}, \bar{x}) = 0$, this inequality can be rewritten in the form

$$w_{t_n}(\bar{x}, \bar{x}, \bar{x}) < \inf\{w_{t_n}(x_1, x_2, x_3) : x_i \in \bar{x} + \varepsilon\, \mathbb{B}_X\} + \frac{\varepsilon^2}{n}. \tag{2}$$

By the Ekeland variational principle applied to the complete metric space S^3, there exist $x_{i,n} \in \bar{x} + \frac{\varepsilon}{n}\mathbb{B}_X$ such that

$$w_{t_n}(x_{1,n}, x_{2,n}, x_{3,n}) \le w_{t_n}(\bar{x}, \bar{x}, \bar{x}) \tag{3}$$

and $(x_{1,n}, x_{2,n}, x_{3,n})$ is a minimum point on S^3 for the function

$$\begin{aligned}(x_1, x_2, x_3) \mapsto f_1(x_1) \quad &+ f_2(x_2) + f_3(x_3) + t_n\Delta(x_1, x_2, x_3) \\ &+\varepsilon(\|x_1 - \bar{x}\| + \|x_2 - \bar{x}\| + \|x_3 - \bar{x}\|).\end{aligned}$$

It is easily seen that for $f : X^3 \to I\!R \cup \{+\infty\}$ with

$$f(x_1, x_2, x_3) = f_1(x_1) + f_2(x_2) + f_3(x_3)$$

one has

$$\partial f(x_1, x_2, x_3) = \partial f_1(x_1) \times \partial f_2(x_2) \times \partial f_3(x_3).$$

So, the continuity of the convex function Δ allows us to apply the Moreau-Rockafellar theorem to obtain $x_i^* \in \partial f_i(x_{i,n})$ such that

$$-(x_{1,n}^*, x_{2,n}^*, x_{3,n}^*) \in t_n \partial\Delta(x_{1,n}, x_{2,n}, x_{3,n}) + \varepsilon\, I\!B.$$

Observing that for any $(u_1^*, u_2^*, u_3^*) \in \partial\Delta(u_1, u_2, u_3)$ one has

$$u_1^* + u_2^* + u_3^* = 0,$$

one gets

$$\|x_{1,n}^* + x_{2,n}^* + x_{3,n}^*\| \leq \varepsilon.$$

By (1) and (3) one has

$$w_{t_n}(x_{1,n}, x_{2,n}, x_{3,n}) \underset{n\to\infty}{\longrightarrow} f_1(\bar{x}) + f_2(\bar{x}) + f_3(\bar{x}) \tag{4}$$

and by definition of w_t one has

$$f_1(x_{1,n}) + f_2(x_{2,n}) + f_3(x_{3,n}) \leq w_{t_n}(x_{1,n}, x_{2,n}, x_{3,n}).$$

As $x_{i,n} \to \bar{x}$ and each function f_i is lower semicontinuous, it follows from the last inequality and from (4) that

$$f_1(x_{1,n}) + f_2(x_{2,n}) + f_3(x_{3,n}) \underset{n\to\infty}{\longrightarrow} f_1(\bar{x}) + f_2(\bar{x}) + f_3(\bar{x}).$$

The latter and the lower semicontinuity of f_i once again imply that $f_i(x_{i,n}) \underset{n\to\infty}{\longrightarrow} f_i(\bar{x})$. Furthemore, according to (2) and (3) one has

$$w_{t_n}(x_{1,n}, x_{2,n}, x_{3,n}) - \inf_{(x_1,x_2,x_3)\in S^3} w_{t_n}(x_1, x_2, x_3) \underset{n\to\infty}{\longrightarrow} 0$$

and hence by Lemma 1 once again one has

$$t_n(\|x_{1,n} - x_{2,n}\|^2 + \|x_{2,n} - x_{3,n}\|^2 + \|x_{3,n} - x_{1,n}\|^2) \underset{n\to\infty}{\longrightarrow} 0. \tag{5}$$

Fix now any $(u_1^*, u_2^*, u_3^*) \in \partial\Delta(u_1, u_2, u_3)$. By definition of the convex subdifferential, one has for any $v \in X$ with $\|v\| = 1$ and any real number $r > 0$ (taking $x_1 = u_1 + rv$, $x_2 = u_2$ and $x_3 = u_3$)

$$\begin{aligned}\langle u_1^*, rv\rangle &\leq \frac{1}{2}(\|u_1 - u_2 + rv\|^2 - \|u_1 - u_2\|^2) + \frac{1}{2}(\|u_1 - u_3 + rv\|^2 \\ &\quad -\|u_1 - u_3\|^2) \\ &= \frac{1}{2}(\|u_1 - u_2 + rv\| - \|u_1 - u_2\|)(\|u_1 - u_2 + rv\| \\ &\quad +\|u_1 - u_2\|) + \frac{1}{2}(\|u_1 - u_3 + rv\| - \|u_1 - u_3\|) \\ &\quad (\|u_1 - u_3 + rv\| + \|u_1 - u_3\|) \\ &\leq \frac{1}{2}r(\|u_1 - u_2 + rv\| + \|u_1 - u_2\|) + \frac{1}{2}r(\|u_1 - u_3 + rv\| \\ &\quad +\|u_1 - u_3\|)\end{aligned}$$

and hence dividing by r and taking the limit for $r \downarrow 0$ one obtains

$$\langle u_1^*, v\rangle \leq \|u_1 - u_2\| + \|u_1 - u_3\|.$$

This implies for $i = 1, 2, 3$ that

$$\|u_i^*\| \leq \|u_1 - u_2\| + \|u_2 - u_3\| + \|u_3 - u_1\|.$$

Using the latter in (5) one obtains

$$\|x_{i,n}^*\|(\|x_{1,n} - x_{2,n}\| + \|x_{2,n} - x_{3,n}\| + \|x_{3,n} - x_{1,n}\|) \underset{n\to\infty}{\longrightarrow} 0.$$

So, it is sufficient to choose some integer n sufficiently large to complete the proof. □

We can now prove the following sequential formulas for the subdifferential of $f + g \circ A$ where f and g are convex functions and A is a continuous linear mapping. The form iii) appeared in our preceding paper [18] and has been derived from the formula in [7] in terms of the approximate ε-subdifferential of convex functions. For the form ii) we refer to [12] and [19, 20]. The proofs in [12] and [20] also use the formula in [7], whereas the proof in [19] employs a convergence result of subgradient set-valued mappings instead, but in reflexive Banach spaces. Note that results in these lines have been established, for reflexive spaces, in our paper [17] with sequences in place of nets and also with the strong convergence of $x_n^* + y_n^* \circ A$ to x^*. Here, we obtain the new form i) and we give a direct proof that does not use the formula in [7]. This form i) introduces a control of the size of the nets $(\|x_j^*\|)_j$ and $(\|y_j^*\|)_j$. The forms ii) and iii) will be easily deduced from i) and we will add two more forms iv) and v) to the list. In section 3, we will provide applications for all the forms. In the proof of the theorem we consider an idea of [3] and [4] in the use of the indicator function ψ_L below.

THEOREM 1. *Let X and Y be two Banach spaces, A be a continuous linear mapping from X into Y, and f and g be two lower semicontinuous convex functions from X and Y into $I\!R \cup \{+\infty\}$ with $f(\bar{x}) < \infty$ and $g(A\bar{x}) < \infty$. Then, $x^* \in \partial(f + g \circ A)(\bar{x})$ if and only if there are nets $x_j^* \in \partial f(x_j)$, $y_j^* \in \partial g(y_j)$ such that*

$$x^* = w^* - \lim(x_j^* + y_j^* \circ A)$$

and such that one of the following assertions holds :

i) $(x_j, f(x_j)) \to (\bar{x}, f(\bar{x}))$, $(y_j, g(y_j)) \to (A\bar{x}, g(A\bar{x}))$, $\langle x_j^*, x_j - \bar{x}\rangle \to 0$, $\langle y_j^*, y_j - A\bar{x}\rangle \to 0$ *and* $(\|x_j^*\| + \|y_j^*\|)(\|y_j - Ax_j\|) \to 0$;

ii) $(x_j, f(x_j)) \to (\bar{x}, f(\bar{x}))$, $(y_j g(y_j)) \to (A\bar{x}, g(A\bar{x}))$, $\langle x_j^*, x_j - \bar{x}\rangle \to 0$ *and* $\langle y_j^*, y_j - A\bar{x}\rangle \to 0$;

iii) $x_j \to \bar{x}$, $y_j \to A\bar{x}$, $f(x_j) - \langle x_j^*, x_j - \bar{x}\rangle \to f(\bar{x})$ *and* $g(y_j) - \langle y_j^*, y_j - A\bar{x}\rangle \to g(A\bar{x})$;

iv) $x_j \to \bar{x}$, $y_j \to A\bar{x}$, $\langle x_j^*, x_j - \bar{x}\rangle \to 0$ *and* $\langle y_j^*, y_j - A\bar{x}\rangle \to 0$;

v) $x_j \to \bar{x}$, $y_j \to A\bar{x}$, $\liminf\langle x_j^*, \bar{x} - x_j\rangle \geq 0$ *and* $\liminf\langle y_j^*, A\bar{x} - y_j\rangle \geq 0$.

Proof. We denote by o) the assertion $x^* \in \partial(f + g \circ A)(\bar{x})$ and we still label i) - v) the corresponding assertions with the additional equality for x^* in the statement of the theorem. The proof will be established by showing that all the assertions o) to v) are equivalent.

o) $\Longrightarrow$ i). We begin with the case $Y = X$ and A is the identity mapping over X, that is $x^* \in \partial(f + g)(\bar{x})$. Denote by $\mathcal{F}$ the set of all finite dimensional subspaces of X and denote by J the directed set $\mathbb{N} \times \mathcal{F}$ with $(n, L) \leq (n', L')$ iff $n \leq n'$ and $L \subset L'$. Fix an integer $n \geq 1$ and a finite dimensional subspace L with $\bar{x} \in L$. The point $\bar{x}$ is a minimum point for the function $(f - x^*) + g + \psi_L$ over X. Putting $j := (n, L)$ it follows from Lemma 2 that there exist $(x_j, x_j^*) \in \partial f$, $(y_j, y_j^*) \in \partial g$ and $(z_j, z_j^*) \in \partial \psi_L$ with

$$\|x_j - \bar{x}\| \leq \frac{1}{n}, \ \|y_j - \bar{x}\| \leq \frac{1}{n}, \ \|z_j - \bar{x}\| \leq \frac{1}{n} \tag{6}$$

$$|f(x_j) - f(\bar{x})| \leq \frac{1}{n}, \ |g(y_j) - g(\bar{x})| \leq \frac{1}{n} \tag{7}$$

and

$$\|x_j^* + y_j^* + z_j^* - x^*\| \leq \frac{1}{n} \tag{8}$$

$$(\|x_j^*\| + \|y_j^*\| + \|z_j^*\|)(\|x_j - y_j\| + \|y_j - z_j\| + \|z_j - x_j\|) \leq \frac{1}{n}. \tag{9}$$

So

$$x_j^* + y_j^* \in x^* - z_j^* + \frac{1}{n}\mathbb{B}_{X^*} \subset x^* + L^\perp + \frac{1}{n}\mathbb{B}_{X^*}$$

and hence

$$x^* = w^* - \lim_j (x_j^* + y_j^*).$$

Moreover, for some $b_j^* \in \mathbb{B}_{X^*}$, one has (because $z_j \in L$ and $z_j^* \in L^\perp$)

$$\begin{aligned} \langle x_j^* + y_j^*, x_j - \bar{x}\rangle &= \langle x^* + \tfrac{1}{n}b_j^*, x_j - \bar{x}\rangle - \langle z_j^*, x_j - \bar{x}\rangle \\ &= \langle x^* + \tfrac{1}{n}b_j^*, x_j - \bar{x}\rangle - \langle z_j^*, z_j - \bar{x}\rangle + \langle z_j^*, z_j - x_j\rangle \\ &= \langle x^* + \tfrac{1}{n}b_j^*, x_j - \bar{x}\rangle + \langle z_j^*, z_j - x_j\rangle \end{aligned}$$

and hence by (6) and (9)

$$\langle x_j^* + y_j^*, x_j - \bar{x}\rangle \to 0. \tag{10}$$

Note that by definition of the subdifferential

$$\langle x_j^*, \bar{x} - x_j\rangle \leq f(\bar{x}) - f(x_j) \quad \text{and} \quad \langle y_j^*, \bar{x} - y_j\rangle \leq g(\bar{x}) - g(y_j)$$

and hence taking (7) into account we obtain

$$\limsup \langle x_j^*, \bar{x} - x_j\rangle \leq 0 \quad \text{and} \quad \limsup \langle y_j^*, \bar{x} - y_j\rangle \leq 0. \tag{11}$$

Therefore, it follows from (10) that

$$0 \leq \liminf \langle x_j^*, \bar{x} - x_j\rangle + \limsup \langle y_j^*, \bar{x} - y_j\rangle$$

and hence using both inequalities in (11) we get

$$0 \leq \liminf \langle x_j^*, \bar{x} - x_j\rangle \leq \limsup \langle x_j^*, \bar{x} - x_j\rangle \leq 0$$

which ensures

$$\lim \langle x_j^*, \bar{x} - x_j\rangle = 0.$$

In the same way

$$\lim \langle y_j^*, \bar{x} - x_j\rangle = 0$$

and this completes the proof of the implication in the case when A is the identity mapping.

Consider now the general case, that is $x^* \in \partial(f+g\circ A)(\bar{x})$. We follow a trick due to Rockafellar (see [16]). Put $\varphi(x,y) = f(x)+g(y)$ and denote by ψ_A the indicator function of the graph of A, that is $\psi_A(x,y) = 0$ if $y = Ax$ and $\psi_A(x,y) = +\infty$ otherwise. It is easily seen that $(x^*,0) \in \partial(\varphi+\psi_A)(\bar{x},A\bar{x})$. Then, according to the particular case above, there are nets $(x_j^*,y_j^*) \in \partial\varphi(x_j,y_j)$, $(a_j^*,b_j^*) \in \partial\psi_A(a_j,Aa_j)$ such that

$$(x_j^*,y_j^*) + (a_j^*,b_j^*) \xrightarrow{w^*} (x^*,0) \tag{12}$$

$$(x_j,y_j) \to (\bar{x},A\bar{x}), a_j \to \bar{x}, f(x_j)+g(y_j) \to f(\bar{x})+g(A\bar{x}) \tag{13}$$

$$(\|(x_j^*,y_j^*)\| + \|(a_j^*,b_j^*)\|).\|(x_j,y_j)-(a_j,Aa_j)\| \longrightarrow 0 \tag{14}$$

$$\langle x_j^*, x_j-\bar{x}\rangle + \langle y_j^*, y_j - A\bar{x}\rangle \to 0. \tag{15}$$

Note first that the lower semicontinuity of f and g and the last convergence in (13) imply that $f(x_j) \to f(\bar{x})$ and $g(y_j) \to g(A\bar{x})$. As

$$\begin{aligned}\|y_j - Ax_j\| &\le \|y_j - Aa_j\| + \|A\|.\|x_j - a_j\| \\ &\le (1+\|A\|)\|(x_j,y_j)-(a_j,Aa_j)\|,\end{aligned}$$

it follows from (14) that

$$(\|x_j^*\| + \|y_j^*\|)\|y_j - Ax_j\| \longrightarrow 0.$$

Note also that the definition of $\partial\psi_A(a_j,Aa_j)$ ensures in a straightforward way that $a_j^* = -b_j^*\circ A$ and hence one can derive from (12) that

$$x^* = w^* - \lim(x_j^* + y_j^*\circ A).$$

So, it remains to show that $\langle x_j^*, x_j - \bar{x}\rangle \to 0$ and $\langle y_j^*, y_j - A\bar{x}\rangle \to 0$. Observe that

$$\limsup\langle x_j^*, \bar{x}-x_j\rangle \le 0 \quad \text{and} \quad \limsup\langle y_j^*, A\bar{x}-y_j\rangle \le 0$$

because

$$\langle x_j^*, \bar{x}-x_j\rangle \le f(\bar{x}) - f(x_j) \quad \text{and} \quad \langle y_j^*, A\bar{x}-y_j\rangle \le g(A\bar{x}) - g(y_j)$$

and hence (as in the first case above) it follows from (15) that

$$\langle x_j^*, \bar{x}-x_j\rangle \to 0 \quad \text{and} \quad \langle y_j^*, A\bar{x}-y_j\rangle \to 0.$$

i) $\Longrightarrow$ ii) and ii) $\Longrightarrow$ iii) are obvious.

iii) $\Longrightarrow$ o). Suppose that $x^* = w^* - \lim(x_j^* + y_j^*\circ A)$ with $x_j^* \in \partial f(x_j)$ and $y_j^* \in \partial g(y_j)$ satisfying iii). Then for any $x \in X$ one has

$$\begin{aligned}&\langle x_j^* + y_j^*\circ A, x-\bar{x}\rangle \\ &= (\langle x_j^*, x-x_j\rangle + \langle x_j^*, x_j-\bar{x}\rangle) + (\langle y_j^*, Ax-y_j\rangle + \langle y_j^*, y_j - A\bar{x}\rangle) \\ &\le (f(x) - (f(x_j) - \langle x_j^*, x_j-\bar{x}\rangle))+ \\ &\quad (g\circ A(x) - (g(y_j) - \langle y_j^*, y_j - A\bar{x}\rangle))\end{aligned} \tag{16}$$

and passing to the limit one obtains

$$\langle x^*, x-\bar{x}\rangle \le (f(x)-f(\bar{x})) + (g\circ A(x) - g\circ A(\bar{x})).$$

As this inequality holds for every $x \in X$ we conclude that

$$x^* \in \partial(f+g\circ A)(\bar{x}).$$

i) $\Longrightarrow$ iv) and iv) $\Longrightarrow$ v) are obvious.

v) $\Longrightarrow$ o). As in (16) one has for any $x \in X$

$$\langle x_j^*, \bar{x} - x_j \rangle + \langle y_j^*, A\bar{x} - y_j \rangle + f(x_j) + g(y_j) \\ \leq f(x) + g \circ A(x) - \langle x_j^* + y_j^* \circ A, x - \bar{x} \rangle$$

and hence

$$\liminf \langle x_j^*, \bar{x} - x_j \rangle + \liminf \langle y_j^*, A\bar{x} - y_j \rangle + \liminf f(x_j) + \liminf g(y_j) \\ \leq f(x) + g \circ A(x) - \langle x^*, x - \bar{x} \rangle.$$

According to the assumptions

$$0 \leq \liminf \langle x_j^*, \bar{x} - x_j \rangle \quad \text{and} \quad 0 \leq \liminf \langle y_j^*, A\bar{x} - y_j \rangle$$

and the lower semicontinuity of f and g, the last inequality above implies that

$$f(\bar{x}) + g(A\bar{x}) \leq f(x) + g \circ A(x) - \langle x^*, x - \bar{x} \rangle.$$

This ensures that $x^* \in \partial(f + g \circ A)(\bar{x})$ and hence the proof of the theorem is complete. □

3. Applications

In this section, we will provide applications of various forms of Theorem 1. Recall that applications of form iii) to sequential Lagrange multipliers for general convex mathematical optimization problems are given in our paper [17]. The form iii) is also applied in our paper [18], showing that any exact subdifferential formula under any qualification condition can be derived from it. The form ii) can be employed as in [17] to get a sequential mean value theorem. Those results and the applications that we will develop below demonstrate the usefullness of all the forms in Theorem 1.

As we pointed out in the introduction, our preceding papers derived generalized sequential calculus rules from Hiriart-Urruty and Phelps formula [7,8]. So, we begin by showing how their formula can be deduced from form iii) (for example) in the setting of Banach spaces (the case of non complete vector spaces will be considered elsewhere). We must mention that the proof in [7,8] holds for any locally convex vector space.

The statement of the result makes use of the notion of approximate ε-subdifferential. For a convex function $f : X \to I\!R \cup \{+\infty\}$ that is finite at $\bar{x} \in X$ and for a real number $\varepsilon > 0$, the ε-subdifferential of f is the subset of X^* given by

$$\partial_\varepsilon f(\bar{x}) := \{x^* \in X^* : \langle x^*, x - \bar{x} \rangle \leq f(x) - f(\bar{x}) + \varepsilon, \ \forall x \in X\}.$$

PROPOSITION 1 (Hiriart-Urruty and Phelps formula). *Let X and Y be Banach spaces, $A : X \to Y$ be a continuous linear mapping, $f : X \to I\!R \cup \{+\infty\}$ and $g : Y \to I\!R \cup \{+\infty\}$ be two proper lower semicontinuous convex functions. Then for any $\bar{x} \in X$ with $f(\bar{x}) < +\infty$ and $g(A\bar{x}) < +\infty$ one has*

$$\partial(f + g \circ A)(\bar{x}) = \bigcap_{\varepsilon > 0} cl_{w^*}(\partial_\varepsilon f(\bar{x}) + \partial_\varepsilon g(A\bar{x}) \circ A)$$

where cl_{w^*} denotes the closure with respect to the weak star topology and $\partial_\varepsilon g(A\bar{x}) \circ A := \{y^* \circ A : y^* \in \partial_\varepsilon g(A\bar{x})\}$.

Proof. It is not difficult to see that the second member is included in the first one. So, we prove the reverse inclusion. Fix any $x^* \in \partial(f + g \circ A)(\bar{x})$ (there is nothing to prove if this set is empty). The form iii) of Theorem 1 gives nets $x_j^* \in \partial f(x_j)$ and $y_j^* \in \partial g(y_j)$ such that $x^* = w^* - \lim(x_j^* + y_j^* \circ A)$ and

$$x_j \to \bar{x}, y_j \to A\bar{x}, \quad f(x_j) - \langle x_j^*, x_j - \bar{x}\rangle \to f(\bar{x})$$
$$g(y_i) - \langle y_i^*, y_j - A\bar{x}\rangle \to g(A\bar{x}).$$

Take any $\varepsilon > 0$. There exists some $j_\varepsilon \in J$ such that for all $j \geq j_\varepsilon$

$$f(x_j) - \langle x_j^*, x_j - \bar{x}\rangle \geq f(\bar{x}) - \varepsilon \quad \text{and} \quad g(y_j) - \langle y_j^*, y_j - A\bar{x}\rangle \geq g(A\bar{x}) - \varepsilon.$$

Fix any $j \geq j_\varepsilon$. The definition of $\partial f(x_j)$ and the first one of the latter inequalities imply that for all $x \in X$

$$\begin{aligned}\langle x_j^*, x - \bar{x}\rangle &= \langle x_j^*, x - x_j\rangle + \langle x_j^*, x_j - \bar{x}\rangle \\ &\leq f(x) - f(x_j) + \langle x_j^*, x_j - \bar{x}\rangle \\ &\leq f(x) - f(\bar{x}) + \varepsilon.\end{aligned}$$

This ensures that $x_j^* \in \partial_\varepsilon f(\bar{x})$ and in the same way $y_j^* \in \partial_\varepsilon g(A\bar{x})$ and hence $x_j^* + y_j^* \circ A \in \partial_\varepsilon f(\bar{x}) + \partial_\varepsilon g(A\bar{x}) \circ A$. As this holds for all $j \in J$ with $j \geq j_\varepsilon$ and as $x^* = w^* - \lim(x_j^* + y_j^* \circ A)$, we obtain

$$x^* \in cl_{w^*}(\partial_\varepsilon f(\bar{x}) + \partial_\varepsilon g(A\bar{x}) \circ A),$$

and the proof is complete. □

The second application is devoted to deriving from i) or ii) the graphical density of points of subdifferentiability. This is a well-known result (see [5] and [2]).

PROPOSITION 2. *Let f be a proper lower semicontinuous convex function from a Banach space X into $\mathbb{R} \cup \{+\infty\}$. Then, the subset* $\operatorname{dom} \partial f$ *is graphically dense in* $\operatorname{dom} f$, *i.e. for any $\bar{x} \in \operatorname{dom} f$ and any $\varepsilon > 0$ there exists $x \in X$ with $\partial f(x) \neq \emptyset$ and with $||x - \bar{x}|| + |f(x) - f(\bar{x})| < \varepsilon$.*

Proof. For $S := \{\bar{x}\}$ one has

$$f(\bar{x}) + \psi_S(\bar{x}) \leq f(x) + \psi_S(x) \quad \text{all } x \in X$$

and hence $0 \in \partial(f + \psi_S)(\bar{x})$. So, it suffices to apply the form ii) in Theorem 1. □

We proceed now to provide an application of the form iv) or v) to the famous Rockafellar integration theorem for the inclusion of subdifferentials of convex functions. All the proofs known so far are deep ones, see the original proof in [13] and [15] and also the proof in [21]. To our knowledge, the proof below is the simplest one in the setting of Banach spaces. (However, we must point out that it does not cover the more general case considered in Theorem 2.1. in [21] that concerns the inclusion $\partial f(x) \subset \partial g(x) + r\,\mathbb{B}_{X^*}$ for all $x \in X$, where r is a fixed non negative real number and only g is assumed to be convex but not f).

PROPOSITION 3 (Rockafellar integration theorem). *Let X be a Banach space and $f, g : X \to \mathbb{R} \cup \{+\infty\}$ be two proper lower semicontinuous convex functions such that $\partial f(x) \subset \partial g(x)$ for all $x \in X$. Then*

$$f = g + \text{Constant}.$$

Proof. The case when X is a one-dimensional space is straightforward with the use of the left or right derivative or with the Rademacher theorem (see also [14]). Let us show that the general case can be reduced to the latter case. We know (see for example, Proposition 2) that $\text{dom}\partial f \neq \emptyset$. Fix $a \in \text{dom}\,\partial f$. Then $f(a) \in \mathbb{R}$ and $g(a) \in \mathbb{R}$. We may assume $a = 0$. For each $b \in X$, consider the one-dimensional vector space $L_b := \mathbb{R}b$ and the continuous linear mapping $A_b : L_b \to X$ with $A_b u = u$ for all $u \in L_b$ (i.e. the embedding mapping).

We begin by showing that $\partial(f \circ A_b)(u) \subset \partial(g \circ A_b)(u)$ for all $u \in L_b$. Fix any $(u, u^*) \in \partial(f \circ A_b)$. By iv) in Theorem 1 one has $u^* = \lim x_j^* \circ A_b$ with

$$x_j^* \in \partial f(x_j), \quad \langle x_j^*, x_j - A_b u\rangle \longrightarrow 0 \quad \text{and} \quad x_j \longrightarrow A_b u = u.$$

But the inclusion assumption of the proposition implies that $x_j^* \in \partial g(x_j)$ and hence by iv) in Theorem 1 once again one obtains $u^* \in \partial(g \circ A_b)(u)$. So, the inclusion $\partial(f \circ A_b)(u) \subset \partial(g \circ A_b)(u)$ holds for all $u \in L_b$.

By the one-dimensional argument (see the comments above), one obtains

$$f(a) + g(b) = g(a) + f(b).$$

On the one hand, this equality says that $f(b) < \infty$ if and only if $g(b) < \infty$, that is $\text{dom}f = \text{dom}g$. On the other hand, for all $x \in \text{dom}f = \text{dom}g$

$$f(a) + g(x) = g(a) + f(x)$$

and hence for all $x \in X$

$$f(x) = g(x) + (f(a) - g(a)).$$

This completes the proof. □

Any one of the five forms of Theorem 1 can be applied to prove in a very simple way the maximal monotonicity of the subdifferential operator. This is a known result due to Minty [10] for (everywhere defined) real-valued continuous convex functions, and for lower semicontinuous convex functions with values in $\mathbb{R} \cup \{+\infty\}$ it is due to Moreau [11] in the Hilbert space setting and to Rockafellar [13] and [15] in the Banach space setting.

Remind that a set-valued operator $M : X \rightrightarrows X^*$ is monotone if $\langle x^* - y^*, x - y\rangle \geq 0$ for all $x^* \in M(x)$ and $y^* \in M(y)$. When, in addition, the graph of M is maximal (with respect to the inclusion) among all the graphs of monotone operators, one says that M is maximal monotone. It is a direct consequence of the definition of subdifferential that ∂f is monotone for any convex function f.

PROPOSITION 4. **(maximal monotonicity theorem [13,15])** *Let X be a Banach space and $f : X \to \mathbb{R} \cup \{+\infty\}$ be a proper lower semicontinuous convex function. Then, the set-valued operator $\partial f : X \rightrightarrows X^*$ is maximal monotone.*

Proof. The result is straightforward in the case when X is a one dimensional space (see also [13]). Consider the general case and suppose that there exists $(a, a^*) \in X \times X^*$ with $a^* \notin \partial f(a)$ and $\langle x^* - a^*, x - a\rangle \geq 0$ for all $(x, x^*) \in \partial f$. We may assume by translation that $a = 0$ and $a^* = 0$, i.e. $0 \notin \partial f(0)$ and $\langle x^*, x\rangle \geq 0$ for all $(x, x^*) \in \partial f$. As $0 \notin \partial f(0)$, there exists some $b \in X$ with $f(b) < f(0)$. Put $L := \mathbb{R}b$ and define $A : L \to X$ with $A(u) = u$ and $f_0 : L \to \mathbb{R} \cup \{+\infty\}$ with $f_0(u) := f(Au)$. Then $f_0(b) < f_0(0)$ and hence $0 \notin \partial f_0(0)$. Fix any $(u, u^*) \in \partial f_0$. By Theorem 1, there exist $u_j \to u$, $x_j^* \in \partial f(Au_j)$ with $x_j^* \circ A \to u^*$. By what precedes, $\langle x_j^* \circ A, u_j\rangle = \langle x_j^*, Au_j\rangle \geq 0$ and hence $\langle u^*, u\rangle \geq 0$. This contradicts the one dimensional argument saying that ∂f_0 is maximal monotone. So, the proof is complete. □

References

[1] **H. Attouch, J.-B. Baillon** and **M. Théra**, Variational sum of monotone operators, J. Convex Analysis 1 (1994), 1–29.
[2] **J.M. Borwein**, A note on ε-subgradients and maximal monotonicity, Pacific J. Math. 103 (1982), 307–314.
[3] **J.M. Borwein** and **A.D. Ioffe**, Proximal analysis in smooth spaces, Set-Valued Analysis 4 (1996), 1–24.
[4] **J.M. Borwein** and **Q.J. Zhu**, Viscosity solutions and viscosity subderivatives in smooth Banach spaces with applications to metric regularity, SIAM J. Control Optim. 34 (1996), 1576–1991.
[5] **A. Brønsted** and **R.T. Rockafellar**, On the subdifferential of convex functions, Proc. Amer. Math. Soc. 16 (1965), 605–611.
[6] **M.G. Crandall, H. Ishii** and **P.-L. Lions**, User's guide to viscosity solutions of second order partial differential equations, Bull. Amer. Math. Soc. 27 (1992), 1–67.
[7] **J.-B. Hiriart-Urruty** and **R.R. Phelps**, Suddifferential calculus using ε-subdifferentials, J. Funct. Analysis 118 (1993), 154–166.
[8] , **J.-B. Hiriart-Urruty, M. Moussaoui, A. Seeger** and **M. Volle**, Subdifferential calculus without qualification conditions, using approximate subdifferentials : a survey, Nonlinear Anal. Th. Meth. Appl. 24 (1995), 1727–1754.
[9] **A.D. Ioffe**, On subdifferentiability spaces, Ann. N.Y. Acad. Sci. 410 (1983), 107–119.
[10] **G.J. Minty**, On the monotonicity of the gradient of a convex function, Pacific J. Math. 14 (1964), 243–247.
[11] **J.-J. Moreau**, Proximité et dualité dans un espace hilbertien, Bull. Soc. Math. France 93 (1965), 273–299.
[12] **J.-P. Penot**, Subdifferential calculus without qualification assumptions, 3 (1996), 207–219.
[13] **R.T. Rockafellar**, Characterization of the subdifferentials of convex functions, Pacific J. Math. 17 (1966), 497–510.
[14] **R.T. Rockafellar**, Convex Analysis, Princeton University Press, Princeton, N.J. (1970)
[15] **R.T. Rockafellar**, On the maximal monotonicity of subdifferential mappings, Pacific J. Math. 33 (1970), 209–216.
[16] **R.T. Rockafellar**, Directionally Lipschitzian functions and subdifferential calculus, Proc. London Math. Soc. 39 (1979), 331–355.
[17] **L. Thibault**, Sequential convex subdifferential calculus and sequential Lagrange multipliers (preprint 1994), SIAM J. Control Optim. (1997), 1434–1444.
[18] **L. Thibault**, A generalized sequential formula for subdifferentials of sums of convex functions defined on Banach spaces, Recent Developments in Optimization (Seventh French-German Conference on Optimization hold in Dijon in June 27 - July 2, 1994) edited by R. Durier and C. Michelot, Lecture Notes in Economics and Mathematical Systems, Springer-Verlag Nr. 429 (1995), 340–345.
[19] **L. Thibault**, A direct proof of a sequential formula for the subdifferential of the sum of two convex functions (unpublished paper, 1994).
[20] **L. Thibault**, A short note on sequential convex subdifferential calculus (unpublished paper, 1994).
[21] **L. Thibault** and **D. Zagrodny**, Integration of subdifferentials of lower semicontinuous functions on Banach spaces, J. Math. Anal. Appl. 189 (1995), 33–58.

Université Montpellier II, Laboratoire d'Analyse Convexe, Case Courrier 051, 34095 Montpellier, France

E-mail address: thibault@math.univ-montp2.fr